The Goods of Design

The Goods of Design

Professional Ethics for Designers

Ariel Guersenzvaig

ROWMAN & LITTLEFIELD
Lanham • Boulder • New York • London

Published by Rowman & Littlefield
An imprint of The Rowman & Littlefield Publishing Group, Inc.
4501 Forbes Boulevard, Suite 200, Lanham, Maryland 20706
https://rowman.com

6 Tinworth Street, London SE11 5AL, United Kingdom

British Library Cataloguing in Publication Information Available

ISBN: HB 978-1-78661-540-4

Library of Congress Cataloging-in-Publication Data
Names: Guersenzvaig, Ariel, 1970– author.
Title: The goods of design : professional ethics for designers / Ariel Guersenzvaig.
Description: Lanham : Rowman & Littlefield, [2021] | Includes bibliographical references and index.
Identifiers: LCCN 2020054910 (print) | LCCN 2020054911 (ebook) | ISBN 9781786615404 (cloth) | ISBN 9781786615428 (epub) | ISBN 9781538179918 (pbk)
Subjects: LCSH: Design—Moral and ethical aspects. | Designers—Professional ethics.
Classification: LCC NK1520 .G84 2021 (print) | LCC NK1520 (ebook) | DDC 745.4—dc23
LC record available at https://lccn.loc.gov/2020054910
LC ebook record available at https://lccn.loc.gov/2020054911

The Goods of Design

Professional Ethics for Designers

Ariel Guersenzvaig

ROWMAN & LITTLEFIELD
Lanham • Boulder • New York • London

Published by Rowman & Littlefield
An imprint of The Rowman & Littlefield Publishing Group, Inc.
4501 Forbes Boulevard, Suite 200, Lanham, Maryland 20706
https://rowman.com

6 Tinworth Street, London SE11 5AL, United Kingdom

British Library Cataloguing in Publication Information Available

ISBN: HB 978-1-78661-540-4

Library of Congress Cataloging-in-Publication Data
Names: Guersenzvaig, Ariel, 1970– author.
Title: The goods of design : professional ethics for designers / Ariel Guersenzvaig.
Description: Lanham : Rowman & Littlefield, [2021] | Includes bibliographical references and index.
Identifiers: LCCN 2020054910 (print) | LCCN 2020054911 (ebook) | ISBN 9781786615404 (cloth) | ISBN 9781786615428 (epub) | ISBN 9781538179918 (pbk)
Subjects: LCSH: Design—Moral and ethical aspects. | Designers—Professional ethics.
Classification: LCC NK1520 .G84 2021 (print) | LCC NK1520 (ebook) | DDC 745.4—dc23
LC record available at https://lccn.loc.gov/2020054910
LC ebook record available at https://lccn.loc.gov/2020054911

Contents

Acknowledgements

Because it is not possible for me to acknowledge all the people who in one way or another participated in the writing of this book, I will start by thanking all the designers, design students, and design educators I have encountered through the years, with whom I have explored and discussed the topics that are included in these pages. This book grew out of these formal and informal encounters, which helped me better frame and refine my ideas about the nature of the design profession and its ethical dimension. Without the inspiration, criticisms, comments, advice, objections, encouragement, and all the teachings I received from students, designers, and educators, this book could never have been written.

I am especially indebted to several people who helped me write this book. I should like to thank some of them in particular.

I want to express my most profound indebtedness to my friend and colleague David Casacuberta, with whom I learn, teach, and write about design, technology, and ethics. Besides reading the book's manuscript at various stages and providing invaluable critical commentary and steady nurturing, he kindly and unselfishly answered the hundreds of specific queries and requests for opinions with which I assaulted him.

I must also especially and heartily thank César Astudillo for his honesty, enormous generosity, and the many insightful and challenging comments and suggestions I have received from him on every draft chapter.

Another significant debt is to Julia Benini and Oier Romillo, who also read the manuscript and provided sharp, critical feedback and a good dose of support as well. Special thanks are also due to Nuria Garuz, Angus Robson, Iain Law, Laura Andina, Antònia Pulido, Dani Armengol, Toni Llácer, Oscar Tomico, Raffaella Perrone, and Mercè Graell, who were kind enough to collaborate and provide valuable feedback on parts of the manuscript. I also want to thank others who read portions of the book and provided useful feedback: Ariana Escobar, Ariadna Ariza, Paula López-Nuño, Sebastián Ribas, Reme Martínez Castillo, Airí Dordas, Sal Atxondo, Cris van der Hoek, Ujue Agudo, Jorge Camacho, Grace Ascuasiati, and Alberto Romero.

I would like to particularly thank Ezio Manzini, who was generous enough to engage with me to discuss matters related to design ethics. I consider myself extremely fortunate to have been able to receive feedback and advice from him.

Parts of these book have been presented at several conferences and events: the 4TU.Ethics Bi-annual Conference 'Ethics of Disruptive Technologies' at the Eindhoven University of Technology; the Conference on Design, Art, Science and Technology at the Universitat Autónoma de Barcelona; and the Design Principles and Practices Conference in Barcelona. I also presented my work at internal events organised by design consultancy firms, among them Fjord, DesignIt, and Biko. I thank all those who offered feedback, challenged my views, and made suggestions for improvement.

I wrote the book directly in English, but given I am a not a native speaker, I was lucky to benefit from the linguistic coaching, advice, and revisions I received from Brian Gallagher and Aodh Byrne. I want to thank them for providing many suggestions and helping me get the right formulation on those occasions when I could not.

I would also like to thank Frankie Mace, Scarlet Furness, Alden Perkins, Dhara Snowden, Isobel Cowper-Coles, and everyone else from Rowman & Littlefield who was involved in making this project a reality, as well as the anonymous readers that were appointed to review and comment on an early draft of the manuscript before the book was accepted for publication. I want to express my gratitude to Daniela Bendall from Fortissimo Films, who helped me obtain permission to use the long quote of Frank Gehry that is discussed in chapter 3.

I am indebted to Javier Peña and Albert Fuster, respectively general director and academic director of ELISAVA, Barcelona School of Design and Engineering, where I am a Professor of Design Theory. They believed in this project from the start, and they allowed me to partially count it as an internal research project. I am thankful to them and ELISAVA for their support. A special thanks goes out to all my other colleagues who pepped me up and encouraged me with kind words and enthusiasm. I want to particularly say thanks to my colleague and friend Rosa Llop for the constancy and integrity that she has taught me through the years.

As I write this, I am in full lockdown in Barcelona because of coronavirus (just like everyone else in Spain and half of the countries on Earth). A year ago, when I prepared the final writing schedule for the book after signing the publication contract, I decided to keep the month of May 2020 free of engagements to be able to give the script its finishing touches without external pressures. I always knew that May 2020 was going to be spent in some sort of confinement. Naturally, I could have never imagined that it was going to be like *this*.

But this too shall pass.

The certainty of my family's love in these days of sorrow and worry are a great source of beauty and consolation, and I am grateful for that. I am thankful that my parents, José and Andrea, and my little sister, Paula, have always being there for me and have given me the opportunity to be there for them. I must give special thanks to my partner's parents, Caterina and Arturo, who have provided encouragement and support throughout this project in the way that only loving 'out-laws' can do. I want to thank my daughters, Abi, Eloisa, and Selba, for teaching me patience and endurance by building me up just to knock me down and build me up again; I want to thank them for being kind and forgiving when I get grumpy, and above all for being an infinite source of *naches* and delight.

My partner, Clara, who supported me with ceaseless fuelling and love throughout the two years this book took to research and write, and managed to bear with me as she only can. I want to also thank her for reading through the drafts and providing feedback; being a professional designer, her comments counted double. And I must thank her also for designing the cover of this book.

Without all of these people, this book would not be what it is. The merits it may contain are shared with them; the errors that undoubtedly remain are only mine.

Barcelona, May 2020

This book was written with the support of a Research Leave grant from ELISAVA Barcelona School of Design and Engineering for the academic years 2018/2019 and 2019/2020.

List of Figures

Introduction and Overview

This philosophical inquiry concerns the professional practice of design and tries to provide an account of its ethical dimension as well as guidelines for action. Though the book is directly focussed on the ethical aspects of being a professional designer, it does not adopt a view that defends a 'professional role' as being fully separated from the person who holds the role. A central idea the book explores is that when a professional designer feels *uneasy* (in an ethical sense) about making a particular design decision (or as consequence of having made one), their uneasiness is often felt in their *whole* selves, and not in some sort of sealed 'compartment' where their professional life resides and is kept separated from the rest of their own 'true' lives. When designers (and most professionals for that matter) are conflicted and worried about decisions they have to make as professionals, they feel worried as persons, not only as professionals. *The Goods of Design* is about this tension and how to navigate it.

Let's explore the contents and approach of the book in a bit more detail.

Designers occupy a prominent role in how products, services, and environments get from abstract idea to concretion. Given the tremendous influence design has in the way people live their lives, the profession is fraught with ethical questions, but unlike engineering, nursing, or journalism, professional design lacks widespread principles and normative frameworks for addressing ethical issues. Granted, institutions and individuals have formulated codes of design ethics, which may be in some cases useful to prompt discussions and reflections, but they rarely go beyond generalities as emphasising the importance of 'preventing harms' and 'respecting human rights'. What is more, they may offer a distorted picture of what ethics is about by disseminating the idea that it is about compliance with a set of rules and norms that one can memorise in order to be 'ethical'. In short, even though design seems

1

to be a fertile ground for the exploration of ethical issues, the ethical discussions around the professional practice of design are still insufficient. Considering design's importance, this calls for further investigation.

I am aware that, given the multifaceted character of the profession, it would be a pointless endeavour to seek to provide a definitive answer on how designers ought to design. I will be satisfied with providing a general and pluralistic normative direction that can guide the profession, while accommodating a broad diversity of views and values that are present in the field of design. In this sense, my arguments and the theoretical apparatus that will be displayed are also a probe for conversations meant to raise questions and counterarguments that might advance design professional ethics.

The first question that needs to be explored is the professional status of design, which is a much-underdeveloped locus of exploration. What is more, the professional status of design is often taken as a given, yet if we want to investigate design professional ethics we need to first reasonably establish that design is indeed a profession. Alas, this step makes the inquiry into design professional ethics somewhat longer because we need to travel the path from the very beginning of the road.

We could avoid this extra step simply by considering design as an occupational activity, thereby taking advantage of the many excellent accounts that are available of *design as a job*. This, however, would undermine our inquiry methodologically because the account I will provide necessitates understanding *design as a profession* for my arguments to be compelling. I must make clear that in treating design as a profession, my intention relates to exploring the implications the ideal of professionalism may have for designers; I do not seek to increase design's social allure by calling it a profession to set it apart from other occupations in the social hierarchy. The goal of engaging with the notions of 'profession' and 'professionalism' is to formulate an understanding of professions as moral projects worth promoting and undertaking.

Once the professional status of design is plausibly established, we shall turn to the ideas of the philosopher Alasdair MacIntyre, a major figure in late-twentieth- and twenty-first-century ethics and political philosophy. He offers a theoretical account of 'practices' that allows us to conceptualise the design profession as a 'practice'. In the MacIntyrean sense, a practice is not simply a 'customary way of doing something', as the word is colloquially understood. A practice in the sense intended is a specific type of social activity with specific intrinsic rewards and standards of excellence, which MacIntyre calls 'internal goods'. Practices are intrinsically integrated in our own narrative projects as human beings by being a constitutive element of who we are. A profession, insofar as it is a special type of practice, is thus centrally linked to one's identity and life plans.

My purpose is to reflect on what can meaningfully guide design activity by drawing on what, at the same time, justifies and legitimises the design profession itself. My central argument will link the design profession to the promotion of others' wellbeing, whereby wellbeing is understood as living a worthwhile and purposeful life (this is an incomplete description, but it should suffice for now).

This book is primarily written for designers and design students who experience the tension between the professional and the personal dimensions of one's life I referred to at the beginning of this introduction, but also for those who do not *always* feel that way. A misguided understanding of professionalism might make some designers wrongly believe that professional life and personal life should be in separated compartments. They might think that they can care about something or somebody *as persons*, but that their professional *persona* should not necessarily care as well. Whereas the designer who feel this way might realise that their designs could contribute to harming people somehow, they will be unsure as to what extent they should feel personally troubled and responsible for that harm as they are simply 'doing their job'. This volume will make the case that viewing matters this way is not only mistaken but also pernicious for both professional and personal growth. Besides being instrumental in the promotion of others' wellbeing, the practice of design is also central to the designer's *own wellbeing*.

The account I present here introduces a rather novel approach for the study of design ethics. Although I naturally draw on the work of many a design scholar, for developing the ethical thrust of my account, other than by MacIntyre's *After Virtue*, I was also guided and inspired by two volumes from philosophers from outside the field of design: George Moore's *Virtue at Work* and Chris Higgins' *The Good Life of Teaching*[1]. Although I quote more extensively from the former, Higgins' account of how and why teaching is not only intrinsically rewarding but a way of self-enactment and self-growth for the teachers themselves played a central role in the development of my own account.

This book aims to be useful for designers as well as for designers in training (that is, design students) who seek to navigate the complexities they encounter in their profession. But this is not a design self-help book, nor is it a bag of tips and tricks for dealing with ethical issues. Instead, it is a book on professional ethics, in which I aim to provide an account of the design profession as a practice that makes a strategic contribution to society and the common good on the one hand, and plays a key role in the designer's life, life-plans, and own wellbeing on the other.

Designers who are absolutely sure that their professional selves and their 'core' selves should be kept apart from one another will perhaps not be persuaded by the arguments presented here; they will find little use for this book. Conversely, designers who are less sure about whether it is okay to do things one rejects as a person, but is 'supposed' to do them because it is 'their job', might find many useful insights.

In this volume, we will steer away from the well-trodden path of codes of ethics and from a widespread yet misdirected inclination to hope for a set of clear rules about how to decide that can be simply applied to any situation. Given the multifaceted and dynamic nature of the design profession, neither is it my intention to provide a definitive and fully instrumental answer on how designers ought to perform their day-to-day tasks. Given the open-ended, contextual nature of design problems, attempting to provide such a guide would be nothing but chimeric. Rather, this account will provide a *general* normative direction for the design profession, but no checklists, step-by-step guides, nor lists of dos and don'ts will be offered.

We will also steer clear from 'quandary ethics'. I am of the view that most designers are seldom confronted with 'dramatic' ethical topics such as nanotechnology, intelligent robots, or driverless cars that 'must choose the lesser of two evils, such as killing two passengers or five pedestrians' as it happens in the 'Moral Machines' project from the Massachusetts Institute of Technology Media Lab. These topics and dilemmas are primarily relevant for philosophers of technology, regulators, and those directly working in the development of emergent and disruptive technologies, but less so to most professional designers. This is not to say these topics would or should not be of interest to designers; the point is these topics are too far removed from their everyday practice. Most designers design everyday artefacts within the paradigm of existing technology. As just indicated, they do not work on designing experimental prototypes for food packages enhanced with nano-robots, but on more ordinary and mundane affairs: an app for a banking service, a waiting area at a train station, a school textbook, and many other objects and services like the ones we will review in chapter 3. Even if it does not involve life-or-death decisions, and is mainly concerned with everyday affairs, the practice of professional design deserves important ethical consideration, and this book is fundamentally about doing that.

ALIGNING OUR VIEW ABOUT DESIGN

It is likely that most of the readers of this book come from the 'design world' at large. Most of us, presumably, will have at least a minimum understanding of what design is. This does not entail, however, that we share an unequivocal understanding of it. Others might come from outside design, and they might need to deepen their understanding of what it is. Because of this, and before we start dealing with other issues, we need to clarify what we should understand by the word 'design', at least for the purposes of this book. Some definitions will eventually be provided, but I do not aim at putting forward a definitive definition to exhaust the topic. My purpose is to outline a theoretical account that will help establish a great deal of what is relevant for our discussion on the profession of design and ethics.

In the English parlance, the word 'design' is used both as a noun and a verb. Generally, when used as a noun, design refers to a 'plan', a 'sketch', or a 'specification' for the construction of an artefact: for instance, for a building, a car, a computer game, a dress, etc. When used as a verb, it means 'to create', 'to plan', or 'to decide' on the look and functioning of an artefact. Although the technical use for the word 'design' does not greatly differ from how we commonly use the word, several further clarifications are in order.

Design historian John Heskett asks us to consider the shifts of meaning of the word 'design' in a 'seemingly nonsensical sentence' (his own words) that includes four different meanings for the word 'design' as both verbs and nouns: 'Design is to design a design to produce a design'[2].

The first meaning refers to design as a field in itself, as in 'Design has business value'. This first sense conveys the meaning of design as an activity, perhaps as a practice, and a discipline. The second meaning refers to design as planning, as in 'Emma designs a new car'. The third communicates a type of representation or specification, as in 'Peter's design for a car is technically unfeasible'. The fourth refers to an artefact or product that is actually made thanks to the plan, as in 'the design of some modern wind turbines mimics features of humpback whales'.

Thus, we see how complex the term is, with different and interrelated meanings. In this book we will be concerned with all of the four understandings of design to different degrees, but primarily with the first one: design as an activity. Design scholar Richard Buchanan explicates the nature of this activity in a way that is worth quoting in full: 'design is the way we plan and create the complex wholes that provide a framework for human culture—the human systems and sub-systems that work either in congress or in conflict with nature to support human fulfilment'[3].

We are surrounded by *the designed*. It is not necessary to provide yet another enumeration of designed things: everywhere we look we find human-made things that have been designed. Most of us live in a designed world, and while it is true that most cities or towns have not been designed in the strict sense, they are marked by design's influence. Even in what we take to be 'pristine' nature, we find the effects of design. For instance, satellite coverage extends over the entire world and can wire up the most remote and virgin parts of the world, which cease to be isolated. Needless to say, the human-made world is made possible more than by designers alone; there is politics and economics, and states, and science, and engineering, and religion, and many other interwoven areas of activity that shape the imprint that humans leave on the world.

Design has revolutionised our lives, improving how we, for instance, communicate, heal, care for others, learn, teach, move around, build, or dwell. But, especially since the start of the industrial revolution, we have also devised new technologies and forms of labour and accessed new forms of energy that have caused pernicious (and often unintended and unexpected) socio-economic and ecological effects. From about 1950 onward a dramatic acceleration occurred in the magnitude and rate of human-driven processes that bring ecological and socio-economic change on a large scale; because of the screeching velocity of change, this phenomenon is known as 'The Great Acceleration'[4]. If we briefly consider the ecological dimension, we see that the 'complex wholes that provide a framework for human culture' have become too heavy and material-intensive. To an unprecedented degree in human history, humans have massively re-shaped lands and landscapes; caused deforestation, desertification, and melting ice sheets; polluted Earth's soil and air . . . in short, humans have brought about a general degradation of the biosphere[5*]. Evidently, not all of this could have been solely caused by designers. Nevertheless, if we accept Buchanan's definition, and the great importance that is implicitly ascribed to design, we have to at least admit that design and designers must have played some role in The Great Transformation. However important, we shall now pause on the consideration of the effects of design; we will resume this discussion in chapter 4.

Just as it happens with most key concepts in the humanities and the social sciences, there is no definitive consensus among design scholars about what design is. Definitions range from 'a product or process of thinking, modelling or problem solving, to all-encompassing visions of design as the transformation of social environments'[6]. Design theorists Bryan Lawson and Kees Dorst argue that there is no use in naming constituent parts for design or providing totalizing definitions, and they prefer to describe the multifaceted and 'polymorphous' nature of design from different viewpoints. For them, design can be seen as 'a mixture of creativity and analysis', as 'problem solving', as 'learning', as 'evolution', as 'the creation of solutions to problems', as 'integrating into a coherent whole', and as 'fundamental human activity'[7].

For the purpose of this book, I am rather lenient in my view of design, which must be understood as an activity primarily associated with the conception and the planning of the human-made world[8*]. It goes far beyond the popular idea of design as stylistic enhancement or form-giving following aesthetic norms. In this book, the usage of the word 'design' is about framing and crafting the plan to bring about material or immaterial artefacts that do not yet exist but could be realised in the future. Objects but also the built environment are examples of the former, whereas digital interactions or services exemplify the latter. The global aim of design is to explore and project potentialities and possibilities that do not exist yet, thus initiating a change in the current artefactual state of affairs.

Buchanan offers yet a second definition of design that adds a necessary layer of purpose to the application of design and the direction of change it aims to enable: 'Design is the human power of conceiving, planning, and making products that serve human beings in the accomplishment of their individual and collective purposes'[9]. Provided we understand 'products' in a broad sense (as artefacts), his definition coexists amicably with and provides a nice summary of what we have been saying so far about design.

This understanding of design also goes beyond a rationalistic understanding of the problem-solution pair, in which problems are well-defined, coherent, and tractable. Design problems are, oftentimes, 'wicked problems'. 'Wicked' must be understood here in a meaning akin to that of 'tricky'; it is not that the problems are evil or ethically deplorable but rather resistant to solution, as they are never solved for good, at best they are re-solved. Buchanan posits that 'Design problems are "indeterminate" and "wicked" because design has no special subject matter of its own apart from what a designer conceives it to be'[10]. Even when they are not 'wicked' in the strict sense, design problems are usually ill-defined, systemic, complex, and somewhat intractable. Perhaps, because of this, it would be more appropriate to speak of 'design proposals' or 'candidate solutions' instead of 'design solutions'. But, alas, this term is too persistent and all too firmly established in design culture, so we shall stick with it.

Moreover, the proposed solutions co-evolve along with the understanding and definition of the problem, which depends on how we subjectively frame it. Solutions and problems are intertwined and the relationship between them is bidirectional, as solution conjectures are used to explore and formulate problems[11]. Designers take problems seldom as given but frame them *reflectively*, constantly generating and testing hypothesis and understandings within a context broadly stipulated by the design brief. This is a plausible explanation for why different designers would come up with different solutions to solve what would initially appear to be the same problem (for example, designing a chair or developing a new admission system for a hospital).

The link between problems and solutions in design activity is a central thread in the writings of many design scholars and the literature on this issue is vast[12]. Despite decades of scholarly attacks and criticism[13]*, the problem-solving view of design is very resilient and has gained new momentum since the late 2000s with the popularization of 'design thinking' in the business world. It must be recognised that this view is useful for design education, because it is accessible and easy to understand for learners of design, but it is also simplistic and unidimensional. More importantly, it does not properly and wholly explain what design does. This is not a call to stop using the words 'design problem' and 'design solution'; it only asks of ourselves that we keep in mind that design is problem solving inasmuch as it is problem finding and sense-making.

Design is transformative. It has effects that affect our 'lifeworld', that is, the state of affairs in which our pre-reflective, immediate subjective experience of everyday life and what surrounds us occur. Along these lines, design can be seen as a producer of meaning[14], as a way of making sense of things and what things should be, do, and mean. Designing a chair is not only shaping its form but also tacitly answering questions about what a *good* chair is, what a *good* posture is, how comfort *could* or *should* affect human experience, and what the symbolic meaning of a chair could be. These questions go thus beyond basic utility and function, and enter the realm of values and social meanings.

ALIGNING OUR VIEW ABOUT ETHICS

We also need to cast light on what we mean by ethics, if only because the word, just like the word 'design', is used in many different ways by both lay people and philosophers. As we will revisit ethical theories in chapter 6 in more detail, I will keep this section to a minimum.

In its everyday sense, ethics is generally related to acting according to norms and values, and it is associated with notions such as 'good' and 'evil', and 'right' and 'wrong'. In this sense, acting ethically is doing the right thing. The word 'ethics' is popularly used to mean the moral principles of a particular individual, as in 'my ethics' or 'Emma Watson's ethics'. In philosophy, however, ethics is generally understood more broadly, and it is primarily defined as 'the philosophical study of morality'. It is also common, but still *highly* contested, not to make a distinction between 'the moral' and 'the ethical', and to use these words interchangeably. I follow this practice here[15*].

Examples might help clarify the difference between the common and the philosophical senses. In everyday language, we might say that 'Eloise is an ethical person' if she, for instance, has empathy for others or has integrity. Similarly, we could say that a particular design is 'ethical' or 'unethical', if it is good or bad for society, respectively. In this sense, 'ethical' and 'unethical' are evaluative terms that respectively express approval and disapproval. In philosophy, however, when we say that a design has an 'ethical dimension' or that a profession is a 'moral project', we do not mean it primarily in the same sense. We mean above all that that particular dimension or project is subject to ethics because we recognise that it is clear that moral issues are involved. We obviously imply, thereby, that an analysis from the vantage point of ethics will help us understand the nature of those issues, and, eventually, make an evaluation of an action or a thing, for instance. So, when we say that cars 'have an ethical dimension', it should not be understood as to imply that cars are good (or bad). That cars have an ethical dimension simply means that they need to be analysed from the perspective of ethics because, for instance, they deeply affect the way we live our lives, and for how long we live them. Similarly, when we talk about the 'ethics of cars', we mean the 'ethical discussion on cars'.

Ethics provides rational ways in the form of ethical theories that enable us to assess human actions and behaviours, and this is how right and wrong (and other notions) enter ethics. We could evaluate Eloise's actions (or the existence of cars) following a particular theory and decide whether her actions were right or wrong. If the evaluation is negative, some philosophers would refer to Eloise as 'unethical', but many would say that her behaviour is 'ethically inadmissible', or 'indefensible' or 'unjustifiable'. Moreover, we could only evaluate Eloise's actions ethically if we consider Eloise to be a *moral agent*. For engineering ethicist Caroline Whitbeck, moral agents are 'those who can and should take account of ethical considerations' [16]. Different *normative ethical theories* may yield different evaluations as each theory has particular conceptions of what is best. So Eloise's action could be ethically acceptable according to one theory and fully indefensible according to a different one.

These different ways of understanding and assessing ethics have to do with the different ethical theories on which they are grounded. In Western philosophy, Kantianism, utilitarianism, and virtue ethics are often presented as the main ethical theories. These three theories share as a starting point that humans think and deliberate about what to do and how to do, and that they often have *reasons* for acting the way they do; in other words, they can explain and possibly justify their actions according to those reasons.

In this volume, when we refer to 'reasons', we will be primarily referring to so-called normative or justifying reasons: 'reasons which, very roughly, favour or justify an action, as judged by a well-informed, impartial observer' [17]. Reasons in this sense are related to but distinct from *motives*, which may explain why a person acted the way they did, and are the *individual* reasons ('motivating' reasons) that a person has for acting in a particular way. When individual motives and normative reasons converge, we say that the person had *good reasons* for acting the way they did, and what we mean here is that we think they their own motives are aligned with normative reasons.

Although the three ethical theories mentioned earlier share this basic premise, they markedly diverge on the object of analysis. The main difference between virtue ethics and the other two is that the former is structured around *character traits and dispositions*, whereas the latter are concerned with *general principles* (such as the moral obligations one has, or the results of one's actions) that serve as a benchmark to evaluate whether an action is ethically acceptable; that is, when an action (or an omission) is backed by normative reasons.

Whereas Kantianism and utilitarianism are structured around objective moral principles that tell us which acts are morally *right*, and which ones are morally *wrong*, virtue ethics is concerned with discerning *what kind of person one should be*. For virtue ethics there is more to ethics than acts: inner traits, motives, dispositions, and character of an individual matter too. Because of this, virtue ethics is said to be person-oriented and not action-oriented. This does not imply that acts are not important—of course they are—but the focus lies in the big picture of how a life is lived. Thus, virtue ethics goes beyond the notions of right and wrong, and although these notions *are* important, for virtue ethicists ethics is, first and foremost, *about the nature of the good life.*

The differences between the theories highlight two understandings of ethics: a *narrow* one, concerned with right and wrong, and a *broad* one, whose focus is on life as a whole and includes the narrow one[18*]. In the narrow sense, the basic ethical questions are: 'What ought I to do?' or 'What ought I to choose?'; these are questions that imply an auxiliary question: 'What principle should be followed in order to decide?' The narrow approach is often presented in terms of specific ethical dilemmas: problems or questions that require taking a side in the dispute and providing objective reasons for being in favour or against the respective sides (the 'quandary ethics' we referred to earlier). We frequently find this approach in the discussion of, for instance, the ethics of abortion, euthanasia, the death penalty, or the rights of the unborn, and more recently in the discussion of, for example, autonomous weapons, autonomous cars, facial recognition, robot workers, genetically modified foods, genetically altered embryos (popularly called 'designer' babies), or surveillance driven by artificial intelligence. Again, the problem with quandary ethics is that it might give the false impression that ethics is only about major, life-or-death decisions, and not about everyday affairs.

In the broad sense of ethics, the questions are not at all primarily centred on isolated issues and dilemmas. The consequences of acts and discussions of right and wrong *are* relevant, but only in a secondary stage. *The good takes precedence over the right*, as famously asserted by novelist and philosopher Iris Murdoch[19]. To briefly clarify, 'something is "right" if it is morally obligatory, whereas it is morally "good" if it is worth having or doing and enhances the life of those who possess it'[20]. So, in the broad understanding of ethics, we do what is right because it enables us to attain what is good. To illustrate with a simple example: telling the truth is generally the right thing to do because it makes us honest and fair, which are 'good' things to be. In a different case, telling the truth *may* not be the right thing to do; imagine somebody cooks a meal for you and you hate it—you may still say that you enjoyed the food to show tact and gratitude for a kind act, and even empathy as you could imagine that the other person would feel awful if you told the truth. Another example: the right thing to do at a pedestrian crossing is to stop, but not merely because it is a legal traffic obligation—it is the right thing because by stopping at a crossing, the good of safety is preserved. And that is precisely why it is a good thing that stopping at a pedestrian crossing or at a red light is enshrined in a law.

In Aristotelian times (500 BCE), the core question of ethics was 'How ought one live?'; modern neo-Aristotelian philosophers pose this key question as 'what sort of person am I to become?'[21] They hasten to add that the question is answered by living it: 'an answer to it is given *in practice* in each human life'[22]. The words of business ethicist Geoff Moore serve to round up this section with a description of the broad approach to ethics that we shall take in this volume: ethics is 'about discerning how we should live and what it means to live a good life'[23].

WHO IS THIS BOOK FOR?

This book is likely to be of interest mainly to those who are directly involved in the practice of design: especially professional designers, advanced students of design (upper-level undergraduate and graduate levels), design educators, and design scholars. These, naturally, are not separate groups; they are fully intertwined, and the same people often move from one role to another: students become professionals, professionals become educators and scholars.

Professional designers who are interested in initiating a deeper reflection on the ethical side of their professional practice will find many discussions on situations and decisions that I trust will resonate with their own experience as professionals. The book is likely to engage junior designers as well as senior designers, and also those occupying more strategic positions such as design directors. The book is for all sorts of professional designers whether they are self-employed, work for design firms, or work in-house within corporations and organisations. Regardless of their level of design expertise and place of work, designers will obtain valuable knowledge and insights on the fundamental ethical questions of the design profession. This knowledge will empower them to discover what it means to be an 'ethical' designer, which will also serve them to assess their current workplace and how they conduct their own design practice.

Undergraduate students from all disciplines of design will benefit from many of the same topics as professional designers, as a preparation for their future professional careers and personal life. To suit the needs of this group (and the former), I have tried to minimise the use of terms from academic philosophy and to provide accessible descriptions when the term was necessary. Though the book is, admittedly, not a 'light read', neither is it daunting; I believe that if the effort is made there is much to be gained from the reflections it offers. The book will make students aware of the fact that they will inexorably encounter ethical challenges in their future work as professional designers, and they will learn about professional ethics in a way that is close to the experience of design a practitioner would have. Moreover, because education is not only about gaining instrumental skills but about learning how and when to apply that knowledge and for what ends, the volume, without being a textbook, still provides students with ground material to develop ethical expertise in conjunction with the experiences they undergo in the university or design school.

The same is valid for *graduate students* who could also use the contents of this book to further their own theories and design practice. It could serve to anchor investigations in the fields of social design, participatory design, sustainable design, service design, interaction design, design-centred innovation, critical design, and more broadly, design ethics, philosophy of design, design theory, and design methodology.

Naturally, this volume will also particularly interest *design educators* who, besides all of the above, can profit from a profound reflection on the purpose of design, which can serve to stimulate and guide a debate about the purpose of design education, and about the curriculum students are learning. For what notion of profession are we educating students? What type of professionals do we envision? What type of capabilities do we as design educators consider worth developing? The book offers teachers a prompt for discussions with students about the purpose of design, and the type of designers and persons they want to become. Educators will also find material in this volume that will enable them to orient and nurture the students' processes of ethical reflection and character development. In the coda, educators will find a specific orientation on teaching ethics with a focus on practical rationality, away from the instrumentalist rationality that plagues education.

This book might be of interest to *design scholars* and *philosophers* working in areas linked to design ethics, ethics of technology, responsible innovation, professional ethics, and, possibly, in the Capability Approach. I have tried to contribute to the ongoing debates by providing an account that is both accessible enough for designers and students, and still, I hope, solid and rigorous enough to be considered a worthy scholarly source.

Taken more broadly, the book could also be of interest to other people who participate as decision makers or stakeholders in design projects such as project managers, or even clients commissioning design projects. I am of the view that policy makers and think-tank analysts working at organisations with a focus on design and innovation could also benefit from reading the book.

Throughout this book I often adopt a 'we' perspective to refer to what 'we' think, say, do, feel, and experience. This is not to suggest that *all* of 'us' *always* think or feel in the *exact* same way, but only to indicate that others ('you') *could* think, say, do, feel, and experience roughly, and at times, in the same way 'I' do. But who are 'we'? There are three different 'we'. The first 'we' refers to human beings in general; I use it when I say things like 'we shop' or 'we meet friends'; that is, things most people generally do. The second 'we' refers to all the readers of the book, especially when I say things like 'we will analyse' or 'we will review'; this is a welcoming 'we' that invites all readers to do some things together. A third type of 'we' refers primarily to designers (I count myself as one). Within each group there is a smaller subset of members who are referred to when I say things like 'we feel', 'we experience', or 'we would think'. I take this liberty to indicate that I believe that what I am recounting might be shared by others as well. All three types of 'we' are inclusive, and you can remove yourself from the group whenever what I am describing does not apply to you or you find yourself disagreeing with claims I make that purport to include you.

PLAN FOR THE BOOK

This book consists of eight chapters structured into three parts.

Part I (chapters 1 through 4) deals with the activity of design and argues for its professional status.

Chapter 1 offers an account of design that seeks to provide a plausible shared understanding of key terms and themes for the purpose of this book. Given the vastness of design as a subject matter, it is necessary to further align views around our main subject of discussion. The chapter provides a discussion of the notions of 'intentionality', 'creativity', and 'appropriateness' as minimal necessary conditions to talk about design. It also reflects on the epistemic insufficiency under which designers operate and provides two complementary perspectives for looking at different levels of design activity: a 'general' view (where everyone has a capacity to design) and a more restricted 'occupational' one that only includes people who we normally call 'designers'. The chapter ends by reflecting on and making a case for the ethical relevance of design by focusing on how design prescribes norms or standards and by doing so influences our behaviour.

Chapter 2 explores concepts and theories in relation to professions as moral projects. Rather than seeking to provide comprehensive definitions that exhaust the matter (which might actually be a futile endeavour as philosophers and sociologists working on the subject only agree that there is no agreed understanding), the chapter aims to provide basic theoretical building blocks and stipulate some criteria from which to approach the analysis of design professional ethics. It offers an introductory section on professional ethics, as well as one on responsibility. The chapter also provides an examination and a rejection of codes of ethics as a primary approach to professional ethics.

Chapter 3 tries to convincingly establish whether design can be indeed called a profession. This is not the first volume to treat design as a profession, but its professional status, I contend, has not yet been plausibly established from an ethical-normative perspective. This chapter is an effort in that direction. To that end, the chapter covers a broad spectrum of design cases that can be seen as exemplary or paradigmatic, and contrasts them with the criteria for professionalism that were stipulated in the previous chapter. The chapter ends with the provisional conclusion that design can indeed be considered a profession.

Chapter 4 challenges that provisional conclusion through a series of objections, which are structured on the one hand around the role design plays in manipulation and consumerism, and, on the other, on the unintended consequences design brings about. A profession-wide inquiry into professional ethics is presented as a viable course of action to overcome the challenges.

Part II (chapters 5 and 6) delineates the inquiry into professional ethics from a virtue ethics perspective.

Chapter 5 opens up the second part by fleshing out the ethical inquiry that was proposed in the previous chapter. A practice-centred programme of inquiry is presented and contextualised against the larger field of design ethics. The chapter also defines key terms for the inquiry as well as some goals and conditions for it.

Chapter 6 introduces virtue ethics as the theoretical philosophical foundation upon which our professional ethics will be grounded and contrasts it with two important alternative ethical theories (Kantianism and utilitarianism). The second half of the chapter engages with the ideas of Alasdair MacIntyre, upon whose analysis of human practices and internal goods we will extensively rely as the conceptual backbone for part III.

Part III (chapters 7 through the coda) fleshes out the account of practice-centred design professional ethics.

Chapter 7 starts by revisiting the design exemplars from chapter 3 (and some new ones) from a MacIntyrean perspective with the objective of providing a *bottom-up*, descriptive analysis of design practice. It develops a scaffold that enables us to search for and formulate a plausible overarching purpose for design, which is inferred and generalised from the analysis of different instantiations of excellent design. In turn, the overarching purpose is linked to the promotion of human 'capabilities'. Lastly, the notion of 'regulative ideals' serves to explain how designers can be guided in their professional activity without the need for codes of ethics and explicit declarative rules or guidelines.

Chapter 8 explores the notion of 'responsibility as a virtue' and what it might mean for a designer to be responsible, developing a view that focuses on care as a central element. It works through the role that empathy plays in both care (to which it is often associated) and responsibility. The chapter proposes a view that, while fully valuing empathy, sees the designer's personal investment in the design profession as a way of self-enactment, and as a sufficient motive for responsible action. It ends by discussing the role of 'practical wisdom' as the virtue that enables the designer to cope and deal with the plural demands that are made on them.

Chapter 9 gathers together and summarises many of the central themes that arise throughout the book. Acknowledging the dynamics of current professional design, which is embedded in for-profit organisations, the conclusion reflects on the many constraining conditions (political, organisational, and economical ones) that a professional designer faces in the context of today's modern capitalism. It also makes modest suggestions on how the readers can navigate through the difficulties they face in their working lives, without missing out on the opportunities for self-growth they encounter.

The coda is mainly directed at design instructors, and it reflects on the need to integrate reflective non-prescriptivist activities into design education with the goal of developing ethical expertise. While eschewing ethical proselytism, this last section touches upon and suggests how this could be achieved.

As I indicated earlier, when a number pointing to an endnote is accompanied by an asterisk, it indicates that substantive content can be found in the endnote besides the bibliographic reference. The symbol is included to lessen the need to having to flip back and forth to the endnotes to check for substantive content.

NOTES

1. Alasdair MacIntyre, *After Virtue: A Study in Moral Theory* (Notre Dame: University of Notre Dame Press, 2007); Geoff Moore, *Virtue at Work: Ethics for Individuals, Managers, and Organizations* (Oxford: Oxford University Press, 2017); Chris Higgins, *The Good Life of Teaching: An Ethics of Professional Practice* (West Sussex: Wiley-Blackwell, 2011).

2. John Heskett, *Design: A Very Short Introduction* (Oxford: Oxford University Press, 2002), 3.

3. Richard Buchanan, 'Human Dignity and Human Rights: Thoughts on the Principles of Human-Centered Design', *Design Issues* 17, no. 3 (2001): 38.

4. J. R. McNeill and Peter Engelke, *The Great Acceleration* (Cambridge: The Belknap Press of Harvard University Press, 2014).

5. For a historical account of the massive change humans have wrought in the world, see J. R. McNeill, *Something New under the Sun: An Environmental History of the Twentieth-Century World* (New York: W. W. Norton & Company, 2000). For a critical history of the Anthropocene, see Christophe Bonneuil and Jean-Baptiste Fressoz, *The Shock of the Anthropocene: The Earth, History and Us* (London: Verso, 2017).

6. Annina Schneller, *Scratching the Surface: 'Appearance' as a Bridging Concept between Design Ontology and Design Aesthetics*, edited by Pieter E. Vermaas and Stéphane Vial, Advancementes in the Philosophy of Design (Dordrecht: Springer, 2018), 35.

7. Bryan Lawson and Kees Dorst, *Design Expertise* (Burlington: Architectural Press, 2009).

8. This view might remind some readers of the canonical definition of design proposed by John Chris Jones: 'designing is to initiate change in man-made things.' In John Chris Jones, *Design Methods*, second edition (Chichester: John Wiley and Sons, 1992), 4. The resemblance is obviously not fortuitous.

9. Richard Buchanan, 'Design Research and the New Learning', *Design Issues* 17, no. 4 (2001): 9.

10. Richard Buchanan, 'Wicked Problems in Design Thinking', *Design Issues* 8, no. 2 (1992): 16.

11. Nigel Cross, *Designerly Ways of Knowing* (Basel: Birkhäuser, 2007), 102–03.

12. For example, ibid.; Nigel Cross, *Engineering Design Methods: Strategies for Product Design*, fourth edition (Chichester: Wiley, 2008); Bryan Lawson, *How Designers Think: The Design Process Demystified*, fourth edition (Oxford: Architectural Press, 2006); Jones, *Design Methods*.

13. The start of the attack can be dated to the publication of the foundational paper on 'wicked problems': Herb Rittel and Melvin M. Webber, 'Dilemmas in a General Theory of Planning', *Policy Sciences* 4, no. 2 (1973).

14. Ezio Manzini, *Design, When Everybody Designs: An Introduction to Design for Social Innovation* (Cambridge: The MIT Press, 2015), 35–37.

15. The standard reading is to take morality as the object of ethics. There are, however, different views on the subject. Because I do not develop on the distinction between ethics and morality in this volume, I will not explore the topic in further detail and refer the interest reader to Gert Bernard and Gert Joshua, 'The Definition of Morality', Stanford University, https://plato.stanford.edu/archives/fall2017/entries/morality-definition/.

16. Caroline Whitbeck, *Ethics in Engineering Practice and Research*, second edition (Cambridge: Cambridge University Press, 2011), 10.

17. Maria Alvarez, 'Reasons for Action: Justification, Motivation, Explanation', Stanford University, https://plato.stanford.edu/archives/win2017/entries/reasons-just-vs-expl.

18. This distinction, albeit in different terms, has been made earlier by influential authors such as Alasdair MacIntyre, Iris Murdoch, Charles Taylor, and Bernard Williams. Charles Taylor adds an important normative meaning to the distinction. He calls narrow ethics 'morality' and broad ethics simply 'ethics'. For him, contemporary 'morality' is only a fraction of what ancient philosophers defined as 'ethics.' Moreover, Taylor sees contemporary moral philosophy as having 'accredited a cramped and truncated view of morality in a narrow sense'. Charles Taylor, *Sources of the Self: The Making of the Modern Identity* (Cambridge: Harvard University Press, 1989), 3.

19. Murdoch, *The Sovereignty of Good*.

20. Charles Larmore, 'Right and Good', (1998),https://www.rep.routledge.com/articles/thematic/right-and-good/v-1.

21. MacIntyre, *After Virtue: A Study in Moral Theory*, 118.

22. Ibid. Italics in the original.

23. Moore, *Virtue at Work*, 5.

Part I

The Design Profession

Chapter One

Design, Designers, and Normativity

In this chapter, we will focus on the notions of intentionality and creativity in order to complement the general account of design that was presented in the introduction. We will also discuss two perspectives for looking at design activity, which are not at all opposing views, but rather only different ways of looking at different levels of action. To end, we will reflect on the ethical relevance of design by looking at how design influences our behaviour by tacitly prescribing norms or standards.

CONDITIONS AND EPISTEMIC BOUNDARIES FOR DESIGN

I pointed out in the introduction that the view of design adopted in this book is rather lenient. For the purposes of this volume, design must be understood as an activity primarily associated with the conception and the planning of the human-made world to empower and enable humans to pursue their individual and collective purposes. Along these lines, design goes beyond the pursuit of beautiful and functional forms. More concretely, design is about framing and crafting the plan to bring about material or immaterial artefacts that do not yet exist but could be realised in the future. This 'framing and crafting' is inscribed within the exploration and projection of potentialities and the envisioning of possibilities that do not exist yet, thus initiating a change in the current artefactual state of affairs.

Design also goes beyond traditional problem solving. In design, problems are seldom well-defined, coherent, and tractable. Design problems can often be wicked problems, which require a different resolution mindset and are never solved for good. Because design problems are only temporarily resolved, design 'solutions' are inexorably provisional and tentative. What is more, contrarily to well-defined problems, solutions for design problems coevolve along with the understanding and definition of the problem that is being tackled, which depends on how we subjectively frame it.

Much has been written about these issues, but to prevent the discussion in the following chapters from being muddled and unproductive, we still need to add a few more words to our understanding of some more key aspects of the nature of design that will bear on the discussion. To do so, we shall explore two necessary conditions for design, that is, conditions without which one cannot say that something has been designed: intentionality and creativity. Then we will reflect on the relation between the uncertain character of design decisions and the notion of *appropriateness*, which is a particular aspect of the creativity condition. To reiterate, design is a vast subject; it is not in the scope of this book to perform a profound enquiry on design ontology, so these two conditions by no means aim to exhaust the issue, and are highlighted here as they are especially relevant to our discussion.

Intentionality

First, there is the necessary condition of *intentionality*, which, for our purposes, must be understood in its everyday sense of 'being deliberate or purposive' or 'consciously deciding on a course of action'. The deliberate, purposive, and solution-oriented nature of design has been explored extensively in the field of design theory and methodology. Herbert Simon, one of design's most influential thinkers, describes design in an oft-quoted dictum as the activity of devising 'courses of action aimed at changing existing situations into preferred ones'[1]. Purpose and intentionality are embedded in Simon's very words.

The arguments that I will present in the following chapters will also rely on the premise that designers have *agency*, that is, that they are capable of performing intentional actions. Stated broadly, 'an agent is a being with the capacity to act, and "agency" denotes the exercise or manifestation of this capacity'[2]. This, however, raises the question of to what extent do designers' intentions really shape the 'preferred situations' they design.

Science and technology scholars Patrick Feng and Andrew Feenberg explore this issue and discuss three general perspectives for design intentionality.[3] First, there is a 'strong intentionality' perspective in which designers are powerful and have a great deal of control, and can steer the shape of artefacts through their knowledge and values. Designers assess various demands and deliver results that are optimised according to those demands.

The second perspective is that of a 'weak intentionality' in which designers are understood as having different degrees of autonomy in translating their goals into artefacts. They are constrained by a variety of factors—economic, political, institutional, social, and cultural—which leave designers different degrees of room for manoeuvring, resisting, and negotiating them.

These constraints may be very strong due to corporate interests and the dynamics of capitalism or somewhat less strong due to norms and expectations of the corporate culture[4*]. Feng and Feenberg cite the work of engineering scholar Louis Bucciarelli, who offers a more optimistic view, and argue that constraints could be worked out by negotiations between designers and co-workers[5].

The third perspective sees 'design as a function of the broader culture'. Here, the very notion of intentionality is thrown into question by problematising the distinction between designer and society-at-large. In this perspective, design is a function of the broader assumptions that play a role in a given culture. To illustrate this with examples, the way in which things appear to be 'intuitive' in an interface, 'useful' in a domestic device, or 'comfortable' in a chair to the designer is informed by socialisation processes and norms and values that are part of a broader culture. To focus too much on individual designers, or even on teams, is to neglect the larger cultural setting where design occurs.

The strong intentionality perspective is implausible as a general account because it ignores the social and institutional contingency of design. Feng and Feenberg concede that designers 'do have a substantial influence on the design process and sometimes control the outcome'[6]. That 'sometimes' in the previous sentence is crucial, as sociologists, historians of technology, and anthropologists have shown that the design, but also the adoption and use, of products is a dynamic social process that globally and generally exceeds the control of the designer[7]. It would be a mistake to base a book on professional ethics in a view of a designer-god that designs artefacts that will always behave as intended[8*].

The other two perspectives seem much more plausible as designers are not isolated individuals who design in a vacuum with absolute control. Both perspectives will be useful for our analyses.

It is worth talking about intentionality because it matters for ethical questions that are especially relevant to design—for instance, discussions on responsibility or on the moral status of side effects and unintended consequences. The issue of the actual degree of control a designer has is of great importance for professional ethics, especially in the context of modern capitalism in which most professional designers operate. This theme will be taken up again in the next chapters. However, the arguments I will make are agnostic in respect to the perspectives one adopts on design intentionality apart from the strong intentionality perspective, which is implausible, as I just pointed out.

Creativity

Another important condition that design must satisfy is *creativity*. To be able to say that something is designed, creativity is a necessary condition that has to be met. This is not particularly surprising nor controversial. After all, design is inherently bonded with creativity by virtue of design being the creation of a plan for a *new* sort of thing[9], which requires creativity. Psychologist Teresa Amabile defines creativity as the 'production of a novel and appropriate response, product, or solution to an open-ended task'[10]. Two terms included in this definition require further explanation; 'novelty' can, in turn, be understood as 'unusual, statistically infrequent, or completely unique', whereas 'appropriateness' refers to being correct 'in the context of the problem or audience to which it was addressed'[11]. It is the presence of the general qualities of novelty and appropriateness that differentiates creative from uncreative products[12].

Some might worry that this definition of creativity, given the importance attributed to appropriateness and novelty, might be too restrictive and will make too few things count as design. This is a valid concern, but this objection could be worked out without giving up these requirements. To do so we should adopt a flexible perspective regarding what counts as appropriate. Appropriateness could be understood as a gradient continuum (from absolutely inappropriate to absolutely appropriate) and not in a binary, dichotomic sense (appropriate versus inappropriate). Considering appropriateness as a continuum allows for different degrees in which a design can meet this criterion, thus avoiding a view of design that is too restrictive.

A similar argument can be made for novelty. It could also be seen in different shades of grey, as it were: from new *to* old and not only as new *versus* old. Also, the novelty criterion need not apply to *all* features of an artefact, which does not have to be a radical innovation. So, an incremental modification of a single aspect, provided it has a key role in the design plan, would satisfy the requirement of novelty. Introducing a new element to an existing product, substituting a material, or, even, recombining existing elements would yield a new (re)configuration of the previously existing design. Along these lines, it could be argued that the more novel aspects a design includes, provided they are also appropriate, the stronger the claim that that design must be regarded as creative and innovative.

In this way, we could relax the understanding of the constraints imposed by the condition of creativity in order to allow for different degrees of novelty and appropriateness. The concerns seem to be sufficiently dispelled; there is no need to worry about too few things counting as design because of creativity.

Another clarification is in order here: design does not refer to the mere creation of new instances of a type of thing that already exists. When designing a new car, designers produce a precise, concrete, and detailed representation of a new type of car that can be later used to produce actual realisations of the design representation. These actual realisations are cars that can actually be ridden in and become part of our lifeworld.

Creativity is relevant for our larger discussion in the sense that, because of the creative nature of the design solution, the designer, in each project, is likely to be confronted with—at least for them—unknown ethical challenges that require situated and contextual ethical reflection. They cannot simply follow a general procedure and apply pre-existing knowledge, for instance in the form of a checklist. They need to discover and articulate their answers to the ethical challenges afresh. We will revisit this issue in chapter 8 when addressing the subject of moral imagination.

Appropriateness and Epistemic Insufficiency

We said previously that there are different degrees to which a design can meet the criterion of appropriateness. So a design can range from being absolutely inappropriate to absolutely appropriate. Presumably, any designer would aim at least at a minimal degree of appropriateness. In trying to meet this threshold, the designer finds ways to evaluate the appropriateness of a design based on the reasonability and plausibility of a solution in relation to the project's high-level objectives, given the available empirical and theoretic evidence. Designers spend a considerable amount of time evaluating design proposals using both objective criteria, as well as personal standards and values.

But can a designer really *know* if theirs is an 'appropriate response' at all? The inferences carried out during the design project are often the result of abductive reasoning, which is a type of reasoning that is different from the 'classical' methods of logical reasoning: induction and deduction. Abduction is the process of facing an unexpected fact, applying an explicit or tacit rule—already known or created ad hoc—and, as a result, proposing or hypothesising a case that *may be*. This hypothesising is a key feature of design reasoning because 'unlike deduction or induction, abduction allows for the creation of new knowledge and insight' [13]. The basic form of this reasoning when applied to design takes the following form: 'If the design is like *this*, then *that* might happen'. For instance, 'if I make this button red, people will know where to tap to go to the next screen' or, a much broader example, 'if we redesign the city closing or limiting certain streets and areas to cars, other forms of transportation will gain more importance and citizens' wellbeing will increase'.

Of course, after implementation the outcomes of abductive reasoning might turn out differently than expected, even if the reasoning was based on valid prior knowledge. Because of the intrinsic nature of design problems, designers operate in what economist and ethicist George DeMartino calls 'epistemic insufficiency'. Under epistemic insufficiency, an agent 'largely cannot be certain of the effects of the interventions they advocate'[14]. Buchanan made a similar point in reference to wicked problems: 'The problem for designers is to conceive and plan what does not yet exist, and this occurs in the context of the indeterminacy of wicked problems, before the final result is known'[15]. The upshot is that design solutions get implemented in the future so the appropriateness of a solution is initially only notional, as the actual, achieved appropriateness can become clear only *after* the implementation of a design. A design can be thought to be appropriate for a problem or user group during the design process, but result in being absolutely inappropriate after implementation.

Examples abound, but the One Laptop per Child (OLPC) project is a paradigmatic example. The OLPC project set out to change education through the design and introduction of cheap laptops (the project was popularly known as 'the 100-dollar laptop') to children in the developing world. Despite some victories such as the distribution of four hundred thousand laptops to most schoolchildren in the nation of Uruguay, the project failed to transform education for schoolchildren in the rest of the developing countries. Some authors attribute the lack of impact to the failure to anticipate and adequately understand the social, cultural, and institutional problems that could arise in the environment in which the OLPC was to be introduced[16]. But being aware of other cultural settings would not have guaranteed per se that the solution would have been a success. Other critics dismiss the whole thing as a 'flashy, clever, and idealistic project that shattered at its first brush with reality'[17]. One could argue that the designers, coming from developed countries, should have been sensitive to other cultural settings, but the point here is a different one: the designers (and the project leaders and sponsors) most probably thought *at that time* that the solution was appropriate and could work in the target setting. But it did not, as we now know[18*].

Let's consider now two different examples to illustrate two ways in which designs can be inappropriate. First, a design for a concept plane made of the new material graphene[19*] would be currently technically unfeasible and economically unviable, but, provided it could theoretically fly, the design would not be inappropriate *in principle*. Its inadequacies are only practical. Conversely, a design for a common screwdriver with the objective of installing and removing screws would immediately be seen as inappropriate if it was designed to be made of regular glass, as physics tells us that the glass would easily break when used. These examples illustrate that designs may be inappropriate for *practical* reasons (such as the graphene plane, which lacks economic viability or technical feasibility) or for *theoretical* reasons (such as the glass screwdriver, which would break simply because of the laws of physics)[20].

It must be clear by now that designers do not generate fully fledged solutions in the way '4' is the solution for the sum '2 + 2'. Naturally, a design solution is different from a mathematical solution in a myriad of ways, but for the purpose of this discussion, the most important difference is the contextual and relational nature of design solutions. This can be illustrated with an example: a fuel-efficiency dashboard display might successfully contribute to reducing fuel consumption and be thusly seen as an adequate solution. But it might at the same time contribute to potentially dangerous driving habits as it might inadvertently encourage drivers wanting to maximise fuel efficiency to drive faster or roll through stop signs as a way to maximise efficiency[21].

The process of design involves the generation of successive design proposals that need to be evaluated in a dialectic relation that the designer establishes. Design scholar Nigel Cross points out that 'the designer, in constructing a design proposal, constructs a particular kind of argument, in which a final conclusion is developed and evaluated as it develops against both known goals and previously unsuspected implications'[22]. The complexity of design is not only that the designer finds new unexpected implications; the main point here is that due to the intrinsic and substantial epistemic insufficiency of design, they may err when estimating appropriateness, as we have seen from the OLPC and the case of the fuel efficiency dashboard display. Evidently, not all cases of errors in estimation will have ethical relevance, and sometimes the ethicality of a design decision will have nothing to do with mistakenly judging appropriateness, but with conscious, purposive actions.

Of course, *some* designs might be known in advance to be inappropriate (for instance, a graphic designer can know that most people over forty have difficulties reading tiny letters). A designer could use established scientific knowledge and other forms of knowledge to navigate uncertainty. Granted, the epistemic insufficiency is not total as many issues could be overcome. But this will get designers only so far. Many difficulties and uncertainties will persist; because of that, generating, evaluating, and choosing among alternative design proposals is a task fraught with ethical complexities from the start—for instance, when defining what actually counts as an objective, when translating high-level objectives and constraints into concrete success indicators, or when making trade-off decisions when choosing between different intermediate design proposals. The backdrop of substantial epistemic insufficiency will not go away completely.

TWO VIEWS OF DESIGN ACTIVITY

It is now time to explore a different aspect of design activity: the people who undertake it. To put it another way, *who designs*? What is the 'we' that Buchanan refers to when he says that design is the way 'we plan and create'?

An answer comes from design theorist Victor Papanek, who argues that design is something humans do all the time with 'the conscious and intuitive effort to impose meaningful order'[23]. It is worth quoting his well-known passage about the nature of design:

> all men[24]* are designers . . . for design is basic to all human activity. The planning and patterning of any act toward a desired, foreseeable end constitutes the design process . . . design is the primary underlying matrix of life[25].

So, according to this view, executing a mural is design, but design is also baking an apple pie or cleaning and reorganizing a desk drawer[26]. Design scholars such as Donald Norman posit, in a similar fashion, that 'we are all designers' and that everyone designs when rearranging objects on their desks or the furniture in their living rooms[27]. Design philosopher Glenn Parsons points out that design 'is, apparently, a part of nearly every sort of activity in which we engage'[28]. Similarly, designer and scholar Ezio Manzini argues that we live in 'a world in which everybody constantly has to design and redesign their existence'[29].

The answer to the question with which we started this section is straight-forward: *everyone designs*; we *all* design at one time or another. But if everything we do is design and everyone designs, does it still make sense at all to talk about 'designers'? If we accept that everyone designs, is then everyone a designer? Or is the word 'designer' just a synonym for human? If this is the case, it seems that some meaning that we give to the word 'designer' is lost.

How can we capture the particular meaning we attach to the word 'designer' when we use it to refer to famous designers like Paula Scher, Dieter Rams, Philippe Stark, or Marianne Brandt? Our intuition is that there is a particular group of people that we call 'designers', and that not all humans belong to this group. Design luminaries are members of this group, but so are all the anonymous individuals who work at design firms around the world or are self-employed or work at design departments of different organisations such as banks, newspapers, government agencies, or non-governmental organisations.

This group can be set apart from the rest because of what they know and do, and how they do it. It seems legitimate to want to reserve the label 'designer' for people belonging to this group. The upshot of all this is that the particularities that exist between the two groups merit making a distinction between two possible perspectives about design: the general view and the occupational view. We will explore these two perspectives next and substantiate the distinction.

The General View of Design

The first view refers to design as a generic term for the planning, shaping, and making of all material culture; here design is viewed as a universal type of cognitive activity: everyone designs, at least from time to time. Let's call this first conception the *general* view of design. This view is connected to viewing design as a general human *capacity*, in the way Papanek, Manzini, or Norman explicated previously. Heskett points out that design 'can be defined as the human capacity to shape and make our environment in ways without precedent in nature'[30].

This might be a good moment to ponder on a crucial question: why do humans design? The philosopher José Ortega y Gasset (1883–1955) offers a compelling explanation of the role of technology in human life that can be extrapolated to design: we design and make things (heating systems, agriculture machinery, or cars) not to directly satisfy vital needs (keeping warm, eating, moving), but rather to free ourselves from ceaselessly having to face those needs. Artefacts bring about 'a suspension of the primary set of actions with which we meet needs directly'[31]. In other words, humans free themselves from urgent concerns by designing and developing artefacts that satisfy those necessities in advance, as it were. A human, unlike other animals, has 'the possibility of disengaging oneself temporarily from the vital urgencies and remaining free for activities which in themselves are not satisfaction of needs'[32]. The goal that is sought is to be able to focus not on *being*, but rather on *wellbeing*. Ortega writes:

> man's desire to live, to be in the world, is inseparable from his desire to live well. Nay more, he conceives of life not as simply being, but as well-being; and he regards the objective conditions of being as necessary only because being is the necessary condition of well-being[33].

A desire to live well is what drives all humans to design; we design to avoid having to deal with immediate needs such as getting warm and satisfying hunger that constantly assault non-human animals; we free up time to occupy ourselves with the task of being ourselves[34]. By conceiving and making artefacts, humans free up time and energy to work on the task of being themselves, 'in the odd pursuit of realizing [their] being in the world'[35].

Ortega's profound insights can be taken to indicate that we design as a road to self-discovery and to become fully human; that is, to transcend our animal nature and not to have to worry about thinking where to sleep, what to eat, how to avoid being cold, and so forth. But there is more than avoiding these worries: one does not merely want to avoid their 'grim, dismal, negative character'[36]. One wants to avoid them to be able to live 'that kind of life which he regards as most human'[37]. Following Ortega, we could argue that it is through design that human societies and individuals attempt to realise their different life plans: 'man endeavors to realize the extranatural program that is himself'[38]. Thus, the artefacts we design are also constitutive of our life and of the type of person we want to be; ethicist Jeroen van den Hoven tellingly illustrates this: 'it is impossible to be a Samurai without a special sword'.

The Occupational View of Design

The second perspective refers to design as a 'social or institutional practice, or profession'[39]. We can call this the *occupational* view of design. Needless to say, this narrower perspective is a subset of the general view.

This more exclusive notion of design involves a particular set of methodologies and approaches that arose in the mid-nineteenth century as a consequence of the breakdown of traditional craft systems, which coincided with the first serial production processes, like printing and textile manufacturing, and the introduction of methods of mass production of goods during the Industrial Revolution. Whereas before industrialisation, a single craftsperson (perhaps with the help of an apprentice) built a product based on pre-existing artefacts, in the new era of mass production a system of division and specialisation of labour was put in place, and the design and manufacture of a product were no longer carried out by one and the same person. The task of manufacturing the product was divided into subtasks, each assigned to different workers that took care of their part; the crucial division that occurred was *the separation of conceiving from making*. One of these new roles for workers was the role of *designer*, who took care of the planning and sketching (that is, designing) of not just one single product but of a generic specification that all the other workers were able to use to build a multitude of equal-looking products from the original designs. It is, thus, only since the age of the Industrial Revolution that we can speak of (industrial) design in the contemporary sense.

The current generic design process, at minimum, goes from defining high-level objectives and constraints based on a design brief, generating and evaluating design proposals, and producing some sort of specification for functional and formal implementation[40*]. The process concludes when a particular design proposal that effectively passes a threshold of likelihood of success according to the high-level objectives of the project is communicated to the commissioning client and/or producer of the artefact.

Since its origins, design as an occupation has undergone remarkable changes. Over time, specialisation in knowledge and fields of application have advanced to a point where one can rightly speak of hyper-specialisation, whereby a designer is responsible for designing one specific part of an arte-fact and no longer for the whole artefact. Arguably, this is the case for most designers working for or at a large company.

Ezio Manzini argues that design has also changed due to the need to widen its 'field of application', as design is covering more ground than the design of textiles or furniture to include other fields such as the design of services and organisations. It also transformed due to the inclusion of 'new actors', as design used to be designer-led, and since the 1970s it is involving experts from other disciplines and end users to inform and participate within the design process. Finally, design changed its 'relationships with time' and moved from being concerned with closed-ended projects to open-ended, on-going processes[41].

Because design is now concerned with larger challenges (such as the design of cities and businesses), the simple, linear process that we described previously is more complex now than in the past. The activities of defining high-level objectives and constraints (and even the very design brief on which a project is based) have become a crucial open-ended task for any design project that aims to go beyond traditional problem solving. These activities are often referred to as the 'fuzzy front end', a term that emphasises its 'messy and chaotic nature'[42]. The results of working in the fuzzy front end are discovered and framed through highly context-dependent explora-tions that serve to define the fundamental nature of the design challenge that is being addressed and to identify actionable opportunities for design.

It is evident, considering present material culture, that the occupational view of design covers a very broad spectrum. Already in 1952, the famous Italian architect Ernesto Rogers affirmed that architects (that is, architectural designers) design 'from the spoon to the town'[43]. Rogers' contention was that, on an average day, a typical designer may work on the design of a broad range of goods: a spoon, a chair, a lamp, and a skyscraper. Due to specialisa-tion (except perhaps in the case of design celebrities) this is no longer the case. But the variety Rogers alluded to is reflected in the many disciplines and subdisciplines present in design: from fashion design to product design, from service design to app design, and so forth. The list of design specialisa-tions seems endless. In connection with this variety, it is worth emphasising that in this book we are not focused on any design discipline in particular, but on the common overarching activity of *designing*. It must also be acknowl-edged that every discipline has a set of particular specific skills, tools, meth-ods, and, of course, their own cultural traditions anchored in their specific fields. These particularities, however, do not contradict the insights of schol-ars such as Bruce Archer or Nigel Cross regarding 'the deep, underlying patterns of how designers think and act'[44], which make it possible to argue that design, regardless of the discipline, is a distinctive, coherent *designerly* activity different from science or art.

Design as an occupational activity can be seen to be what is known as a 'community of practice'; that is, 'typically a group of professionally qualified people in the same discipline. All of whom negotiate with, and participate in, a mutually understood discourse'[45]. Design scholar Klaus Krippendorff suggests that designers, 'by their very professional commitments to each other, cannot escape and derive their identity from being part of a design discourse community'[46]. It seems reasonable and justified, then, to reserve the role of designer for somebody who is engaged in the second, more specific conception of design; that is, in design as an occupation.

Even though everyone designs from time to time, it seems plausible to claim that not everyone is a designer: being a designer means belonging to a community one enters into by being accultured to and by sharing common goals related to the practice itself, by recognising—and perhaps challenging—the tradition of the community, by having specific and generic skills, by interacting with peers to share knowledge and experiences; in short, by sharing a practice. There is another point that can be made: self-recognition; Krippendorff posits that 'not everyone who acts to make the world a better place calls him or herself a designer'[47]. It seems clear that when people design in the general way, they do not engage in these beings and doings the way *practitioners* do and do not think of themselves as designers. This must sufficiently and plausibly support the claim that not everyone is a designer. However, there is much more that needs to be said about practices; they are a key element of the main argument about professional ethics I will try to put forward. Because of that we will revisit the topic in more detail in chapter 6.

Because this is a book on design professional ethics, one might wonder if this second perspective should not be better called the *professional* perspective instead of the *occupational* perspective. This is an apt concern, but it is too soon to call design a profession. At this point, calling it professional will not add much substance to the distinction we made between the general and the occupational perspectives. A profession is a specific type of occupation, and for design to count as one, it must meet several conditions that we will examine in chapter 2. The reader will not be much surprised to see that it will be argued that design is indeed a profession. We will call it that in due time, but first we need to explore the ethical relevance of design.

THE NORMATIVE DIMENSION OF DESIGN OUTCOMES

Previously, we maintained that the literal aim of a design process is producing a specification for a functional and formal implementation. In this section, we turn to the outcomes of design: what comes after a design has been created and a specification delivered to a producer or manufacturer. My goal is to argue that designed artefacts have ethical relevance. I will do so by looking at their normative dimension of design outcomes, which will, in turn, illuminate the ethical relevance of design activity. The 'ethical' in this sense does not primarily refer to what is 'right' or what is 'wrong', but to the relevance of those concepts. Nor am I concerned in this section with determining if a particular design is ethical in the sense of being or not being, for instance, 'harmful'. My interest in this section is modest: I only aim to convincingly show that design can be a legitimate subject for ethical reflection. To do so, we will explore the notion of normativity.

Something is normative when its 'basic uses involve prescribing norms or standards, explicitly or implicitly'[48]. A norm is a 'socially embedded directive concerning what people should (or should not) say or do'[49]. For example, saying 'I'm sorry' after bumping into someone by accident is a norm. We conform to norms, but not always and not to all norms. Whether a norm applies depends on the context: if a collapsing escalator causes an avalanche of people, it would be awkward to say 'I'm sorry' instead of asking 'Are you OK?' or calling for an ambulance if necessary. So, norms 'pertain to those actions and assertions which are considered desirable (or undesirable)'[50].

The Normative Power of Artefacts

Artefacts can have normative power. An obvious example is a red traffic light: it tells us to stop; stopping is the desirable (and legally mandatory) thing to do. A washing machine could be a less obvious example; it tells us what do or not do: not to mix different fabrics or to set it to wash very dirty clothes at a specific temperature and for a specific duration of time in order to achieve a desirable level of cleanliness. What counts as a desirable level of cleanliness is notably also a result of the interactions of technology and societal norms and values related to gender or class, for instance. Historian of technology Ruth Schwartz Cowan argues that the introduction of running water and household appliances such as vacuum cleaners and washing machines contributed to create new chores and new standards of cleanliness. Also, these new time-saving devices meant that even more was expected of women than before, as higher standards of homemaking arose: 'she needed to select, manage, and, if necessary, repair, the equipment that replaced her labor'[51].

The norms and values of designers—and many others, we must quickly add—involved in the design process are embedded in artefacts. Although this may not always occur willingly, as designers may be unaware of the norms they are embedding, designed artefacts and environments can be seen as carriers of norms and values; these, in turn, often get enacted in the use of said artefacts, against the backdrop of broader societal norms and values.

Sometimes the norms and values that are embedded cause the rejection of the use of the artefacts in which they are embedded. An example of this could be the old Jewish custom of covering mirrors in a house during *shiva*, the week-long mourning period that mourners from across the religious spectrum hold for first-degree relatives' deaths. Mirrors are meant to give us a reflection of ourselves, but shiva is about paying respect to the deceased, so a mirror distracts mourners. For similar reasons, mourners also avoid shaving or using cosmetics—which are paradigmatic instances of the embedding of social norms and values in artefacts and activities—as a focus on oneself prevents them from concentrating on the deceased.

To summarise, the previous examples illustrate how artefacts are a mate-rialisation of ideas about the right and the good that are presented to the user in a normative way.

Embedding Values and Guiding Behaviour

Let's now turn to *choice architecture* to explore more in detail how products of design can have a normative dimension. Choice architecture is the design of the flow of choices that are presented to users or consumers in products or services, as well as the features, noticed and unnoticed, that can influence their decisions[52].

A prime and very subtle example could be found in the men's toilets at Schiphol Airport in Amsterdam, where a picture of a black housefly was etched into the urinals, enticing people to aim at it while urinating. This type of encouragement is known in behavioural economics as a 'nudge'. Nudges are 'any aspect of the choice architecture that alters people's behaviour in a predictable way without forbidding any options or significantly changing their economic incentives'[53]. According to some reports, the fly on the urinal resulted in reducing spillage by 80 percent[54].

Designing this visual encouragement presupposes an idea of what is de-sirable, that is, what is 'right' or 'good': the less spillage the better. The fly in the urinal has thus an important normative dimension that encourages airport visitors to urinate in a particular way. Although most people presumably share the view regarding reducing spillage, many people seem not to pay much attention to where they aim as they urinate, leaving behind a mess. The influence this design exerts is not at the same level of a sign saying 'DO NOT URINATE OUTSIDE THE URINAL' or 'REDUCE SPILLAGE'. The design steers behaviour, but differently from a sign, in this case helping—nudging—people to do what they already consider the desirable thing to do.

Researcher Nynke Tromp and her associates offer a typology of four influence strategies in which design can sway people and steer behaviour[55]. These four types can exert different degrees of force (from weak to strong) and have different levels of salience (from implicit to explicit). The first type is 'coercive design', which explicitly and strongly influences behaviour—for example, placing visible cameras to discourage shoplifting, or installing full-height turnstiles in entrances to stadiums, subways, or other facilities to prevent people from sneaking under or jumping over a traditional turnstile. Second, 'persuasive design', which is still explicit but weak—for example, billboard or television campaigns to reduce drink-driving. Third, 'seductive design', which is implicit and weak, meaning that the users often do not realise that they are being influenced. This is what happens with Youtube's (or Netflix's) 'autoplay' feature. Users end up in a 'rabbit hole' watching many more videos or series episodes than they wanted to watch initially. Eventually, they glance at the clock and ask themselves 'Can that be right? Has it really been that long?'[56]* We find another example of this at 'Big Bin Little Bin' bin collection scheme in Bournemouth (United Kingdom)[57], whereby the big bin is for recycling and the little bin is for general rubbish; the difference in size is supposed to encourage recycling. The fourth type is 'decisive design': here the designer implicitly but strongly regulates the user behaviour. An example would be a park bench fitted with vertical seat dividers that make it impossible for a person to lie down on the bench.

It is interesting to note that the typology is not an exact classification scheme; design can discourage or encourage particular behaviours by triggering different psychological processes that depend on how the user experiences the interaction with the product. For illustrative purposes, we placed the 'Big Bin Little Bin' in the seductive type as the message it sends can be read as a tacit reminder of the need of recycling out of care for the environment. But the example could also have been placed in the decisive type, as the design could be read as a message saying that one ought to recycle more than one puts in the waste bin.

In all the cases illustrated here, normative views were intentionally imparted to the designs: the products are designed with the clear intention of influencing behaviour in a particular direction (for instance, toward recycling or reducing spillage), with a specific conception of the good. However, although most of us would agree that recycling or reducing spillage in public toilets is a good thing, not all of us would agree that it is good—or even acceptable—to fit armrests onto public benches so that it becomes impossible to lie there, which particularly affects homeless persons who have to sleep rough. When designers intentionally intervene through their designs by discouraging sleeping rough, they also embed in their designs the conception of what is or is not 'desirable' behaviour regarding sleeping rough. But the normative view that these benches carry is far from being widely defended and evidently raises substantial ethical concerns. Philosopher of technology Robert Rosenberger makes a powerful argument on how unjust this so-called hostile architecture is. He does so by showing how anti-homeless laws and everyday public-space designs, such as benches, litter bins, or ledges, aim at pushing the homeless people out of public space and into further danger and disadvantage[58].

Decisive design is also used in the design of search engines and their result pages. Search engines, by design, prioritise some content while throttling or relegating other. After a web search, the user sees ranked results split over many pages of what the search engine's algorithms deem to be relevant content. Research shows that clickthrough rates on search engine result pages follow a power law distribution: the first result of the page significantly attracts much more traffic than the second result, while the second attracts significantly higher traffic than the rest of the results[59]. This means that the top two results—only the top one in mobile search—have tremendous advantages over the rest. Here, too, we see a clear example of normativity embedded in interface design. The ordinal presentation of the results greatly determines how a user goes through the results and what specific content they access when searching for information on current political affairs, health, job offers, or suggestions regarding entertainment or shopping. It is because of this user behaviour that a whole new industry appeared on the internet called 'search engine optimization' that aims at increasing the traffic to a web service through the optimisation of the positioning of said web service in the results page of a search engine. A similar phenomenon occurs in social networks such as Facebook or Twitter, where users are shown content served by algorithms that assume what a user would want to see. To make these assumptions the algorithms compute many factors such as previous browsing behaviour, the time a user spends on a given piece of content, likes given, comments, location, etc. These algorithms—which are instantiations of decisive design—create what activist and writer Eli Parisier has termed a 'filter bubble': 'a unique universe of information for each of us . . . which fundamentally alters the way we encounter ideas and information'[60].

Yet another example of decisive design can be found in the choice for the default settings in products. Default settings force people to use a product in specific ways until the settings are changed (if they can be changed at all). Even a thing as simple as defining the default search engine in a web browser or a smartphone can have important ethical implications. In their anti-trust case against Google, the European Commission found evidence that users 'who find search and browser apps pre-installed on their devices are likely to stick to these apps'[61]*. This would be not surprising to interaction designers, who know that most users do not change the defaults settings[62]. Users act this way due to an uncontested phenomenon known as 'status-quo' bias, which indicates that most individuals have a strong preference for the current state of affairs. So, with this in mind, we can see that choosing a default search engine in the design of a product has normative power and thus ethical relevance. Of course, the very presence of a search engine in a product is normative itself.

How Things Ought to Be

The normative dimension of design we have been exemplifying here is suc-cinctly put forward by Herbert Simon: '[design] is concerned with how things ought to be'[63]. Being concerned with how things ought to be presup-poses having at least a tacit and approximate idea of what is desirable or undesirable. Defining how things ought to be is thus a value-laden activity. Along this line, philosopher of technology Carl Mitcham points out that 'different designs embody (implicitly or explicitly) distinct socio-political assumptions and visions of life, designing itself constitutes a new way of leading, or a leading into, different technological lifeworlds'[64].

We know from Feng and Feenberg's analysis previously discussed that designers do not have absolute control over the design outcomes. Believing so would be incurring in what philosopher of technology Don Ihde has termed the 'designer fallacy': 'the notion that a designer can design into a technology, its purposes and uses'[65]. Moreover, the consumption, adoption, and use of products is known to be not a passive act but a dynamic process through which people engage with products in ways other than those in-tended—and foreseen—by designers. Design outcomes always exceed the intent of the designer, as they display many more possibilities than those originally intended. Ihde also coined the notion of 'multistability'[66] to refer to the different trajectories of use any design product can have: one can use a hammer to hit a nail (which is presumably its intended use), but a hammer can also become a piece of art when used by an artist as expressive material. Evidently, it could also be used as a murder weapon or to crack nuts. These trajectories are often neither fully determined by the designer or others in-volved, nor by the properties of the product itself, but by multiple factors and practices embedded in specific cultural and political contexts.

To sum up, designers, among others involved in the development of material culture, have great influence on how people will live as they get to embed theirs or someone else's normative views in the things they design. Even if the normative views are someone else's, it is designers who have the special skills to embed those normative views in their designs. The ethical relevance of this should be clear by now. Multistability does not mean that designers *never* determine *at least one way* in which a product will be used or experienced; it only means that there is a complex interplay between design-ers, users, and other parties involved in specific contexts. Sometimes, the designer gets it their way and the intended use or experience is the primary use or experience. Other times it is one among many other uses, and on other occasions it fails to achieve its intended use at all. Also, even if the actual use of a product or a service is not necessarily fully determined in advance by the designer, and accepting also that their capacity for control can be con-strained, designers, simply by planning 'how things ought to be', aspire to prefigure the use or experience their designs will bring about. This necessary task of prefiguration is fraught with ethical tensions between the possible different design trajectories that may be brought into existence.

To summarise, in this chapter we have discussed two necessary conditions for design (intentionality and creativity), and the role that epistemic insufficiency plays in evaluating designs. We have also considered two perspectives for thinking about the meaning of design: first a general view that involves all people, and second an occupational view that specifically relates to designers. We ended by reflecting on the ethical relevance of design by considering a key dimension of design, that is, its normative dimension. Normativity affects both the designed artefacts (design as a thing) themselves which carry normative power as well as to the process of conceiving these artefacts (design as an activity) which aims at defining how things ought to be.

This last section provides an unfinished justification for the inquiry conducted in this book; it is unfinished in so far as it only justifies the 'ethics' part in 'design professional ethics'. The other part that needs justification is the usage of the term 'profession' in connection to design. But this will have to wait until chapter 3. First, in chapter 2, we will explore concepts and theories in relation to professions and the idea of professionalism as a moral project.

NOTES

1. Herbert Simon, *The Sciences of the Artificial*, third edition (Cambridge: The MIT Press, 1996), 111.
2. Markus Schlosser, 'Agency', Stanford University, https://plato.stanford.edu/archives/fall2015/entries/agency/.
3. Patrick Feng and Andrew Feenberg, 'Thinking About Design: Critical Theory of Technology and the Design Process', in *Philosophy and Design: From Engineering to Architecture*, edited by Pieter E. Vermaas, Peter Kroes, Andrew Light, and Steven A Moore (Dordrecht: Springer, 2008), 106–10.
4. For the strong view on constraints, Feng and Feenberg cite David Noble, *America by Design: Science, Technology, and the Rise of Corporate Capitalism* (New York: Alfred A. Knopf, 1977). For the less-strong view on constraints, see Gideon Kunda, *Engineering Culture: Control and Commitment in a High Tech Culture* (Philadelphia: Temple University Press, 1993).
5. Louis L. Bucciarelli, *Designing Engineers* (Cambridge: The MIT Press, 1994).
6. Feng and Feenberg, 'Thinking About Design', 117.
7. Daniel Miller, *Material Culture and Mass Consumption* (Oxford: Basil Blackwell, 1987); Bruno Latour, *Pandora's Hope* (Cambridge: Harvard University Press, 1999); Wiebe E. Bijker, Thomas Parke Hughes, Trevor Pinch, and Deborah G. Douglas, eds., *The Social Construction of Technological Systems: New Directions in the Sociology and History of Technology*, anniversary edition (Cambridge: The MIT Press, 2012).
8. Mike Monteiro's book on design ethics presents such a view. See my critical review: Ariel Guersenzvaig, 'Book Review', *Journal of Design Research* 17, no. 1 (2019).
9. Glenn Parsons, *The Philosophy of Design* (Cambridge: Polity, 2016), 11.

10. Teresa M. Amabile, 'Componential Theory of Creativity', in *Encyclopedia of Management Theory*, edited by Eric H. Kessler (Los Angeles: Sage, 2013), 134.

11. Teresa M. Amabile, 'Social Psychology of Creativity: A Consensual Assessment Technique', *Journal of Personality and Social Psychology* 43, no. 5 (1982): 999.

12. Ibid.

13. Jon Kolko, *Exposing the Magic of Design: A Practitioner's Guide to the Methods and Theory of Synthesis* (Oxford: Oxford University Press, 2010), 24.

14. George DeMartino, '"Econogenic Harm": On the Nature of and Responsibility for the Harm Economists Do as They Try to Do Good', in *The Oxford Handbook of Professional Economics Ethics*, edited by George DeMartino and Deirdre McCloskey (New York: Oxford University Press, 2016), 78. DeMartino refers to professional economists, but the dynamics of the situation he describes can be extrapolated to designers.

15. Richard Buchanan, 'Wicked Problems in Design Thinking', *Design Issues* 8, no. 2 (1992): 18. Italics in the original.

16. Kenneth L. Kraemer, Jason Dedrick, and Prakul Sharma, 'One Laptop Per Child: Vision vs. Reality', *Communications of the ACM* 52, no. 6 (2009).

17. Adi Robertson, 'OLPC's $100 Laptop Was Going to Change the World—Then It All Went Wrong',https://www.theverge.com/2018/4/16/17233946/olpcs-100-laptop-education-where-is-it-now.

18. My colleague Julia Benini, who worked on the project in Brazil, commented on a draft of this chapter that there might be a bright side to this story as the OLPC arguably paved the way for the rise of the 'Netbook', small, inexpensive mini laptop computers that were introduced in the late 2000s. Netbooks did play a key instrumental role in digital inclusion initiatives, such as the programme 'Conectar Igualdad' (Connect Equality). This programme, which was launched in 2010 by the Argentinian National Government to foster basic digital education, has distributed millions of netbooks to schoolchildren from all levels of education across Argentina; see SITEAL, 'Programa Conectar Igualdad',http://www.tic.siteal.iipe.unesco.org/politicas/859/programa-conectar-igualdad. Other people involved in the OLPC project share the view that the OLPC computer was important for the development of the netbook market; see Charles Kane, Walter Bender, Jody Cornish, and Neal Donahue, *Learning to Change the World: The Social Impact of One Laptop Per Child* (New York: Palgrave Macmillan, 2012), 50.

19. A new carbon-based material with a long list of traits such as extreme hardness and efficient conductivity that could have radical implications for the future of design and engineering.

20. This is a variation of an argument developed by Parsons, which aims to help establish whether something can count as design or not, *The Philosophy of Design*, 11.,

21. James W. Jenness, Jeremiah Singer, Jeremy Walrath, and Elisha Lubar, *Fuel Economy Driver Interfaces: Design Range and Driver Opinions* (Washington, DC: U.S. National Highway Traffic Safety Administration, 2009).

22. Nigel Cross, *Designerly Ways of Knowing* (Basel: Birkhäuser, 2007), 51.

23. Victor Papanek, *Design for the Real World*, second revised edition (Chicago: Academy Chicago Publishers, 1984), 4.

24. Evidently, women too. Scattered across the book we will encounter a few quotations that include instances of gendered, non-inclusive language; in all of these cases, 'men', clearly, does not mean 'males', and should be simply read as 'people', 'persons', or 'humans'. The same applies to other gendered nouns present in the included quotations.

25. Papanek, *Design for the Real World*, 3.

26. Ibid.

27. Donald A. Norman, *Emotional Design: Why We Love (or Hate) Everyday Things* (New York: Basic Books, 2004), 224–25.

28. Parsons, *The Philosophy of Design*, 19.

29. Ezio Manzini, *Design, When Everybody Designs: An Introduction to Design for Social Innovation* (Cambridge: The MIT Press, 2015), 1.

30. John Heskett, *Design: A Very Short Introduction* (Oxford: Oxford University Press, 2002), 5.

31. José Ortega y Gasset, *Toward a Philosophy of History*, translated by Helene Weyl (New York: W. W. Norton & Company, 1941), 92.

32. Ibid., 92–93.

33. Ibid., 98–99.

34. Ibid., 118.

35. Ibid., 117.

36. Ibid., 95.

37. Ibid., 93.

38. Ibid., 122–23.

39. Greg Bamford cited in Parsons, *The Philosophy of Design*, 21.

40. This is obviously a very simplified model of the design process. Since the beginning of the design methods movement in the early 1960s, design models have been a central thread in design methodology, and many models have been defined. For an historical inventory of over eighty models of the design process see Hugh Dubberly, 'How Do You Design: A Comparison of Models', Dubberly Design Office, 2005, http://www.dubberly.com/articles/how-do-you-design.html. For a more holistic take on the design process, see Harold G. Nelson and Erilk Stolterman, *The Design Way: Intentional Change in an Unpredictable World* (Cambridge: The MIT Press, 2014).

41. Manzini, *Design, When Everybody Designs*, 53.

42. Liz Sanders, 'Is Sustainable Innovation an Oxymoron?', in *Changing Paradigms: Designing for a Sustainable Future*, edited by Peter Stebbing and Ursula Tischner (Aalto: Aalto University School of Arts, Design and Architecture, 2015), 296.

43. Victor Petit and Bertrand Guillaume, 'Scales of Design: Ecodesign and the Anthropocene', in *Advancements in the Philosophy of Design*, edited by Pieter E. Vermaas and Stéphane Vial (Cham: Springer International Publishing, 2018), 475.

44. Cross, *Designerly Ways of Knowing*, 11.

45. Michael Tovey, 'The Passport to Practice', in *Design and Designing*, edited by Steve Garner and Chris Evans (London: 2012), 5.

46. Klaus Krippendorff, *The Semantic Turn: A New Foundation for Design* (Boca Raton: CRC Press, 2006), 11.

47. Ibid., 31.

48. Michael Proudfoot and A. R. Lacey, 'Normative', in *The Routledge Dictionary of Philosophy* (London: Routledge, 2010), 277.

49. Hans Radder, 'Why Technologies Are Inherently Normative', in *Handbook of the Philosophy of Science*, edited by Dov Gabbay, Paul Thagard, and John Woods (Amsterdam: Elsevier, 2009), 893.

50. Ibid.

51. Ruth Schwartz Cowan, *More Work for Mother: The Ironies of Household Technology from the Open Hearth to the Microwave* (London: Free Association Books, 1989).

52. Richard H. Thaler, Cass R. Sunstein, and John P. Balz, 'Choice Architecture',https://ssrn.com/abstract=1583509.

53. Richard Thaler and Carl Sunstein, *Nudge: Improving Decisions About Health, Wealth, and Happiness* (New Haven: Yale University Press, 2008), 6.

54. Ibid., 3–4.

55. Nynke Tromp, Paul Hekkert, and Peter-Paul Verbeek, 'Design for Socially Responsible Behavior: A Classification of Influence Based on Intended User Experience', *Design Issues* 21, no. 3 (2011).

56. For a fun and very light read on the different stages of falling down a rabbit hole, see Kate Drozynski, 'The 10 Stages of Falling Down a Youtube Rabbit Hole',http://www.mtv.com/news/2283473/youtube-rabit-hole/. For a more distressing view on how YouTube leads viewers down a rabbit hole of extremism, see Zeynep Tufekci, 'Youtube, the Great Radicalizer',https://www.nytimes.com/2018/03/10/opinion/sunday/youtube-politics-radical.html.

57. Bournemouth Borough Council, 'Bins and Recycling', Bournemouth Borough Council,https://www.bournemouth.gov.uk/BinsRecycling/BinsandRecycling.aspx.

58. Robert Rosenberger, *Callous Objects: Designs against the Homeless*, third edition (Minneapolis: University Of Minnesota Press, 2017).

59. Bing Pan, 'The Power of Search Engine Ranking for Tourist Destinations', *Tourism Management* 47 (2014).

60. Eli Parisier, *The Filter Bubble: What the Internet Is Hiding from You* (London: Penguin Books, 2011), 9.

61. European Commission, 'Antitrust: Commission Fines Google €4.34 Billion for Illegal Practices Regarding Android Mobile Devices to Strengthen Dominance of Google's Search Engine', European Commission,http://europa.eu/rapid/press-release_IP-18-4581_en.htm. According to the European Commission, 95 per cent of users who had Google as the default search engine search through it. Conversely, on Windows Mobile devices (where Google Search is not the default) fewer than 25 per cent of all used it; the other 75 per cent of search queries happened on Microsoft's Bing search engine, which is pre-installed.

62. Jared Spool, 'Do Users Change Their Settings?', UIE,https://archive.uie.com/brainsparks/2011/09/14/do-users-change-their-settings/; Nick Babich, 'The Power of Defaults', UX Planet,https://uxplanet.org/the-power-of-defaults-992d50b73968.

63. Simon, *The Sciences of the Artificial*, 114.

64. Carl Mitcham, 'Ethics into Design', in *Discovering Design: Explorations in Design Studies*, edited by Richard Buchanan and Victor Margolin (Chicago: The University of Chicago Press, 1995), 179.

65. Don Ihde, 'The Designer Fallacy and Technological Imagination', in *Philosophy and Design: From Engineering to Architecture*, edited by Pieter E. Vermaas, Peter Kroes, Andrew Light and Steven A Moore (Dordrecht: Springer Netherlands, 2008), 51.

66. Don Ihde, *Postphenomenology and Technoscience: The Peking University Lectures* (Albany: State University of New York Press, 2009), 47; Peter-Paul Verbeek, *Moralizing Technology: Understanding and Designing the Morality of Things* (Chicago: The University of Chicago Press, 2011), 20–22.

Chapter Two

Professions as Moral Projects

A volume on design professional ethics carries two assumptions from the start. The first is that design is a professional activity; the second is that it merits specific ethical reflexion. In the previous chapter, we discussed the latter assumption and argued for the ethical relevance of design basing this claim on the normative character of both designed artefacts and design activity. The other assumption, namely, that design is a profession, will be substantiated in chapter 3. Yet before we are ready to do that, in this chapter we need to explore the notions of 'profession' and 'professionalism' because it is important to determine which understandings of professions and professionalism are worth promoting and undertaking. The chapter also offers an introductory section on professional ethics to prepare the ground for the next chapters. It also offers a treatment of responsibility, which will be a crucial notion for our account. The chapter ends by providing an examination and a rejection of codes of ethics as a primary approach to professional ethics.

OCCUPATIONS AND PROFESSIONS

The words 'profession' and 'professional', just like the word 'design', have many different meanings. There are two everyday meanings for the word 'professional' that we better discard right away as they do not concur with the intended senses that the word will have in this inquiry; this is only meant as a necessary clarification and not at all to attempt to prescribe preferred uses for the word in common language. The first sense of 'professional' we will discard is when it is used to refer to a person that earns a living for what they do. Understood in this manner, 'profession' is simply synonymous with any paid occupation or job. The emphasis here is on the earning of a living, in getting paid for undertaking an activity or being part of a paid occupation.

In this sense a professional is the opposite of an amateur. For a professional chess or poker player, for instance, playing chess or poker is their occupation. An amateur, on the contrary, does not get paid for playing chess or poker, and they play it for fun or for other, multiple reasons.

Another everyday sense in which the term 'professional' is often used is to characterise a very good performance. People use it to refer to a high level of skill demonstrated by a person undertaking an activity. Somebody can say, for instance, that a tour guide has done a professional job if they were knowledgeable and diligent. Similarly, this adjective is also frequently applied to tradespeople such as builders or tailors, to denote a reputation for technical knowledge, fair pricing, and general reliability.

Although not fully unrelated to paid work or quality of skill, the intended use of the word 'profession' in this volume will have to do with a set of specific characteristics that are present in *some but not all* occupations. Under this understanding, a 'professional' is a practitioner of one of those occupations that can be considered a profession. When used as an adjective it refers to the type of roles, activities, aims, or standards that are related to professions. Nevertheless, it is uncontroversial to claim that the line between professions and non-professions is fuzzy and contingent on the way we understand the notion. With one set of criteria, some occupations might count as professions, whereas with a different set, the same occupation may be viewed differently. Some authors go beyond this relativistic conception of professions and argue that there is no intrinsic difference at all between professions and occupations, other than professions having higher social status, and more political and economic power than other occupations. We will consider this critical view shortly, but first, we need to better clarify the meaning of the term.

The Essentialist Account

Given the multiple meanings of the word, it is not surprising that no generally accepted definition of the term 'profession' is to be found. A convenient way to start is with an influential account provided by ethicist Michael Bayles in the 1980s. For him, professions are often characterised by three necessary features that have been identified by almost all scholars who prior to him had studied the matter: 1) extensive training, often requiring academic degrees; 2) although physical skill may be involved, the intellectual component is predominant in the training and in the practice; and 3) the trained ability enables the professional to provide an important service to society, which in the case of technologically complex modern societies is indispensable, as they rely on the application of specialised knowledge to function[1]. For Bayles, other features are commonly found in professions, but they are not as important as the previous three. These are certification or licensing,

professional organisation of members, a considerable degree of self-regulation, and autonomy within the work context. Many similarly overlapping
accounts can be found that emphasise the central role knowledge plays, the
promotion of basic social values, or the addressing of vital needs[2].

This type of *essentialist* account implies that in order to be counted as a
profession, an occupation must possess these key characteristics. Yet the
question still remains as to what degree they must be possessed. Bayles
seems to argue for a discrete 'yes/no' answer, whereas other influential
scholars such as Bernard Barber propose a continuum range of professionalism in which some occupations are more professional than others[3]. In this
view, medicine is more 'professional' than nursing; similarly, a surgeon who
routinely performs operations is more professional than a doctor whose main
task is to perform routine medical examination of, for instance, commercial
pilots or air controllers at a medical examination centre. Barber also distinguishes established professions from emerging professions, which are 'quasi
profession'; writing in the 1960s he mentions management, pharmacy, and
social work as examples. An emerging profession has an uncertain status in
relation to the criteria we mentioned previously, that is knowledge and community orientation. When these quasi professional occupational groups become more professional they can claim recognition as such[4]. This characterisation is not without criticism. Engineering ethicist Caroline Whitbeck
argues that the social status of an occupation is irrelevant from the point of
view of professional ethics[5]. She cites the long and rich history of nursing
ethics to support her view. Implicit in her claim is that an internal view
regarding the commitment to the community is more important than the
external recognition, which might be contingent on many other factors than
the actual community orientation and knowledge a profession has. Nursing
may have been negated full professional status because of the power and
gender dynamics in the medical professions[6].

The essentialist account we have reviewed has been superseded in sociological investigations of professional practice, but it is still present in the
literature on professional ethics[7] as it operates as a plausible ideal type and as
a useful benchmark to establish whether one is dealing with a profession.
Because of its descriptive capacity, we will retain this set of criteria, albeit
with some modifications, and we will use it later to decide whether design
can plausibly count as a profession. This account, however, still lacks normative power as it fails to provide an adequate level of detail about what professionals *ought* to be and do. In the next section, we shall refine and deepen
this issue by discussing professionalism.

TWO KEY ELEMENTS BEYOND COMPETENCE

We should understand the notion of professionalism as the normative requirements an occupation must meet to qualify as a profession. These requirements justify the special status of professions and have particularly to do with the nature of professionals' aims, and the qualities that characterise professionals in their capacity of professionals. Some commentators have signalled that the criteria for professionalism boil down to two types of elements: *cognitive* and *normative*[8]. The cognitive element is related to training and to the theoretical and technical knowledge and expertise that a professional must possess and that separates them from non-professionals. The normative element relates to the general service ideals or aims of professions, and the derivative ideals regarding the attitudes a professional must have toward their own profession itself. To avoid confusion with other instances in which we used the word 'normative' (as in the 'normative dimension of design'), we will replace 'normative element' with 'public service element', which also conveniently emphasises the orientation toward community.

In any case, the cognitive and the public service elements can be seen as individually necessary and jointly sufficient conditions for professionalism, which means that *both* conditions have to be met in order to call an occupation a profession. The condition of training and knowledge can be seen as a requirement for the provision of expert service, given the type of issues, situations, and problems a professional will normally face. But knowledge and expertise are not sufficient in themselves. In this understanding, being maximally competent qualifies a person as a professional if and only if that knowledge and expertise is put to the service of the public good. This orientation toward community is what differentiates the professional from, for instance, the commercial entrepreneur, whose social role 'positively sanctions self-interested behaviour'[9]. Naturally, having a conscious and explicit focus on serving the public does not in the least mean that this must be the *one and only* goal of professionals—it only indicates that serving the public must be primary and is indispensable.

Ethicists Justin Oakley and Dean Cocking propose that a way of distinguishing between professions and occupations is to assess the strength of their connection to a key human good; they suggest that 'the more an occupation's body of special expertise deals with a *key* human good, the greater claim that occupation has to be properly regarded as a *profession*'[10]. Arguably, although nurses and doctors deal with human goods that are evidently more crucial to people than the goods provided by journalists or teachers, it is also straightforward that the contribution made by the latter can still be seen as central to human wellbeing in a larger sense. Other occupations, such as sales manager, cook, gardener, or welder, although important for other reasons, cannot be considered professions as their connection to *key* human goods is much weaker and non-specific.

Other additional features of professions, such as mandatory certification, licensing, or professional organisation of members (that is, the 'machinery of professionalism')[11] are undoubtedly relevant from a sociological point of view, but are less relevant from an ethical perspective. Besides, as ethicist David Carr posits: 'there might be purely contingent historical reasons why such features have failed to achieve full institutional recognition or embodiment'[12]. In the case of design, we could plausibly claim that because design is a recent occupational activity, it has not yet managed to achieve these other secondary features.

For a working definition of 'profession' that fits our understanding of professionalism, we can follow Whitbeck's definition. She concisely words the two elements we just introduced: 'professions are those occupations that both require advanced study and mastery of a specialised body of knowledge, and undertake to promote, ensure, or safeguard some aspect of others' well-being'[13].

The Public Service Element of Professions

The ideal of providing an important service to society has a long tradition, and the need for specific knowledge and expertise is a derivative of this ideal. It arose in ancient times and endured through medieval times, when there were three professions: priesthood, medicine, and law. Of course, it is rather anachronistic to speak of professions as the term has a much more recent origin in the thirteenth century. In its current sense it first appeared in the sixteenth century, until then these professions were described as sciences or studies[14]. These three 'classical professions' took care of the most important dimensions of human existence: the soul (priests), the body (doctors), and the social relations (jurists). Because of the greatness of the goods they served, these professions became separated from the rest of trades and occupations, gained privileges and social status, and established stringent access and training requirements for new members[15].

In modern times, with the Reformation and the birth of the professions as we understand them today, the ideal changed to accommodate new views promulgated by bourgeois-capitalistic entrepreneurs. These new views, deeply influenced by Calvinism as Max Weber has taught us, made it admissible (and even desirable) to be concerned with obtaining wealth[16]. Modern professions ultimately lost their connection with religion after the Enlightenment, but the service ideal of going beyond self-interest persists.

The public service element requires professionals to develop a strong sense of dedication to their profession and professional identity[17]. Also, and possibly more importantly, it prescribes commitment to others' welfare, and a strong sense of responsibility toward society, broadly understood. I have already mentioned Bayles' global aim for professions, which was 'to provide an important service to society'. Similarly, business ethicist Norman Bowie emphasises the proper goal of professional skill and the service orientation of professions: 'the professional skills are service skills, specifically skills that benefit humankind'[18]. Other authors similarly speak of providing 'an important service to society'[19], addressing 'vital [human] needs'[20], having a 'primary orientation to the community interest'[21], a 'commitment to the public'[22], and of assuring that professional competence will be put to 'socially responsible uses'[23]. Along these lines, every profession has a distinct focus for its service efforts: nurses promote the wellbeing of patients by providing care and assisting doctors, teachers do so by delivering knowledge and promoting student autonomy, and so forth. The rationale is that because these goods are so strategic to human life, society designates a specific professional group to deal with these matters.

This is the classical perspective on professionalism, and it is known as the 'social trustee' account. The need for trust in the professions arises from the complexity of professional knowledge and practice, whereby laypeople are not always able to assess the quality or efficacy of professional practice[24]. In this account, prestige, autonomy, and sometimes monopolistic power are granted to professionals (or to professional organisations) in exchange for commitment to the public and society's interests in the form of professional service, for which highly specialised practical and theoretical knowledge is necessary. Business ethicist Robert Solomon bases his claims for the recognition of businesspeople as professionals *precisely* on the intimate connection between professionalism, expertise, and serving the public: 'one of the essential features of a profession is the enforced qualifications and competence of its practitioners where the public good is concerned'[25]. The responsibility professionals have toward society stems from the trust bestowed upon them.

PROFESSIONALISM AND ITS DISCONTENTS

In passing, I have mentioned that several authors have challenged the standard account of professionalism. This 'radical critique', as it is often referred to, argues their case on the apparent apolitical nature of professions that ignores what the professions have done in order to secure that trust and the special position it affords. Thus, the critics see professionals as essentially self-serving. They are also concerned with the way the relationship is structured between professionals and the public and how this relationship affects the public's capability to recognise and act on their interests[26*].

Power Through Scarcity

One critic is sociologist Magali Sarfatti Larson. She performs a historical analysis of how professions have been established since the nineteenth century and concludes that professionalisation is the struggle of some occupational groups for an advantageous economic market position through the mobilisation of the notions of competence and service ideals. Professionals organised themselves to attain power through a process that enabled them to constitute and control a market for their own field of professional expertise. 'Professionalisation is thus an attempt to translate one order of scarce resources—special knowledge and skills—into another—social and economic rewards'[27]. The monopoly positions many professions enjoy were and are a necessary tactic to maintain scarcity.

In her analysis, Sarfatti Larson convincingly shows how claims to service, as well as knowledge, helped professions to secure their positions. However, this does not entail that the claims of competence and service ideal were (or are, for that matter) void of plausibility. Political scientist Albert Dzur concedes that making those claims indeed helped the professions secure a special position, but challenges Sarfatti Larson's view: merely rhetorical claims to competence and service 'were unlikely to have been enough to secure and maintain privileges and autonomy without the "actual results." [It] is clear that some of the social functions heralded by the social trustee account were served as well'[28].

The Disabling Professions

Ivan Illich is another of the prominent critics that challenges the social trustee account. He argues that professions actually *disable* the very people they pretend to help. He calls professions 'dominant, authoritative, monopolistic, legalised—and, at the same time, debilitating and effectively disabling the individual'[29]. Illich charges against the professions with great sarcasm and sharpness, this passage that summarises his view is worth quoting in length:

> Merchants sell you the goods they stock. Guildsmen guarantee quality. Some craftspeople tailor their product to your measure or fancy. Professionals tell you what you need and claim the power to prescribe. They not only recommend what is good, but actually ordain what is right. Neither income, long training, delicate tasks nor social standing is the mark of the professional. Rather, it is his authority to define a person as client, to determine that person's need and to hand the person a prescription[30].

Illich was particularly virulent against the professions of school teaching and medicine as practised in modern societies[31]. But his critique may also be illustrated with the example of experts in parenting or alimentation such as child psychologists and nutritionists. Illich would say that we, humans, have been able to pursue both activities since the beginning of our species and developed huge amounts of knowledge around them, but we are now less able to do so because these professionals obtained the 'power to dictate'[32] what *good* parenting or eating is, thus rendering non-professional knowledge invalid or deviant.

Conversely, Dzur argues that professions are not necessarily disabling. For him, it is the prescriptive power professionals have that effectively disables people by taking tasks away from them, which in turn causes laypeople to be less confident in their own abilities. Simultaneously, because professionals are acknowledged by the public to perform better than non-professionals, the general public loses confidence in the competence of laypeople[33].

The social trustee model is based on an 'economy of trust', whereby the public trusts the professional to self-regulate, determine standards, and perform according to these criteria; laypeople are thus excluded whereas professionals get to define the meaning of public service[34]. Illich's critique points to a serious problem with the classic account of professionalism. Because professionals self-regulate, even if they aim to serve the public through their expertise, they do so in a way that is disabling as it excludes from these deliberations the very public they purport to serve.

For Illich, the solution for these pernicious power dynamics is deprofessionalisation: 'The time has come to take the syringe out of the hand of the doctor, as the pen was taken out of the hand of the scribe during the Reformation in Europe'. Right after this call to action, he makes an implausible claim: 'most curable sickness can now be diagnosed and treated by laymen'[35]. And if this strikes us as difficult to accept it is, in his view, because the 'medical ritual' is deceiving us by creating artificial complexity. Can he be right?

Against Deprofessionalisation

Illich—and Sarfatti Larson as well—seem to neglect or undervalue an important factor: the drive professionals often feel for their professions. It is a powerful personal force that can be channelled for goals other than securing a market position, achieving status, and exerting power on others. This motivation affects at least two important dimensions of work: first, the work in itself, which is marshalling sophisticated and cognitively demanding strategies to perform skilfully the core activities of a profession[36]. Second, the internal and external struggles professionals engage in to maintain moral and ethical standards and to be able to do work that is socially responsible and ethically sound even when facing market and organisational pressures. Our everyday experience may sufficiently provide evidence that this actually happens; it's possible that many of us might know people who exemplify this behaviour. I do not want to linger on this issue much longer, but we have all seen the pictures of the bruised, exhausted faces of countless nurses, doctors, and other health-care professionals around the world caused by spending long hours at work wearing protective masks and goggles during the worst days of the COVID-19 pandemic. Naturally, not all professionals exhibit *this* level of commitment to their profession and to others. Still, some sceptics might remain unconvinced by an account based on subjective experience; in that case, there is also empirical evidence showing instances of professionals being able to maintain integrity and do good work in difficult situations[37].

This aspect of the personal drive professionals feel to do their work will be crucial to the arguments I will develop in the second part of the book.

Again, Illich has shed light on crucial negative aspects of professionalism, but is deprofessionalisation the answer? Dzur thinks it is a mistake and calls for a different kind of professionalism, one that is oriented toward 'public capabilities'. He grounds his disagreement with the critics on three reasons: first, professionals can function as barriers and as disablers, but they can also remove barriers and enable the public; second, there are growing movements of professional reform within professions; and third, the critics overestimate the power exerted by contemporary professions given that they are in fact being undermined by both the market and the state[38].

Indeed, even medicine, one of the paradigmatic classical professions, yielded to the need for external regulation and abandoned the idea of full self-regulation and autonomy as a defining characteristic of professionalism, accepting also the necessity of interdisciplinary cooperation and a focus on patient autonomy, not full professional autonomy[39]. This is translated into the ethical requirement to obtain informed consent for medical treatment, for participation in medical research, and for participation in teaching exercises that is affirmed by most medical organisations around the world[40]. Due to the complex technical nature of some medical procedures and the high emotions that can be involved, obtaining fully informed consent might still prove difficult in practice; however, the notion that the patient controls their own fate in medical affairs is the dominant doctrine in the modern physician-patient relationship[41].

New Dynamics in the Professions

Another development that must be addressed is the market trend of changing professional practice from single or autonomous practitioners paid on a fee-for-service basis to one of team practice and paid by salary that was signalled by Barber in the early 1960s and has become the norm in recent years[42]. An important implication of this trend is that professionals are increasingly governed by managers[43]. This has raised concerns over professionalism being compromised by the employee status of many professionals, who might be less able to uphold professional values[44]. These concerns seem to be ungrounded; there are some professions with a long history of commitment to service ideals, like nursing or professionals in the public sector, in which most practitioners have historically been non-autonomous employees, as is happening now with engineers, lawyers, and medical doctors; expecting self-employment from professionals to qualify as such might be anachronistic[45].

There are still unequivocal honorific connotations in the usage of the words 'profession' and 'professional'. But due to the developments we just reviewed, the connotations of social prestige and power associated with being a professional are plausibly less intense now than before, as being a professional does not automatically suggest the allure of self-employment and a high income that was the norm in the liberal professions of the past.

The critics of the social trustee account have persuasively shown that the model fails to involve laypeople through an open process where their true interests can be defined; they have also made us aware that this failure is intrinsically inimical to the public's real needs. The classic account needs to be revised keeping in mind that the professions *can* be marshalled for the public good. Albert Dzur nicely sums up how the professions could process the potent challenges rightly made by Illich, Sarfatti Larson, and others we have not covered here: 'seeking the public good *with* and not merely *for* the public . . . in a way that is tightly connected to the empowerment of laypeople'[46]. Calling for a new professionalism is not new, and many efforts have been undertaken to reformulate it[47*]; needless to say, this volume too is a step in that direction.

PROFESSIONAL ETHICS IN A NUTSHELL

Having clarified what we mean by 'profession', it is time to consider our topic at large: professional ethics. We could start by clarifying that in this volume, when we speak of professional ethics, we refer to it in philosophical sense; thus, not in the sense of the moral principles of a particular professional, as in 'my doctor's ethics' or 'Bill Gates' ethics'. A quick working description might be that professional ethics is a field of ethical inquiry whose purpose is to study the decisions, reasoning, and actions of professionals.

Professions, at least in the way we understand them, are inevitably fraught with ethical issues; after all, professions seek to promote, ensure, or safeguard some aspect of others' wellbeing and to address vital needs. It is because professions are 'moral projects' that professionals are from the outset involved in matters of genuine ethical controversy, Carr argues. For him 'appreciation of the ethical must lie at the heart of any professional understanding and deliberation worthy of the name'[48].

Ethicists working in professional ethics seek to understand the ethical dimension of a professional activity and of the professionals involved. At the same time, they aim to devise guiding principles or ideals for them. The way philosophers pursue these goals is usually by choosing a normative ethical theory (such as Kantianism or virtue ethics, which will be introduced in chapter 6) to analyse and reflect on the implications of the theory for a range of issues in the profession. The findings and conclusions of this analysis become 'normative'; that is, they 'allow us to say how professional life *should* be conducted' [49].

Depending on the ethical theory upon which the ethicist decides to base their approach, professional ethics is concerned with a set of specific interests. In practice, it all comes down to what particular authors, guided by the ethical theory, understand as necessary for good professional practice. So, for instance, we find a frequent focus on questions around autonomy, obligations, responsibility, integrity, trustworthiness, confidentiality, privacy, justice, informed consent, non-paternalism, beneficence, non-maleficence, and competency.

Often, professional ethics is a term that evokes images of solemn oaths like the Hippocratic Oath, which is an oath of ethics historically taken by physicians. Other times, professional ethics is associated with codes of conduct or of ethics, which many professions and occupations have and to which members are supposedly bound. This is not surprising as a code of ethics is indeed a common way to attempt to integrate ethics in a profession, but professional ethics should not be conflated with codes of ethics or codes of conduct. This conflation rests in a misunderstanding as both seek to promote good professional behaviour. Professional ethics is, however, a larger rational endeavour that is open to substantiated disagreements emanating from the multiple perspectives that may participate in the discipline. It is not a set of definitive norms and rules as one finds in a code of ethics given their importance and popularity. The nature and value of codes will be discussed in the section titled 'Doesn't Design Need a Code of Ethics?' but first we need to the explore the notion of responsibility.

A Primer to Responsibility

A particular sphere of interest that is central to our inquiry is that of *responsibility*, which is a key concept in professional ethics. Responsibility, just like 'design' or 'ethics', can be understood in a plurality of ways, and it is thus necessary to briefly provide a few insights and clarification on the topic.

We should start by noting the difference between legal and moral responsibility. A designer acquires legal responsibility toward a client, for example, when they enter a contract; similarly, an employee has, for instance, the legal responsibility not to disclose confidential information or to accept bribes. If one does not comply with a legal obligation (a law or a binding contract, for instance) one might be fined or be held 'liable' to pay damages (or even sent to prison by a judge, in extreme cases). Legal responsibility, however, is not the topic we will be considering here. It is responsibility in an ethical sense that interests us. We are particularly interested in the notion of responsibility when it is connected to moral norms, principles, and values.

On some occasions, legal and moral responsibility might overlap, rendering an action both morally and legally inadmissible. This, however, is not always necessarily so; for example, it is not illegal to design (or to use) clothes that are produced under the so-called sweatshop model, which involves reducing the production costs by offshoring production to poor countries and resorting to exploitative and unhygienic working conditions (especially for women and children). Although these practices are not illegal in the (rich) countries where the producing brands are based, their ethical admissibility is highly contested, to say the least[50*].

Keeping in mind that we are not interested in the legal understanding of responsibility, a further internal distinction can be made within the notion of moral responsibility. Three different meanings for the notion in an ethical sense are frequently encountered in the literature: as *obligation*, as *attribution*, and as *virtue*[51*]. These three different meanings are not unrelated to one another as we see in the following example: 'Peter was responsible (obligation) for writing the design proposal for that new project; he was very busy with other projects, but he is a very responsible person (virtue), and did a great job. He is responsible (attribution) for winning the project'.

The first meaning of responsibility refers to the *obligations* that are inherent in a job or duties associated with a particular position; for example, an interface designer is responsible for developing visual elements such as input controls, navigational and informational components (for example, buttons, icons, search fields, etc.). In some cases, this type of responsibility can be assigned or self-assumed on a case-per-case basis; for example, when we say that 'Sophie is responsible for the communication with the client during this project'. Examples can be found outside the professional realm: for example, a 'designated driver' has the obligation to abstain from alcohol as they are in charge of driving their friends or family members home safely.

The second sense for responsibility is related to the *attribution* of responsibility. We attribute responsibility if we say, for example, that Facebook is responsible for the infamous Cambridge Analytica data scandal[52*]. We say so because we believe they should have designed and implemented better security mechanisms, and they failed to adequately consider the risks of third-party data use. The evaluation and the subsequent praise or blame of a person's actions (or omissions) to act takes place in terms of 'good' and 'bad', or 'right' and 'wrong'. In the negative sense, responsibility is thusly associated with blameworthiness, accountability, and liability[53].

The third understanding of responsibility is as a *virtue* or character trait. Here a responsible person is one who deliberates about their actions and consequences, showing moral maturity. For example, if we say that 'Sophie is a responsible interface designer', we might be saying that they are, for instance, reliable, meticulous, careful, and trustworthy. A responsible person does not act (necessarily) out of duty, but because that is how that person is or how they would like to be (that is, what they aspire to be). A responsible person cares, and being caring, meticulous, and reliable is constitutive of who they are. Sophie has an obligation to the client (for instance, delivering sufficiently good work), but she seeks to submit high-quality work because that is the type of designer she wants to be, not only because she has to, and she would do so even if she knew that the client has low evaluative standards to judge the submitted work.

These examples illustrate a further characterisation that can be made: responsibility can be forward- or backward-looking[54]. Responsibility is forward looking when it relates to actions or events that have to occur in the future; for example, when going the extra mile to exceed a client's expectations and seek excellence in the results (responsibility as virtue) or when deliver the agreed upon results on time (as obligation). Conversely, responsibility can be backward-looking when it refers to an evaluation of past acts or omissions (responsibility as attribution); for example, if a design team delivers substandard work the team or an individual can be held responsible for the delay in the project and the work will probably have to be done over again.

Attribution need not be related to a blameworthy action, but to the failure to act despite the obligation to do so. This is called an 'omission'. An omission can be fully deliberate, but it can also arise out of *negligence* or *recklessness*. For example, one can deliberately fail to disclose to a potential client that one is already working for a direct competitor, which could result in a conflict of interest. On the other hand, an omission can also be motivated by carelessness or lack of consciousness of the obligation. Responsibility can thus still be attributed, even without intent, depending on the gravity of the effects provoked by the omission. The upshot is that one can be held responsible for things one did not directly or explicitly intended to do.

These notions are useful to evaluate Facebook's responsibility in the Cambridge Analytica data scandal. Even accepting for the sake of argument that Facebook acted *without* premeditation or malice, the omission to adequately guarantee an acceptable threshold of safety and privacy might indicate that Facebook was negligent and reckless. This, in turn, would make them morally responsible, albeit in a different form than Cambridge Analytica. According to most reports, Facebook could feasibly have done something to prevent the data from being harvested as they had the means to do so, but they did not. They had the ability to act responsibly, but they failed to do so.

They can be held responsible because if they had acted responsibly, *the data misuse would have not taken place*[55].

Somebody could argue that Facebook was unaware of what was going on. Claiming ignorance could diminish Facebook's responsibility, but only slightly. A strong case could be made that they had the effective capacity to reasonably foresee the illicit usage of the data and were invested with sufficient effective power to prevent those risks. Again, this is precisely what grounds the attribution of responsibility. Facebook was (and still is) entrusted with people's private data, and it has a crucial role in how people access the news, develop views on politics and social affairs, and keep contact with friends and family[56*]. The values and goods at stake were so great that the responsibility to take appropriate measures to safeguard them was also great[57*]. And still is.

In chapter 8, we will resume the discussion on responsibility, approaching the notion from the perspective of virtue.

DOESN'T DESIGN NEED A CODE OF ETHICS?

Every discussion of professional ethics and responsibility commonly and immediately brings the idea of codes of ethics to our minds, but these notions should not be conflated. Codes of ethics and codes of conduct are documents that lay down guidelines for recommended, required, or forbidden behaviour, but they are not the same as professional ethics. Also, although both codes of ethics and of conduct aim at guiding behaviour, they are not exactly the same. Codes of ethics tend to more generally enunciate values and ethical principles, whereas codes of conduct are more practical and specific in regard to their prescriptions. This is an important difference that needs to be duly noted, but it does not have great implications for our discussion.

Codes are usually promulgated by professional organisations, companies, or other institutions such as government agencies and non-governmental organisations. Engineering ethicists Ibo Van de Poel and Lambèr Royakkers distinguish two main types of codes: 1) professional codes, promulgated by a professional organisation, that address all professionals practising a given profession; 2) corporate or institutional codes that aim at guiding the behaviour of a company's or an institution's employees. For them, professional codes could be boiled down to the discussion of three main elements: 1) how the profession ought to be conducted, 2) obligations toward employers and clients, and 3) responsibility toward society. Corporate codes include primarily mission statements, core values, and the responsibilities toward stakeholders[58].

Codes could have three different functions: regulatory, aspirational, and educational[59]. Regulatory codes are mandatory and aim at enforcing behaviour. Aspirational codes seek to shape individual values and actions but allow the practitioner to decide for themselves. Lastly, educational codes aim at important values that need to be considered in a given situation.

Chartered or licenced professionals such as doctors or engineers have such codes, to which they are subjected to different degrees. Although it is not required to be a member of a professional organisation to practise design, nor is a general license to practice design mandatory, many design professional organisations have formulated codes of ethics. Evidently, because design organisations lack enforcement power, the codes they formulate are aspirational or educational.

Codes can also be formulated by prominent individuals within a profession, in this case their aim at fostering conversations and discussion between peers. In the field of design, the code formulated by designer Mike Monteiro serves as one example of this kind of code[60*].

Let's briefly introduce other institutional examples from the field of design. The first is AIGA, the professional association for design, which has a code of conduct that provides ethical and behavioural standards for professional designer[61*]. Among its norms, AIGA expects that 'a professional designer shall not knowingly accept instructions from a client or employer that involve infringement of another person's or group's human rights'[62]. A second example comes from the 'Code of professional ethics', developed by the World Design Organisation (WDO), a large organisation in the field of industrial design[63*]. Article 1 of the WDO code reads: 'Industrial designers' ultimate responsibility to their clients shall be realized by providing appropriate and original designs, which represent both value and benefit to their clients, clients' customers and the general public, while meeting the clients' ethical, business objectives'[64]. It is outside the scope to analyse these codes in detail, but we will return to them shortly.

Pros and Cons of Codes of Ethics

Several scholars believe that codes of ethics may provide useful ethical guidance on professional issues and could be a starting point for an integrated ethics program. For Caroline Whitbeck, they are a 'guide to the moral problems, temptations, and pitfalls [that are common in practice] and [offer] guidance on how to respond well to them'[65]. Ethicist Michael Davis, for instance, writing on the importance of codes of ethics for engineering, argues that codes are central for educational and evaluative purposes, and serve to understanding engineering itself as a profession[66].

On the other hand, a frequent charge made against codes is that they are vague and inconsistent. The problematic character of its application becomes evident when integrating the many principles that are included in a code. For example, the WDO code of ethics reads: 'Benefit the client' (article 1),

'Benefit the user' (article 2), 'Protect the earth's ecosystem' (article 3), 'Enrich cultural identity' (article 4), and 'Benefit the profession' (article 5)[67]. Of course, these are only the titles, but it is easy to see how difficult it would be to apply the code to practical situations. How should these articles be ranked in importance? According to what principle? No ranking criterion is provided. And remember article 1? Providing designs that represent 'both value and benefit to their clients, clients' customers and the general public, while meeting the clients' ethical, business objectives' is easier said than done. How is one supposed to integrate the many possibly conflicting interests of all these parties to reach a sound decision? Codes of ethics undoubtedly can serve as pointers to important issues that designers should be thinking about, but, alas, they are no panacea.

A notable critic of codes of ethics is philosopher John Ladd, who called the whole idea of a code of ethics an 'intellectual and moral absurdity'[68]. For him, among other ills, codes of ethics would encourage a practitioner to deliver the minimum a code requires instead of the best that they can do. He also claims that 'the attempt to impose such principles on others in the guise of ethics contradicts the notion of ethics itself, which presumes that persons are autonomous moral agents'[69].

In defence of codes, it is hard to see how a moral agent would lose their autonomy because of a code of ethics; after all, the vagueness of the codes *will* require ethical judgement as it cannot be followed blindly. Indeed, applying a code of ethics requires a great deal of critical thinking. A practitioner cannot follow rules algorithmically in order to obtain an ethical outcome. Codes of ethics do not seem to contradict ethics itself as Ladd argued. They might indeed be vague, but this very vagueness guarantees ethical autonomy. As Van de Poel and Royakkers argue, 'the code is maybe better considered as a set of guidelines that is helpful in judging cases than as a set of strict prescriptive rules'[70]. One could easily imagine that codes of ethics could be useful as a guide for dealing with and responding to the moral issues that arise in practice. This liberates codes from the charge of being detrimental to autonomy but makes them more ambiguous.

Alas, we are back at square one: in order to apply a code, a professional will still have to deal with the complexity of open-ended ethical considerations. So, in order to decide to apply the code, a professional will have to first recognise the moral saliency of the situation, which is no easy feat in itself, and then they will have to decide how to apply the code, which is only a guideline and requires a lot of judgement and reasoning. It is doubtful that a professional capable of these complex, open-ended ethical considerations will truly need a code of ethics at all.

Ladd might be onto something more substantive when he claims that the 'most mischievous' side effect of codes is that they divert attention from the macro-ethical problems of a profession to its micro-ethical problems. For him, codes of ethics shift the locus of attention from the collective discussion on the role of a profession in society and its effect on the public interest to the individual decisions on particular issues.

But do codes work at all? A 2018 behavioural ethics study conducted on sixty-three software engineering students and 105 professional software developers shows that even when given explicit instructions to consider the code of ethics of the Association for Computing Machinery, reading said code had no observed effect on the participants' decision making[71]. Software developers are a different professional group than designers, but this is not an isolated finding; a considerable amount of literature points in the direction of codes tending to be routinely ignored by professionals and being ineffective as guidance because people fail to recognise the ethical issues present in a situation, which inadvertently prevents them from following the code prescriptions[72]. George DeMartino makes the compelling claim that the ineffectiveness occurs when the codes are 'not embedded in and do not arise from a robust field of professional ethics'[73].

This indicates that one needs a robust professional ethics if one wants to formulate an effective code that goes beyond merely formulating general points such as respecting human rights or not engaging in racism. These are points with which, on the other hand, most professionals would agree. To codify proscriptions and prescriptions that exceed the obvious, a prior recognition of the ethical saliency of relevant issues is indispensable, as well as a careful consideration of them, which is precisely what a broad professional ethics is about. In the current context whereby design does not have a robust professional ethics, crafting a code of ethics would be putting the cart before the horse.

Having gained a clearer understanding of professionalism and the associated issues that we are grappling with, the next challenge is to 'bite the bullet' and attempt to convincingly establish whether design can be indeed called a profession. In the following chapter, we will analyse several features of design activity and contrast them with our criteria for professionalism.

NOTES

1. Michael D. Bayles, *Professional Ethics* (Belmont: Wadsworth Publishing Company, 1981), 7–8.
2. Daniel E. Wueste, ed. *Professional Ethics and Social Responsibility* (Lanham: Rowman & Littlefield Publishers, 1994), 11 ; Christopher Meyers, *The Professional Ethics Toolkit* (Hoboken: Willey Blackwell, 2018), 24–25; David Carr, *Professionalism and Ethics in Teaching* (London: Routledge, 2000), 23.
3. Bernard Barber, 'Some Problems in the Sociology of the Professions', *Daedalus* 92, no. 4 (1963): 671–72.
4. Ibid., 676–77.

5. Caroline Whitbeck, *Ethics in Engineering Practice and Research*, second edition (Cambridge: Cambridge University Press, 2011), 79.

6. Andrew Edgar, 'Professionalism in Health Care', in *Handbook of the Philosophy of Medicine*, edited by Thomas Schramme and Steven Edwards (Dordrecht: Springer, 2017), 688.

7. Sarah Banks and Ann Gallagher, *Ethics in Professional Life: Virtues for Health and Social Care* (London: Palgrave Macmillan, 2009), 14.

8. Lisa Newton, 'Professionalization: The Intractable Plurality of Values', in *Profits and Professions: Essays in Business and Professional Ethics*, edited by Wade L. Robison, Michael S. Pritchard, and Joseph Ellin (Clifton: Humana Press, 1983), 23; Melanie Walker and Monica McLean, *Professional Education, Capabilities and the Public Good: The Role of Universities in Promoting Human Development* (Oxon: Routledge, 2013), 24.

9. Edgar, 'Professionalism in Health Care,' 684.

10. Italics in the original. Justin Oakley and Dean Cocking, *Virtue Ethics and Professional Roles* (Cambridge: Cambridge University Press, 2001), 80

11. Carr, *Professionalism and Ethics in Teaching*, 44.

12. Ibid., 43.

13. Whitbeck, *Ethics in Engineering Practice and Research*, 77.

14. John J. Ahern, *An Historical Study of the Professions and Professional Education in the United States*, volume: Doctor of Philosophy (Chicago: Loyola University, 1971), 74.

15. Adela Cortina, 'El Sentido De Las Profesiones', in *10 Palabras Clave En Ética De Las Profesiones*, edited by Adela Cortina and Jesus Conill (Pamplona: Verbo Divino, 2000), 17.

16. Max Weber, *The Protestant Ethic and the Spirit of Capitalism* (London: Routledge, 1992).

17. Tang Joyce, 'Professionalization', in *International Encyclopedia of the Social Sciences*, edited by William A. Darity (Farmington Hills: Macmillan Reference, 2008), 516.

18. Norman E. Bowie, 'The Profit Seeking Paradox', in *Papers on the Ethics of Administration*, edited by N. Dale Wright (Provo, Utah: Brigham Young University, 1988), 99.

19. Carr, *Professionalism and Ethics in Teaching*, 23.

20. Meyers, *The Professional Ethics Toolkit*, 24.

21. Barber, 'Some Problems in the Sociology of the Professions,' 672.

22. Joyce, 'Professionalization,' 516.

23. Talcott Parsons, 'Professions', in *International Encyclopedia of the Social Sciences*, edited by David L. Sills (New York: Macmillan, 1968), 536.

24. Edgar, 'Professionalism in Health Care,' 679.

25. Robert C. Solomon, *Ethics and Excellence: Cooperation and Integrity in Business* (Oxford: Oxford University Press, 1992), 139.

26. For a general discussion of the critique, see Albert W. Dzur, *Democratic Professionalism: Citizen Participation and the Reconstruction of Professional Ethics, Identity, and Practice* (University Park: The Pennsilvanya State University Press, 2008), 79–104. For a discussion of the critique against educational professionalism, see Carr, *Professionalism and Ethics in Teaching*, 49–54.

27. Magali Sarfatti Larson, *The Rise of Professionalism: A Sociological Analysis* (Berkeley: University of California Press, 1977), xv1–xv11.

28. Dzur, *Democratic Professionalism*, 70.

29. Ivan Illich, 'Disabling Professions', in *Disabling Professions*, edited by Ivan Illich, John McKnight, Irving K. Zola, Jonathan Caplan and Harley Shaiken (London: Marion Boyars, 1977), 19.

30. Ibid., 17.

31. Ivan Illich, *Medical Nemesis: The Expropriation of Health* (New York: Pantheon Books, 1976); Ivan Illich, *Deschooling Society* (London: Marion Boyars, 2002).

32. Illich, 'Disabling Professions,' 18.

33. Dzur, *Democratic Professionalism*, 102.

34. Ibid., 98–99.

35. Ivan Illich, *Tools for Conviviality* (New York: Fontana/Collins, 1973), 48.

36. For an influential overview of such strategies, see Donald Schön, *The Reflective Practitioner: How Professionals Think in Action* (New York: York: Basic Books, 1983).

37. Howard Gardner, Mihaly Csikszentmihaly, and William Damon, *Good Work: When Excellence and Ethics Meet* (New York: Basic Books, 2001).

38. Dzur, *Democratic Professionalism*, 99–100.

39. Edgar, 'Professionalism in Health Care,' 693–94.

40. See, for example, Committee on Ethics of the American College of Obstetricians and Gynecologists, 'Informed Consent',https://www.acog.org/Clinical-Guidance-and-Publications/ Committee-Opinions/Committee-on-Ethics/Informed-Consent.

41. Hongjie Man, 'Informed Consent and Medical Law', in *Legal and Forensic Medicine*, edited by Roy G. Beran (Berlin, Heidelberg: Springer Berlin Heidelberg, 2013), 865.

42. Barber, 'Some Problems in the Sociology of the Professions,' 683; Magali Sarfatti Larson, 'Looking Back and a Little Forward: Reflections on Professionalism and Teaching as a Profession', *Radical Teacher* 99, no. Spring 2014 (2014): 11.

43. Sarfatti Larson, 'Looking Back and a Little Forward', 11.

44. Whitbeck, *Ethics in Engineering Practice and Research*, 99.

45. Ibid., 100.

46. Dzur, *Democratic Professionalism*, 129–30.

47. For a general overview with a focus on medicine, see Edgar, 'Professionalism in Health Care', 690–94. For new approaches to professionalism in the fields of journalism, restorative justice, and bioethics, see Dzur, *Democratic Professionalism*, 135–243.

48. Carr, *Professionalism and Ethics in Teaching*, 28.

49. Timo Airaksinen, 'Professional Ethics', in *Encyclopedia of Applied Ethics (Second Edition)*, edited by Ruth Chadwick (San Diego: Academic Press, 2012), 617.

50. This issue is thornier than it seems at first glance. It cannot simply be resolved by marshalling regulation banning manufacturers, for example, from using child labour. An immediate full-fledged ban on child labour could 'could be devastating for the children involved', as they and their families do not have other means for survival. Jessica Nihlén Fahlquist, 'Responsibility as a Virtue and the Problem of Many Hands', in *Moral Responsibility and the Problem of Many Hands*, edited by Ibo Van de Poel, Lambèr Royakkers, and Sjoerd D. Zwart (New York: Routledge, 2015).

51. Domènec Melé, *Business Ethics in Action* (London: Red Globe Press, 2020), 60. Ibo Van de Poel, 'Moral Responsibility', in *Moral Responsibility and the Problem of Many Hands*, edited by Ibo Van de Poel, Lambèr Royakkers, and Sjoerd D. Zwart (New York: Routledge, 2015), 14. I follow Melé's classification here. Van de Poel's is roughly the same, but he disaggregates the third type (as attribution) into three types and ends up with five main types.

52. In early 2018, Cambridge Analytica, a British political consulting firm, illicitly harvested the personal data of millions of Facebook users without their consent and used it for political advertising purposes.

53. Melé, *Business Ethics in Action*, 60; Van de Poel, 'Moral Responsibility', 14–15.

54. Van de Poel, 'Moral Responsibility'.

55. For an insider's account of the scandal and its widespread implications, see Brittany Kaiser, *Targeted: The Cambridge Analytica Whistleblower's inside Story of How Big Data, Trump, and Facebook Broke Democracy and How It Can Happen Again* (New York: Harper, 2019).

56. A strong case can be made that Facebook had the moral responsibility of preventing any third party from eroding democracy through the distribution of fake news and disinformation. Also, Facebook economically benefitted from the nefarious advertising that was placed on its own platform, a profit that is evidently tainted by how the data was gathered and by the ends of the advertising (political manipulation). Due to space reasons, I am not considering these issues in further detail as they are not central to the explanation of the notion of omissions and recklessness.

57. This, some of you must have already realised, is related to the moral injunction widely known today as the 'Peter Parker principle' ('With great power comes great responsibility') which was popularised through the Spider-Man comic books written by Stan Lee.

58. Ibo Van de Poel and Lambèr Royakkers, *Ethics, Technology, and Engineering: An Introduction* (Malden: Wiley-Blackwell, 2011), 33–43.

59. Greg Wood and Malcolm Rimmer, 'Codes of Ethics: What Are They Really and What Should They Be?' *International Journal of Value - Based Management* 16, no. 2 (2003): 184–85.

60. Mike Monteiro, *How Designers Destroyed the World, and What We Can Do to Fix It* (San Francisco: Mule Design, 2019). Similarly, designers Samantha Dempsey and Clara Taylor have initiated a project that aims to help design teams define the ethical guidelines of their professional engagement in the form of a designer's oath. Samantha Dempsey and Clara Taylor, 'Designer's Oath: Collaboratively Defining a Code of Ethics for Design', *Touchpoint: The Journal of Service Design* 7, no. 1 (2015).

61. In their own words AIGA, is 'the profession's oldest and largest professional membership organization for design—with more than 70 chapters and over 18,000 members'. AIGA, 'About AIGA',https://www.aiga.org/about/.

62. AIGA, *Design Business and Ethics*, third edition (New York: AIGA | the professional association for design, 2009), 35.

63. The World Design Organization, formerly the International Council of Societies of Industrial Design (Icsid), is an international non-governmental organisation founded in 1957 to promote the profession of industrial design. WDO, 'Who We Are',https://wdo.org/.

64. World Design Organisation, 'Code of Professional Ethics' (Montreal: World Design Organization, n/a), 2.

65. Whitbeck, *Ethics in Engineering Practice and Research*, 96.

66. Michael Davis, 'Thinking Like an Engineer: The Place of a Code of Ethics in the Practice of a Profession,' *Philosophy and Public Affairs* 20, no. 2 (1991): 150–67. See also Wood and Rimmer, 'Codes of Ethics'.

67. WDO, 'Code of Professional Ethics'.

68. John Ladd, 'The Quest for a Code of Professional Ethics: An Intellectual and Moral Confusion', in *Engineering, Ethics, and the Environment*, edited by P. Aarne Vesilind and Alastair S. Gunn (Cambridge: Cambridge University Press, 1998), 21.

69. Ibid., 211.

70. Van de Poel and Royakkers, *Ethics, Technology, and Engineering*, 61.

71. Andrew McNamara, Justin Smith, and Emerson Murphy-Hill, 'Does ACM's Code of Ethics Change Ethical Decision Making in Software Development?' (paper presented at the Proceedings of the 2018 26th ACM Joint Meeting on European Software Engineering Conference and Symposium on the Foundations of Software Engineering, Lake Buena Vista, FL, USA, 2018).

72. See, for example, John R. Boatright, 'Swearing to Be Virtuous: The Prospects of a Banker's Oath', *Review of Social Economy* 71, no. 2 (2013): 150.

73. George DeMartino, 'Epistemic Aspects of Economic Practice and the Need for Professional Economic Ethics', *Review of Social Economy* 71, no. 2 (2013): 168.

Chapter Three

Is Design a Profession?

In the introduction, I pointed out that the professional status of design was commonly assumed, and unless we want to develop an inquiry into professional ethics on a shaky foundation, it is necessary to go beyond this taken-for-grantedness. A design professional ethics only makes sense insofar as design is a profession. In this chapter, we will make a proper case for viewing design as such.

This will undoubtedly not be the first-ever account of design as a profession. The difference between this account and previous ones is that this one does not primarily seek to describe what professional design is nor stipulate what it should be *as a whole*. We will concentrate on analysing several features of design professional practice and contrast them to the normative notions that were introduced in the previous chapter, where we described what professions *are* and defined what they *ought* to be.

Rich and sophisticated accounts of the design profession can be, for instance, found in Klaus Krippendorff's *The Semantic Turn*, in Harold G. Nelson and Erik Stolterman's *The Design Way*, and in Mike Press and Rachel Cooper's *The Design Experience*[1].

Although very different from one another, each of these volumes deals with topics that I consider here. For example, in his influential account, Krippendorff discusses being part of a community of practice with a focus on methods and ways of 'languaging'. In their multitextured philosophical analysis of the nature of design, Nelson and Stolterman offer a significant reflection on professional design, and view 'service' as 'a defining element' of design activities. Press and Cooper pay attention to professionalism (in a chapter tellingly titled 'The Design Profession'), providing methodological, technical, and attitudinal considerations. In these works, however, the professional status of design is taken as a point of departure, but it is not argued.

This does not in the least diminish their value, but it highlights the way in which my treatment is distinct and where the value of the present account lies.

This account will add, thus, a new layer of analysis to existing ones. It will juxtapose insights from the field of design studies (some of which were presented in the introduction and in chapter 1), as well as my own reflections on practical cases, against a normative conception of professions. The overall goal of this chapter is to substantiate the claim that design can rightly be viewed as a profession, but the findings we make and the insights we gather will also have bearing on the discussion about professional ethics that this book is about.

To facilitate the analysis, we will use a simplified version of the three main criteria we reviewed in the previous chapter. The conditions of 'extensive training' and the 'intellectual component' will be boiled down to the *cognitive element* of professions. Instead of speaking of 'providing a service to society', we will refer to the *public service element* of professions. So let's now take on the challenge of ascertaining if the assumption has enough substance to justifiably call design a profession.

THE COGNITIVE ELEMENT

To reiterate, the *cognitive element* consists of several components: extensive training (often academic), theoretical and technical knowledge, skills, and expertise. Following Bayles, the cognitive element presupposes that professions have a predominant intellectual component, while allowing for the performance of manual or physical skill as well. Importantly, what is defining for professionalism is not the intellectual difficulty of the activity or the intellectual nature of the training per se. The defining mark is in the capacity to exercise high-quality intellectual judgement in applying the knowledge that was gained through training. In this section, I will argue that design exhibits this type of behaviour in its core activities and that there is a vast network of professional and academic institutions that generate genuine theoretical and technical knowledge around design and prepares future practitioners to exercise it.

Design Education

As we have repeatedly said, one important factor in the cognitive element is training. Perhaps in 1952 the same designer might have designed 'from the spoon to the town', but design is currently a practice that has been highly specialised into different fields, each with specific training curricula. Even a staunch defender of the view that 'everyone designs', such as Ezio Manzini, argues that even though 'design capability is a widespread human capacity, to be usable it must be cultivated'[2]. Training is necessary because profession-al design requires more than the general human capacity for design can offer. A designer needs to learn how to frame problems; generate and evaluate solutions that will be technologically, functionally, socially, and aesthetically appropriate; and make detailed functional and formal specifications of their designs. For some influential authors, a designer is concerned with creating 'meaning' rather than with only defining formal and functional dimension of artefacts[3].

We will start with a brief overview of academic training in design. Be-cause there were no design schools at the time design came about, early designers were often trained in art, architecture, or engineering. From the Arts and Crafts movement of William Morris and John Ruskin grew the Central School of Art and Crafts (now known as Central Saint Martin's College of Art and Design), which was established in 1896 in London, Unit-ed Kingdom[4]. In 1919 the Staatliche Bauhaus was opened in Weimar, Ger-many[5]. The Bauhaus is generally considered the most important school in design history as it influenced the subsequent development of design theory, practice, and teaching as no other school. In 1936, Maud Bowers became the first person in history to receive a bachelor's degree in industrial design; the degree was granted by the Carnegie Institute of Technology (later known as Carnegie Mellon University) in Pittsburgh, United States[6]. Throughout the 1930s and 1940s, the Royal College of Art (London, United Kingdom) began the teaching of product design and the provision of specialised professional instruction including graphic and industrial design. In 1967, it was granted a Royal Charter, endowing it with university status and the power to grant its own degrees[7].

Since the 1950s, many design schools have been opened and departments of design established in existing institutions. The very influential Ulm School of Design was founded in the mid-1950s in Ulm, Germany, offering four-year programmes in product design and visual communication, among oth-ers[8]. During the 1960s and 1970s, design internationally became a valid degree-level discipline, underpinned by a growing body of theory[9]. Since the 1990s, academic education in design has grown around the world, especially in Europe. Now, a great number of universities, including many of the most prestigious ones, offer academic degrees in different areas of design at under-graduate and postgraduate level, offering doctoral studies too. The Italian magazine Domus publishes a ranking of Europe's Top 100 Schools of Archi-tecture and Design[10]. This indicates the sheer number of academic options that are available in Europe alone, where there are strict mechanisms in place (such as the so-called Bologna Process) for assessing the quality or education and research outputs. The growth and institutionalisation of design as a legit-imate and respected academic discipline alongside the rest of academic fields underscore design as a discipline with genuine intellectual underpinnings.

Knowledge Transfer

Besides schools, professions have found several ways of transferring and generating specialised knowledge that is relevant to every particular professional community: we can especially mention scholarly and non-scholarly journals, and academic and non-academic conferences. This is the case in design, too, where we find several indexed, peer-reviewed scholarly journals such as *Design Studies*[11] and *She Ji: The Journal of Design, Economics, and Innovation*[12], which are published by Elsevier; *Design Issues*[13], which is published by The MIT Press; and the *International Journal of Design*[14], which is open access and independently run by a dozen scholars from around the world. Besides these generalist journals, one can find a myriad of academic journals with a narrower focus, such as those devoted to design history, design education, or to the specific disciplines of design. Also, many non-scholarly high-quality design magazines are available around the world, such as *Interactions magazine*[15]. The catalogues of professional and scholarly publishing houses feature many volumes on design—one can simply consult this volume's bibliography and further reading sections to confirm that this assertion is not exaggerated.

Neither is it an exaggeration to say that there are thousands of design conferences around the world that highlight and foster academic and professional discussions on design. Design Indaba[16], OFFF Festival[17], Il Salone del Mobile,[18]and Interaction (IxDA)[19]are some examples of large yearly professional gatherings. The Design Research Society organises well-attended, highly influential biennial academic conferences[20]. Many other yearly or biennial conferences are organised by institutions such as the Design History Society[21], the Design Management Institute[22], or Cumulus, the International Association of Universities and Colleges of Art, Design and Media[23]. These are only some examples, but they sufficiently show that around design there is a rich and established academic and professional milieu of production and dissemination of knowledge that is broadly similar in kind to the ones existing in other professions.

Informal Training Is Training

Back to training. It should be mentioned that it is a well-known fact that many designers have degrees in different areas other than design. It is also not uncommon for a designer to have a degree in architecture, computer science, or engineering, but we must quickly add that design is a standard part of the current curriculum of those studies. It is also true that a designer can have a degree in a completely unrelated area or no degree at all. In fact, many designers have no formal training in design and engage in design activity coming from other academic or professional disciplines. We could cite the example of famous graphic designer David Carson, who has a degree in sociology and briefly worked as a lecturer before entering the field of design[24]. That a designer can lack a formal degree in design does not entail, however, that serious training is not necessary to *become* a designer. Carson acquired analytic and observational skills during his training as a sociologist that could be transferred to design, but Carson had to master new techniques and methods by self-learning and through experimentation, as well as acquiring the necessary theoretical knowledge. Also, by entering design practice, he joined a social activity in which he could acquire both tacit and explicit knowledge and develop design expertise through interaction with other designers, with whom he shared and discussed the aims and standards of design activity. He became thus socialised and accultured into design.

Given design is not a licenced or chartered profession, specific training in design is not an entrance-level requirement in the same sense that training as a nurse is necessary to practice as a nurse. However, without training in design (either formal or informal), it is virtually impossible to *function* in today's design world. Designers deal with large, complex processes, product-service systems, businesses, and organisations, and due to the complex nature of the technical, aesthetical, and functional demands posed to them, they need to acquire and develop specific skills and expertise that exceed the general design competencies that all human beings possess.

Situated High-Quality Reasoning and Judgement

As we stated at the beginning of this section, being capable of exercising high-quality judgement in practical professional situations is the mark of the professional. To put it differently, to ask whether a given occupation exhibits the general features of professional engagement we should look at the 'qualities of reflection and judgement required' by the issues and problems it raises[25]. It could be argued that since design is about crafting plans and abstract specifications, which is a sophisticated intellectual task, it is self-evident that the required level of reflection and judgement is likely to be high.

In the introduction and in chapter 1, we characterised the occupational view of design as an intellectually demanding activity underpinned by a set of rich core cognitive and metacognitive processes. Summarising these, design scholar Nigel Cross argues that 'design ability requires steering a path through the uncertainty from an ill-defined problem situation to the end-point of a satisfactory and high-quality resolution. Key to that ability is the designer's careful and often apparently tacit selection of strategies and tactics'[26].

For many, this would be sufficient to recognise that design meets the conditions stemming from the cognitive element of professionalism. But a sceptic might argue that designers base their work on an individual talent developed through on-the-job experience. They could, for instance, focus on the time designers spend drawing and sketching. An implicit assumption in the sceptics' view is that these activities are purely routinised behaviour that is carried out automatically without the need for cognitive or intellectual effort.

Drawing and sketching are certainly driven by procedural memory, which is the part of our long-term memory that is responsible for 'how to' instructions for (motor) skills and tasks that coordinate how to perform everyday procedures, such as walking or brushing our teeth. But procedural memory also drives more specific and sophisticated behaviours such as playing the piano or playing golf. Importantly, procedural memory often draws on hard-to-acquire tacit knowledge to steer the application of what is stored in memory. Also, as it does not involve conscious thought, our abilities just seem to 'flow' as we perform them, without much effort in figuring out what to do[27].

Naturally, drawing does involve fine motor skills that are applied 'automatically' when executing the necessary movement of hands and other parts of the body, but the sceptics are wrong if they view this activity as wholly non-intellectual because of the lack of conscious behaviour. Design activity is reflective, and the automatic mode (moving the hand in order to draw) and the evaluative mode (judging the result) are deeply intertwined: one does not make sense without the other. What is more, design could hardly be driven only by procedural memory alone, if only because the problems that the designer tries to solve are different every time and need to be framed and reframed anew requiring thus novel solutions to be generated. We could say that even though drawing can become second nature for many designers, drawing is only a *part* of design. It would be a serious conceptual mistake to equate drawing with design.

What designers do when they draw a sketch is akin to what different medical professionals (for instance, surgeons, dentists, or physiotherapist) do with a client or patient: their performance is driven by motor skills and procedural memory but they do so activating at the same time an extensive declarative knowledge of the human body. For example, a dentist needs practice to train their motor skill to competently use their dental drill, and a surgeon needs training to use a surgical knife, but properly using a dental drill or a surgical knife requires much more than training motor skills. (Needless to say, there is a lot more to dentistry and surgery than using tools skilfully.) Using a drill may become intuitive for a dentist after many years of training, but, for instance, a dentist treating severe dental decay needs to intellectually discern what is going on, where they are going to drill, where the drilling is going, and when to stop, plausibly switching between analytical and intuitive reasoning modes.

To summarise, designers engage in drawing and sketching in order to explore problems and generate solutions. These activities are effective and sophisticated ways to explore the solution space within a constrained setting in a very conscious way. Designers use drawing in combination with their analytic and synthetic planning skills to face their design challenge. The act of drawing is thus not only a way of simply drafting a result. Drawing is a key instrumental cognitive strategy that designers *use* to achieve those results; it is a tool to think with. Ample empirical evidence supports this assertion[28].

A similar claim can be extended to prototyping and model making; we see a very illustrative instantiation of using a model to think with in the documentary *Sketches of Frank Gehry*, about the Canadian American architect Frank Gehry[29]. Here we see how Gehry cognitively engages with a model of a building he is designing with his team. At one particular moment in the film, Gehry is clearly not happy with the model and, without losing sight of it, he utters, 'Pretty funny, . . . it's weird', and then lets out a sigh of frustration. After a few seconds he continues, 'Well, let's look at it for a while, be irritated by it and we'll figure out what to do'. Sidney Pollack, the director of the documentary, realises Gehry's obvious discomfort and asks him, 'What don't you like?' Gehry's reply is, 'I don't know yet. It seems a little pompous, a little pretentious'. After some more dialogue between Gehry and Pollack, Gehry's reflection on the model seems to have paid off and he finds a solution he is happy with, but this is somewhat secondary. Then, and here comes the example that is more relevant to us, Gehry turns the model around and clearly not happy with the model declares to a team member, 'This side, I still don't like this side, Craig, I still don't like it'. He shuffles in his chair and claims: 'I know why I don't like it, you know. I'll tell you why I don't like it'. As he is going to tell us why he does not like it, he puts a crumpled piece of thin cardboard against the empty side of the model he does not like and says, 'This has to get crankier'. Visibly disgruntled, he tosses the crumbled cardboard away. Then Pollack shows us several seconds of a pensive Gehry and a team member; both are visibly stuck. All of a sudden, Gehry says, 'I know how to do it. Just corrugate it [the cardboard]'. Then his team member modifies the model according to Gehry's directives. Gehry is pleased with the result, 'See how it works?'

The sudden realisation Gehry had is what design theorists call a 'creative leap'[30]. It is a way of bridging problem ('pompous and pretentious') and solution ('needs to be crankier'), and synthesising a variety of goals ('not being pompous and pretentious') and constraints ('crankiness needs to match the building'). In retrospect, we can see that the final idea of corrugating the cardboard draws upon the discarded idea of crumpling it. The crumpled element was cranky but not in the right way for the type of building he was designing; that idea was a very important stepping stone, though. This case exemplifies a type of expertise that might be sufficient to persuade some sceptics into agreeing that expert design behaviour exhibits the high qualities of reflection and judgement that are required of an occupation to count as a profession.

Reflection-on-Action

Because the cognitive element is of such crucial import for professionalism, a further argument will be provided to support the claim that design meets this criterion. This argument relies on a phenomenon that is cognitively very sophisticated: that of reflecting on one's work activity. In *The Reflective Practitioner*, his highly influential study of professionals, Donald Schön describes reflections of two types: *in*-action and *on*-action[31].

Reflection-in-action refers to the process that allows the designer to shape and reshape their work while they are working on it. It is more than mere trial and error because it is performed with intent—sometimes even before an error or problem occurs—and it is called upon when the designer detects that something is not going according to plan or when a surprising event is detected. Through this reflection, the designer reassesses the understanding of the problem and implements changes in their design. The case with Gehry epitomises this type of reflection.

Conversely, reflection-on-action is performed after a design situation. It consists in the ability to scrutinise one's own design processes and one's actions and thoughts. This activity enables the designer to enquire into the reasons why they acted as they did, in order to evaluate and improve their own process, thusly increasing their levels of skill and expertise.

Reflective practice, rather than mere routinised, instrumental problem solving, is what professionals engage in to deal with the uncertainty, instability, uniqueness, and value conflicts that are characteristic of design problems. Schön characterises design as a 'reflective conversation with the situation'[32]. Because of design's inherent complexity, the designer generates intermediary solutions other than those initially intended; this makes the designer reflect and form new appreciations and understandings that allow them to take new courses of action. He describes this 'conversation' as follows:

> He shapes the situation in accordance with his initial appreciation of it, the situation 'talks back', and he responds to the situation's back-talk. In a good process of design, this conversation with the situation is reflective. In answer to the situation's back-talk, the designer reflects-in-action on the construction of the problem, the strategies of action, or the model of the phenomena, which have been implicit in his moves[33].

After reading this fragment, perhaps some critics might still reply that situations do not 'talk back'. We should concede that this is true in the strict sense. Nonetheless, using this metaphor does not make the activity less cognitively sophisticated. It is the designer's strategy to act *as if* the situation were talking back in order to engage in this dialectical activity.

Before we end this section, a last consideration is in order. There is no canon and no seal of legitimacy for professional techniques. Or as David Carr puts it, 'there cannot even be any uncontroversial account of what empirical evidence, or which technical skills, are professionally relevant to . . . practices'[34]. It would be grossly missing the point to try to decide if an occupation is a profession by evaluating every specific method and skill and compare it to some sort of canonical list of professional skills and techniques. What is important is to determine if the cognitive element of design *taken as a whole* can be seen as analogous to that of other professions in regard to the level of sophistication required and attained in professional judgement and reasoning. *Designerly* approaches such as sketching and prototyping are technical skills that are adequate for design; they may not be adequate other professionals. Just as proficiency at cross-examination may be indispensable for a lawyer defending a client but is of no use to a radiologist, some design techniques *are* profession-specific after all. This may sound all too obvious, but it is also pertinent to note that 'design thinking', a framework for complex problem solving using design methods, is being applied across many different fields, also in the classical professions such as law[35]or medicine[36]*. However, this adoption of design by other disciplines has been received critically by some design researchers as it could contribute to a trivialisation of design knowledge[37].

To sum up, a strong case has been made in favour of counting design as a profession on the grounds of its cognitive element. I trust that most sceptics will be inclined to see that design requires and exhibits cognitive behaviour that goes *beyond* the technical application of empirical craft knowledge, and thus attains a level of cognitive sophistication that is in line with what is expected in other professions. The same can be concluded regarding the available options for higher education and professional training in design.

THE PUBLIC SERVICE ELEMENT

In this section we will examine design in relation to the *public service element* of professions, which is primarily concerned with the general aim of providing an important service to society. This service ideal is the flagship of professions in the view of professionalism we have put forward. In the following, I will defend the claim that design is indeed in the possession of the public service element and that the provision of said service is not merely incidental but at the core of design activity. Considering that design meets the cognitive element condition, accepting that design has a central public service element would mean that, at least at first sight, design meets both conditions to be rightly considered a profession. Possible objections to the claim of design having a public service element may arise as we advance through the examples, but I will postpone their consideration until chapter 4, where they will be dealt with more thoroughly.

This might be a good moment to restate that in our understanding, design is not merely the outer layer of things, it is the first step in the way abstract ideas get transformed into the reality that surrounds us to assist and enable people in the accomplishment of their 'individual and collective purposes', as posited by Richard Buchanan. Because of this, it is not an exaggeration to say that design has a major impact in all areas of life: 'from the spoon to the town' claimed Rogers, and he is right!

Most of the things and spaces around us are the product of some kind of design activity. From the cereal box somebody buys at a supermarket to the bus seats that withstand vandalism (or spills!). From government-built social housing to expensive cars, from school textbooks to medical equipment: everything has been designed. In the human-made world we are surrounded by design; everywhere we look, we see designed things. And in the developed countries, these things have, more often than not, been designed by professional designers, who, together with non-designers, shape the human-made world.

Of course, not everything in the human-made world was designed by professional designers; many artefacts and environments were designed by non-designers (we explored this issue in chapter 1). Along these lines, some sceptics could also rightly claim that a lot of objects are not designed *at all* in the strict sense, as they are simply new instantiations of existing patterns. That is, repetitions of existing objects stemming from vernacular traditions. This may very well be true, but it is irrelevant and need not be a problem for the putative professional status of design. After all, the professional status of registered nurses is not questioned just because people with health problems are sometimes looked after by family members. Similarly, disease is on some occasions not treated by doctors at all but by shamans, healers, or others who also claim knowledge and healing powers. The same happens with interpersonal disputes or agreements, which are, on perhaps most occasions, solved and reached without the need of lawyers or judges.

In chapter 1, we also defined and described design in a particular way— all definitions serve a strategic role—that emphasises the ubiquitous nature of design and its key role in human life. It has direct and indirect effects on our immediate subjective experience of everyday life and what surrounds us. The case for accepting the presence of a central public service element in design is straightforward: design provides the backdrop against which everybody's lifeworld occurs. In the following, to substantiate this assertion, we will review and discuss several examples of design projects that I believe show particularly well the nature of the service that design provides society and what positive value it adds. These projects, however, must not be read as a succession of isolated instances of design activity; they need to be juxtaposed with the theory we covered regarding the nature of design.

My claim is not that design provides an important service to society because such-and-such design project serves society in such-and-such a way or adds such-and-such value. This would be a weak claim as it would be grounded on particular *design projects*. The following examples aim to illustrate a stronger claim. It is because of *the very nature of design* that it should be clear that it provides an important service to society: which is namely being concerned with the conceiving, planning, and making of the material and immaterial infrastructure for our existence.

To gain a deeper understanding of how design is interwoven into our lives, in discussing examples I propose following Buchanan's 'Four Orders of Design'[38], which reflect the broadening of design's range of action and inquiry, and its successive stages of transformation, whereby the previous orders are not superseded but maintained and incorporated. The Four Orders are:

First order: symbols and visual communications
Second order: objects
Third order: interactions and services
Fourth order: environments and systems

It is important to note that each order is not an exact category but rather 'a place for rethinking and reconceiving the nature of design'[39]. The orders are not arranged in a ladder of quality or importance: the first order is not worse or less important than the fourth order; only of lower complexity. Also, orders overlap and designed artefacts may belong to one or more orders; for example, a dishwasher has a symbolic order (for instance, in the pictograms of the control panel), an objectual order (the device itself as a thing), a service order (the customer care service), and, hopefully, a systemic order. We will start with the design of symbols, and finish with the design of systems.

Symbols and Visual Communications

We encounter this order in the design of printed and digital media that surrounds us. Some examples could be newspapers, magazines, books, schoolbooks, learning materials, copybooks, informational brochures, brands, packages of goods, banknotes, static and animated advertising, and signage.

Visual information is a good place to start. Let's first focus on transit information. Whether in the form of traditional maps and signage or as journey planning apps, millions of people would get lost or be unable to find

their destination when using the underground in cities like New York, Beij-
ing, Sao Paulo, London, or Berlin. Maps and networked transit information
help passengers find the way to their destination easily and with minimal
cognitive effort, enabling them to move around with pleasure and convenience. But besides providing information and reducing commuting times,
maps, signs, and apps constitute an important part of the visual design of a
city's mass transit system, and together with a strong visual identity can
make citizens of all social classes proud to rely on public transportation to
move around their city[40]*.

Design also plays a key role in the design of communications for other
public contexts where people from different countries or with different languages are expected to be found. Otl Aicher's signage design for the 1972
Munich Olympics is an example that has been widely imitated in the signage
systems of airports, hospitals, or city halls. Also, good signage for pedestrians as well as road and traffic signs for vehicles is imperative in modern
cities, as it has a big impact on how people move around: directing them,
ensuring them they are on the right route, and confirming they have arrived at
their destination.

But information design is more than about effective way finding; good
information design is also closely related to safety. Imagine a motorist driving on a motorway under heavy rain and low visibility conditions. Motorway
exit signs are needed that are highly readable from afar, so that they can
switch lanes calmly and properly indicate their exit from the motorway.
Nobody likes having to make a dangerous last-minute jostle to the exit lane,
let alone being behind or next to the car that makes that move. Design takes
care of that. What is more, good information design can be a matter of life
and death in extreme situations such as a serious fire. The brutal fires that
raged through Kings Cross Underground Station in London (1988) and
Düsseldorf airport (1996) lead to loss of life as people were unable to locate
emergency exits[41]. Better design could have helped people get to safety.

Visual communications design can play a crucial role in people's lives by
contributing to civic participation. The design of a voting ballot is an evident
manifestation of the way design can provide a service to democracy. This
time, the issue can be better illustrated with an example of how a bad design
had serious implications on the 2000 U.S. presidential election. In this election, thousands of voters in Palm Beach County, Florida, marked their ballots
for another candidate other than the one intended or simply spoiled their
ballots by voting twice because their graphic layout was so confusing; this
poorly designed ballot is known as the 'butterfly ballot'. According to some
political scientists the 'butterfly ballot' 'caused more than two thousand
Democratic voters to vote by mistake for Reform Party candidate Pat Buchanan, a number larger than George W. Bush's certified margin of victory in
Florida[42]. In the United States, researchers found a myriad of design prob-

lems with voting ballots in general, not limited to the 2000 election[43]. The researchers from the Brennan Center for Justice at New York University School of Law have made an inventory of these problems and concluded that 'poor ballot design frustrates voters, undermines confidence in the electoral process, and contributes to related Election Day problems'[44]. Good design would have increased usability and prevented these grave problems in the service of citizens and democracy.

Clearly, there is much more to civic engagement than voting. National and local governments around the world produce a wide array of forms and official documentation that are key instruments in the communication and implementation of laws and in the promotion of civil, social, or cultural rights. To give an example, design (or the lack thereof) often has a huge impact on whether, for instance, a person is able to fill an official form to receive benefits they are entitled to or to complete their annual tax returns.

Another case worth mentioning is the design of state symbols for the newly formed nation of the Republic of South Sudan, a country that became independent from Sudan in 2011. In order to forge a collective national identity, all graphic representations for the country (national flag, coat of arms, banknotes, passports, letterhead, and stationery) had to be designed from scratch in six months[45]. This shows that even in a country still ripped by bloody internal antagonisms and with a devasted economy, design can be marshalled to build up a sense of shared identity and pride, so that the citizens of a new nation are able to see their citizenship in the nation as more important than ethnic nationalities.

The symbolic order of design has clear effects in the creation of communities of mutual interest, urban tribes, lifestyles, and subcultures. The skater and surfer subcultures that emerged in California in the late 1980s and early 1990s formed around magazines such as *Transworld Skateboarding* and *Beach Culture*, with David Carson, who we mentioned in chapter 1, as art director. These magazines featured work by artists and photographers as well as interviews with skaters and surfers. All of this heavily influenced the outward appearance as well as the attitudes and beliefs of people identifying with these subcultures. Although *Transworld Skateboarding* aimed at promoting a respectable image for skateboarding, *Trasher* magazine had a more 'skate and destroy' ethos of cultural resistance[46]. In this way skate magazines became a key factor in the social processes that shaped the meaning of the very act of skateboarding, which had and still has great influence in youth culture across the world.

Objects

The design of objects is the design of 'instruments of action'[47]. Almost everything we do, we do with the support of something that, more often than

not, is designed. Think of the objects we—and I assume we all live in designed environments—encounter and use when waking up and getting ready
to go out: alarm clock or phone; toothbrush; toothpaste; mouthwash; sink;
towel; toilet, in some countries also a toilet shower or a jug or a bidet; shower
room or cabin; shower head; towel again. And then, perhaps, comb or hair
brush, and certainly clothes and shoes. And we did not even mention the
utensils we need for breakfast. Naturally, some of you have breakfast before
having a shower, and some of you have a shower not in the morning but
before going to bed or wash yourselves differently—but the point remains:
very few things are done without designed things.

   Everyday objects, or, as the curator Paola Antonelli calls them, 'humble'
objects[48], serve to illustrate a related point. We often do not even notice
them. They are almost transparent and do not call attention to themselves, but
we still use them every day and for important tasks. One such object is the
ubiquitous ballpoint pen. We owe its current design to the French company
Bic, which dramatically improved the quality and reliability of existing ballpoint pens in the late 1950s by using new materials and innovating in the
design of its flagship pen, the Bic Cristal. Ballpoint pens changed writing
forever; students did not have to dip their pens in ink any longer or use ink
cartridges for fountain pens, which were more expensive. People could write
wherever they wanted. For the first time people could write with permanent
ink as easily and quickly as with a lead pencil. Although we live in a world
where many families still cannot afford to buy their children even a ballpoint
for school, the ballpoint pen has arguably contributed to a democratisation of
writing given its very low price in comparison to other writing technologies.

   Many other everyday objects could be mentioned here: lamps, chairs,
kitchen bowls, calculators, garbage bins. Designers Naoto Fukasawa and
Jasper Morrison call these objects 'normal'. Refining the normal core existence of objects (the *chairness* of a chair, for instance) 'so that it fits in with
our lives today'[49], is what design does. When this happens, these objects
become 'super normal'. Something is super normal when it is 'good to have
around that you use in a completely satisfactory way without having to think
about its shape or decipher any hidden message or trickiness'[50]. But being
super normal depends on becoming embedded in everyday life. It is actually
contingent on lacking something that is often associated with a mistaken or
naive conception of design. According to design curator Silvana Annicchiari-
co, super normal objects do not have 'style, identity, originality, remarkable-
ness', and not every object has the 'capacity to conceal its features until they
become virtually invisible'[51], and simply remain normal. But a no-frills
shopping basket, like the ones we find at most supermarkets, or a Bic pen are
simply super normal; they are so unremarkable that they enable us to focus
on the shopping or on the writing, and this too is a major contribution of
design.

Every day of our lives is connected to objects, and whether they are meant for everyday or for exceptional use, some of them end up mattering very much to us. Probably we all have a cup, a spoon, a chair, or other object we cherish especially. We form especial relationships with our things: doodle on the fogged mirror after a shower, have a preferred seat at the table, or have a shirt that we never wear but refuse to throw away. These relationships have a clear anthropological dimension. According to anthropologist Daniel Miller:

> Material culture matters because objects create subjects much more than the other way around. It is the order of relationship to objects and between objects that creates people through socialisation whom we then take to exemplify social categories such as Catalan or Bengali, but also working class, male, or young[52].

Objects are constitutive elements for social and individual identity; they are much more than mere instruments we use as they contribute to shaping and structuring our lives. Almost three decades before Miller's findings, sociologists Mihaly Csikszentmihaly and Eugene Rochberg-Halton had studied the significance of material possessions in contemporary life and shown how people invest their domestic environments with meaning and the things included therein. For carving meaning into materiality people need to use their own imagination. The authors posit that 'each person can discover and cultivate a network of meanings out of the experiences of his or her own life'[53]. They also make a very important distinction between 'objects of action' such as musical instruments or radios, and 'objects of contemplation' such as books or photographs.

Until the 1980s the design of objects took place within what can be termed the functionalist paradigm, which was governed by architect Louis Sullivan's famous dictum: 'form ever follows function'. The shape, and even the very existence, of objects had to be defined according to some utilitarian principle. In these years, a *semantic turn* occurred that broke with that tradition. This turn was summarized by Bruce Archer as the conviction that 'humans do not respond to the physical properties of things—to their form, structure and function—but to their individual and cultural meanings'[54]. Although in this book I do not focus on the debates around form and function, it is convenient to keep in mind that a very compelling case has been made in favour of regarding objects having a symbolic dimension too. Clothing and furniture are prime examples of the embedding of a symbolic order in objects. At the same time, however, objects invested with meaning are not only symbols that convey something about us to the world—they can be, and are, used instrumentally as well.

Because of all this, design has become a key part of the development of the consumption society, whereby consumption drives economic growth but can also be seen as a source of meaning and identity. It is, above all, much more than a way to simply fulfil basic functional needs, but involves the satisfaction of emotional, sensorial, and expressive experiences[55]. Previously, we said that designers codify meanings in artefacts, but that is only a part of the story of consumption. In decodifying those meanings, people will find and make new meanings and uses. This explains why one of the major concerns of human-centred approaches is understanding how people use objects in their daily life and how they create new individual and social meanings. For Press and Cooper, designers become thus 'cultural intermediaries' within the framework of consumer culture that play a key role in helping people find meaning and self-identity in a very complex world[56]. The dissemination of different levels of comfort that a large part of the world has come to attain in the last hundred years is due to consumer society. The spread of durable goods such as transportation vehicles, domestic furniture, home appliances, and non-durable consumer goods such as clothing or cosmetics has enormously contributed to people's pleasure, comfort, and living standards.

Although the issue of consumerism and its pernicious effects on society as a whole will be addressed in greater detail in the next chapter when we will consider objections, I feel a brief clarification may be pertinent. The more obvious point I try to make is that design has the capacity for improving the quality of people's living conditions through reducing drudgery or by, for instance, improving safety, hygiene, or pleasure. But I am also trying to emphasise that design can provide an important service to society by facilitating economic exchange and contributing to the interests of business organisations as well. This assertion may sound contradictory and even preposterous to some, but that depends on how we understand economic practice. It is neither contradictory of preposterous if we think that economic activity *can* have goals that go beyond mere economic growth or profits, but rather in the direction of obtaining meaning in its actions and social legitimacy by empowering people to pursue a life that they have reasons to value. This view echoes the words of Amartya Sen, who has argued that 'the overall success of a modem enterprise is, in a very real sense, a public good'[57].

Now, having made this clarification, we can resume the discussion of examples the third order.

Interactions and Services

Of course, no designed object or symbol was ever a fully stand-alone artefact as it was necessarily embedded in a social environment. However, this third order makes that embeddedness more explicit as it has a more integrative focus than the previous orders, strategically involving the situated experience of the human beings that use them.

In the early 1980s, design started to become heavily influenced by a series of applied research programmes such as cognitive psychology or ethnography of workplaces. Due to digitalisation and the increased complexity and automation of work environments, the study of human performance in complex socio-technological interactions became essential. The multidisciplinary field of human-computer interaction (HCI) grew out of these developments and with it the field of interaction design and interface design, which became established design disciplines in the mid-1990s. From an early focus on users of a computer system and its usability (the extent to which it allows users to complete their intended tasks) and accessibility (aimed at guaranteeing access to a system to all people), these approaches moved toward a more holistic focus integrating not only users but also other important stakeholders. Out of these new approaches emerged the disciplines of user experience design and service design in the late 1990s and early 2000s. Also, in the early and mid-2000s, with new approaches that resulted in what is broadly known as the design thinking framework, designers started to get involved at a more strategic level and in a new set of design challenges, which became more complex. Design researcher Jane Fulton Suri posits:

> We are designing integrated and dynamic interactions with objects, spaces and services and helping companies with more strategic decisions. Expanded opportunities have spawned developments in traditional design practice. [There] are developments relating to awareness of people's experience. . . . Combinations of projective techniques and empathic exercises are more holistic in scope and yield results that can be more viscerally understood[58].

There is a vast literature making a case for the benefits of good design of computers based on usability. A canonical volume of the field of HCI, Jenny Preece's *Human-Computer Interaction*, already in 1994 cited prior research indicating that with adequately designed technology, costs were reduced, work levels improved, and absenteeism reduced[59]. This has not changed since then, and considering computers and interfaces in general have become more and more ubiquitous, the need for good technology design has only increased. It is then not surprising that improving interface design to avoid errors in data entry is just another way design can dramatically benefit society. An example from more recent research: 'skilled nurse drug dosing errors can be reduced by a factor of over 6 by improved design of user interfaces'[60]. It is common knowledge in HCI that poor interface design causes human errors (or we could also say error induced by bad design), whereas good design prevents it[61*].

We observe the beneficial effects of design even in areas not commonly seen as the purview of design, such as services. The Oslo University Hospital, the largest hospital in Scandinavia, hired global design firm Designit to rethink the entire referral and diagnostic process for breast cancer patients.

Informed by qualitative user research involving workshops with employees across the hospital's departments, as well as in-depth interviews with patients, design contributed to facilitating collaboration between the project team, the hospital staff, and top management. This human-centred approach resulted in a redesigned service solution that reduced the waiting time from referral to diagnosis from twelve weeks to a total of seven days. The project positively impacted patients' and their relatives' lives, as well as the hospital workflow; as a result of this success, it has become a precursor for a Norwegian national standard for breast cancer procedures[62].

Another example related to health care is the Design Bugs Out project, which was launched by the British Department of Health, the National Health System, and the U.K. Design Council. This project set out to fight health-care-associated infections, that is infections that are acquired in hospitals or due to health-care interventions. Designers and manufacturers were brought together with clinical specialists, patients, and frontline staff. Based on research with users and collaborative work with stakeholders, the project led to the development and implementation of new designs for products such as commodes, bedside cabinets, patient chairs, and overbed tables specifically designed to reduce the incidence of infections by making hospital furniture and equipment easier and quicker to clean and by eliminating dirt traps[63].

Design scholars Anna Meroni and Daniela Sangiorgi present several illustrative case studies of design for service around a variety of issues: health services, e-learning platforms, business, human resources, immigration, the welfare state, or digital connectivity. They summarise the many contributions that service design can provide 1) engagement of users through codesign (in which users actively participate) when designing or redesigning services that are consistent with their needs and behaviours; 2) reaching deeper into an organisation, setting in motion 'deeper transformation processes' and conceiving new business and service models; and 3) discovering and conceiving collaborative solutions where users are not only users or cocreators but become coproducers of their own services[64].

The public sector is gaining increasing understanding of the convenience of integrating design for innovation in the public sector into mainstream practice. Research carried out by a network of eleven European partners supported by the European Commission shows that 'design thinking is the way to overcome common structural flaws in service provision and policy-making'[65].

Similarly, many studies emphasise the business value of design[66]. To cite a recent example, this value has been shown in a cross-industry study of the design practice of three hundred publicly listed companies over a five-year period in multiple countries conducted by the global management consulting firm McKinsey & Company; the study indicates a strong correlation between the integration of design within a company and business and financial performance[67]. For business organisations, design fosters the performance of cross-functional teams, a high degree of specialisation and interdisciplinarity, and an internal innovation culture based on prototyping and experimentation, among other important aspects[68]. This and similar points have been made during the last decade by many influential authors who lay a bridge between design and business, such as Tim Brown, Roger Martin, Lucy Kimbell, and Roberto Verganti. Many organisations around the world, such as the U.K. Design Council, are dedicated to raising awareness about the value that design can add to business, as well as to non-profit organisations, and the public sector. Similar institutions exist in, for instance, Spain, Singapore, Denmark, and Germany.

Environments and Systems

The fourth order integrates all the previous orders into systems. We should clarify upfront that the word 'system' here does not refer to material systems like the storage systems that, for instance, IKEA sells. Here we refer to much larger societal systems, that 'integrate information, physical artefacts, and interactions in environments of living, working, playing, and learning'[69], into what Buchanan called the 'framework for human culture'[70]. These systems integrate the orders of symbols, objects, and services to provide the very infrastructure of our existence.

Previously, we touched on the economic and social value that design provides to a wide range of areas of application. One area we have left unexplored is public space, which could be conceptualised as a system connecting the different facets of our lives through our common-built environment. Most of us encounter public space on a daily basis, and those who do not, due to health, work, or other reasons, miss it a lot. Public space surrounds us: the streets and squares we cross when we go to school or work, or to buy groceries or to go to the cinema. We play with our children in parks or playgrounds, we picnic, we celebrate birthdays, we run and cycle and skate. In public space, we find cultural and commercial exchange in the form of markets and fairs, we take rests on public benches, and we wait at bus stops. When public space is well designed, it is and feels safe and inviting for all types of people regardless of their age, gender, or colour of skin. Well-designed public space allows for many activities that people find important such as wandering around, playing, celebrating, relaxing, being with friends or strangers, demonstrating and protesting for a political or social cause, and engaging in religious or cultural celebrations.

[*Author's note:* The previous paragraph was written several months before the coronavirus disease outbreak of early 2020. Now, as I revise it in early May, and with more than half of the world's population under home lockdown due to the pandemic, I am overcome with a poignant feeling of

longing. What is fascinating and gut-wrenching at the same time is the certainty that other three billion people feel exactly the same way and long to do
those things for which public space is essential. Some analysts and pundits
argue that 'the world as we know it will change for good', whereas on the
other corner others posit that after a while everything will go 'back to normal'. At this time nobody can *really* know how our relation with the public
space and its infrastructures will be in a post-pandemic world. What it is
certain is that public space *is* key and will remain key and not only because
we long for the delightful sensation of being in the open air without wearing
a face mask, but because public space is an indispensable location for communities to function as such in the pursuit of their common goals. A prime
example of this is the Black Lives Matter protests that were ignited by the
killings of Breonna Taylor and George Floyd. The protests took place in
cities all across the United States during May and June 2020 when hundreds
of thousands of people, even in the midst of the COVID crisis, went out on
the streets to protest and march together against inequality, racially motivated violence, and structural racism.]

Besides our subjective experience, there is much evidence of the economic and social value of well-designed public space in a wide range of areas
including health care, educational environments, housing, civic pride and
cultural activity, business, and crime prevention. Research shows that cities
in the United Kingdom and around the world have received far-reaching
economic, health, and social benefits from making the best of their public
spaces[71]. But even if design can in principle improve our quality of life
through delivering better public spaces, due to the importance of the car for
the city, and the urban realm in general, people in cities have fewer opportunities than necessary to be outdoors and enjoy public space. In what some
authors call 'the era of neoliberal urbanism', a spectrum of unevenly distributed developed public space facilities appear that privilege wealthy dis-
tricts[72]. Admittedly, design by itself cannot possibly be able to remediate
this, but when operationalised strategically, it can be a helpful resource to
help tackle these negative trends.

The Superblocks Project provides an example of how design can be mobilised for improving both the availability and quality of the public space for
pedestrian traffic. This project started in the city of Barcelona, Spain, and
rapidly extended to other important Spanish cities. Barcelona, just like many
cities around the world, suffers from a lack of green spaces, has high levels
of pollution and environmental noise, and high accident rates. In short, a
superblock is a repurposing of blocks from the existing strict grid layout
characteristic of Barcelona to form a conglomerate of several blocks (usually
nine); seen from above, a superblock would look like a Rubik's Cube. General traffic is permitted only on the perimeters, and internal traffic is primarily
restricted to residents and at very low speeds[73]*. The project is currently

being deployed across the city and aims at achieving a more sustainable mobility, revitalizing public spaces, promoting biodiversity and urban green spaces, and promoting the urban social fabric, social cohesion, and self-sufficiency in the use of resources. Finally, it aims to integrate governance processes to involve citizens in the definition of projects and the development of actions [74].

The Superblocks Project illustrates how designers are collaborating with public organisations at a strategic level to tackle wicked problems, that is, complex, intractable social problems whose intrinsic complexity is due to their systemic nature. To put it differently, in a big city the problem of traffic cannot be approached from the perspective of mobility alone. To reach an improved state of affairs, one needs to include other important issues besides mobility: dwelling, commerce, work, leisure, safety, environment, and others perhaps.

A project like the Superblocks is more than just the sum of its parts. Organizational theorist Russel L. Ackoff has taught us that each part of a system could affect the behaviour of the whole, what makes the system's constitutive elements interdependent [75]. This is why holistic approaches based on participation and active engagement of users and stakeholders are not only useful but necessary. There are of course other environments and systems besides public space: homes, workplaces, schools, transport stations, and hospitals are some examples that illustrate the systemic order in which design operates.

Relatedly, in a more general reading, although it is still an emergent approach within both design and policy making, design has much to contribute to policy making, as has been already alluded to in a narrower perspective. Design can be used to address many of the current multidimensional societal challenges for which traditional problem solving approaches are not suitable. New approaches are necessary, and design is one of them. Policy makers can and do benefit from design in many ways, as it is a way to meaningfully address difficult social problems, framing issues for action and for discovering new challenges that need to be addressed, thus improving the quality of life for citizens and society as a whole. Designers can bring many skills to the table. Design can be useful for immersive research that provides deeper understanding of the issues at stake and serves to empathise with citizens and stakeholders. It can also deliver functional and meaningful answers to these challenges, from the early insight generation stages that are concerned with deciding what can be done, to the executional stages. It can foster collaboration between policy makers and internal and external stakeholders. Several authors conclude that design offers discovery, visualisation, and prototyping techniques that are powerful means of conceiving, consolidating, and testing emergent solutions that could put citizens back at the heart of policy making [76].

We could consider the use of design as a way of making sense of the future, not in the sense of predicting it but in the sense of bringing desirable futures to cognitive reality[77]. The goal of design here is not necessarily to envision a short-term design solution that will be implemented after a design project is finished, but to gain insights about trends and drivers of change and to foster complex social debates through participation. This neatly fits our understanding of design as being concerned with 'how things ought to be': in this case design helps people create shared images of possible futures for the world. One way this is done is through explorations of future scenarios aided by visualisation and prototyping techniques that bring these abstract scenarios to the realm of what can be perceived. Engaging in these explorations could transform people's activities and lifestyles as they are used to catalyse wider and more systemic transformations. Here, design is operationalised by engaging communities to conceive new possible realities by facilitating stakeholder participation through storytelling and the creation of visual narratives that can engage the public and serve to reach consensus around complex issues such as housing, sustainability, public space, food services, or mobility.

As an example of this approach, we could consider the Victorian Eco-Innovation Lab. This public-funded project aims at bringing a strategic perspective into the reshaping of food services in the city of Melbourne, which currently has carbon-intensive food systems in a country, Australia, that is already experiencing the effects of climate change. The Victorian Eco-Innovation Lab project developed scenarios based in the year 2032, which were turned into design briefs to be further explored through design methods; it has enabled collaborative conversations with industry, policy makers, and the wider public audience to identify possible trajectories of change that can lead to lifestyles and economies high in wellbeing and low in environmental impacts. The project's design outcomes served, in the words of its leaders, to 'create visions that can be easily understood, reinterpreted and used by public sectors to stimulate and drive new social innovations opportunities that lead to more sustainable lifestyles'[78].

To end, we could also mention, albeit briefly, the so-called critical design practices such as 'critical design', 'speculative design', or 'design fiction', which seek to promote constructive reflection through design. For Daniela Sangiorgi and Kakee Scoot, the unifying characteristic of these critical approaches 'is their methodological use of designed objects and systems to elicit critical reflection among users, observers and the designers themselves'[79]. These approaches are primarily concerned with exploration and reflexion as a way of knowing, and their results are often disseminated through exhibitions or publications. This feature contrasts somewhat with the more applied programmes such as Manzini's 'design for social innovation', which, although still exploratory, aims more directly and purposely at enabling, implementing, and replicating design solutions that can contribute to social change[80].

Different approaches to the design of systems are regularly covered in the scholarly and professional literature. Authors emphasise how designerly approaches involving the defining capacities of design, that is, the ability to prefigure and develop new futures in the form of symbols, objects, interactions, services, environments, and systems, serve to foster communications between multiple actors and stakeholders with seemingly opposed or conflicting interests[81].

To sum up, I have shown ample empirical evidence indicating that design makes a positive contribution by improving quality of life, access to opportunities, and economic value. In this section, we have reviewed many cases that illustrate the ways in which design provides an important service to society. The upshot is that design is key in the conception, development, and implementation of the very material and immaterial infrastructure of our existence. I believe that a strong case has been made in favour of the claim that design has a public service element that sufficiently meets the second condition of professionalism. If my arguments so far have been persuasive, design could thus be considered to meet the second key criteria for counting as a profession. Considering that the intellectual element condition has also been met (which was argued in the section titled 'The Cognitive Element'), we could therefore say that design can be provisionally considered a profession, because it meets the two key criteria stipulated in our normative account.

Accepting, albeit provisionally, that the occupation of design is a profession has an obvious implication for the status of designers: if the occupation of design can be called a profession, its practitioners (designers) can be considered professionals.

However, the claim regarding the public service element of design, which is undoubtedly the most ethically laden of the two conditions we reviewed, will certainly not go uncontested. Hence the provisional of design's professional status. Several counterexamples to what I have been arguing can be put forward, which could indicate that design may not only fail to provide a service to society but also be outright detrimental to the public interest. This is a serious caveat for the professional status of design, and it must be addressed. In the next chapter, we will entertain several objections and attempt to discern if the argument in favour of considering design as a profession can overcome these challenges and attain a more consolidated professional status. Before we get to that, we need to explore a terminological issue in the next section.

DESIGN PROFESSION OR DESIGN PROFESSIONS?

Considering design a profession allows us to introduce a small, but significant, modification to the terminology we used when we made the distinction between the general view of design and the occupational view of design in the section titled 'Two Views of Design Activity' in chapter 1. The modification is simply to replace the term 'occupational' with the term 'professional'. I do not think this point requires further arguments; the demonstration of its validity is contingent on the extent to which the arguments in favour of design as a profession are cogent.

So far, we have been treating design as a *singular* design profession, but should we rather treat it as a *group* of related professions? The purview of design is very broad and includes many different disciplines. Let's briefly consider if we should stick to the singular or if we better should use the plural form. If some of you find this question uninteresting, I invite you to skip this section altogether.

Some design scholars prefer to use the plural form and speak of 'design professions'[82], perhaps to emphasise the variety in disciplines. An argument could be made that the difference between some disciplines is so big that it might warrant considering them different professions. Notwithstanding this possible objection, I will carry on using the singular form on two grounds: first, the different design traditions and disciplines share an important common ground anchored in the *designerly* approach that binds the disciplines together, which was reviewed in chapter 1. Besides, and increasingly so, due to the complexity of current design problems, the different disciplines interact on conceiving the human-made world through projects that cross the boundaries of disciplines, organisations, stakeholders, and users.

Second, the singular form makes clearer that we are referring to one specific type of professional (designers in our case). We also see this with the singular 'medical profession', which refers to doctors of medicine, whereas the plural 'medical professions' includes other professions such as nurses, pharmacists, or therapists. In a similar manner, the plural 'design professions' could conceivably be used to refer to other professionals or technicians that work in the realm of design, but are not properly designers such as illustrators, photographers, model makers, project managers, three-dimensional rendering specialists, etc.

The singular form does not in the least mean that design has succeeded already in becoming a cohesive profession like architecture, medicine, or engineering, the latter two of which manage to function and be perceived as a self-contained profession, despite a high degree of internal specialisations. Having clarified this terminological issue, we can move on to the next chapter.

NOTES

1. Klaus Krippendorff, *The Semantic Turn: A New Foundation for Design* (Boca Raton: CRC Press, 2006); Harold G. Nelson and Erilk Stolterman, *The Design Way: Intentional Change in an Unpredictable World* (Cambridge: The MIT Press, 2014); Mike Press and Rachel Cooper, *The Design Experience: The Role of Design and Designers in the Twenty-First Century* (London: Routledge, 2003).
2. Ezio Manzini, *Design, When Everybody Designs: An Introduction to Design for Social Innovation* (Cambridge: The MIT Press, 2015), 1.
3. Krippendorff, *The Semantic Turn*.
4. Jonathan Woodham, 'Central School of Art and Design', in *A Dictionary of Modern Design* (Oxford: Oxford University Press, 2016).
5. Bernhard E. Bürdek, *Design: History, Theory and Practice of Product Design* (Basel: Birkhäuser, 2005), 27.
6. Jim Lesko, 'Industrial Design at Carnegie Institute of Technology, 1934-1967', *Journal of Design History* 10, no. 3 (1997): 274.
7. Royal College of Art, 'Our History',https://www.rca.ac.uk/more/about-rca/our-history/.
8. Bürdek, *Design*, 37.
9. Rachel Cooper, 'Design Research—Its 50-Year Transformation', *Design Studies* 65 (2019): 7–8.
10. Domus Guide 2017:https://www.domusweb.it/en/news/2016/12/06/europe_s_top_100_schools_of_architecture_and_design.html.
11. *Design Studies*:https://www.journals.elsevier.com/design-studies.
12. *She Ji*:https://www.journals.elsevier.com/she-ji-the-journal-of-design-economics-and-innovation.
13. *Design Issues*:https://www.mitpressjournals.org/loi/desi.
14. *International Journal of Design*:http://www.ijdesign.org/.
15. *Interactions*:https://interactions.acm.org.
16. Design Indaba | a better world through creativity:https://www.designindaba.com/
17. Offf Barcelona: Board: https://offf.barcelona/
18. Salone del Mobile Milano: https://www.salonemilano.it.
19. Interaction 20:https://interaction20.ixda.org.
20. DRS Conferences:https://www.designresearchsociety.org/cpages/conferences.
21. DHS Conferences: https://www.designhistorysociety.org/conferences.
22. DMI Conferences: https://www.dmi.org/page/Conferences.
23. Cumulus Conferences: https://www.cumulusassociation.org/homepage/cumulusconferences.
24. Lewis Blackwell, *The End of Print: The Graphic Design of David Carson* (London: Lawrence King Publishing, 1995).
25. David Carr, *Professionalism and Ethics in Teaching* (London: Routledge, 2000), 45.
26. Nigel Cross, 'Editorial: Design as a Discipline', *Design Studies* 65 (2019): 3.
27. Dawn M. McBride and J. Cooper Cutting, *Cognitive Psychology: Theory, Process, and Methodology*, second edition (Thousand Oaks: Sage, 2019), 117.
28. See, for example, Bryan Lawson, *How Designers Think: The Design Process Demystified*, fourth edition (Oxford: Architectural Press, 2006); Nigel Cross, *Designerly Ways of Knowing* (Basel: Birkhäuser, 2007); Gabriela Goldschmidt and William L. Porter, eds., *Design Representation* (London: Springer-Verlag, 2004).
29. Sydney Pollack, 'Sketches of Frank Gehry' (Artificial Eye, 2007).
30. Cross, *Designerly Ways of Knowing*, 66–70.
31. Donald Schön, *The Reflective Practitioner: How Professionals Think in Action* (New York: York: Basic Books, 1983), 49–69.
32. Ibid., 76.
33. Ibid., 79.
34. Carr, *Professionalism and Ethics in Teaching*, 49.
35. Stanford Law School, 'The Legal Design Lab', Stanford Law School,https://law.stanford.edu/organizations/pages/legal-design-lab.

36. For opposing views regarding the integration of design thinking in medical education, see Basil Badwan, Roshit Bothara, Mieke Latijnhouwers, Alisdair Smithies, and John Sandars, 'The Importance of Design Thinking in Medical Education', *Medical Teacher* 40, no. 4 (2018); Jacqueline E. McLaughlin, Michael D. Wolcott, Devin Hubbard, Kelly Umstead, and Traci R. Rider, 'A Qualitative Review of the Design Thinking Framework in Health Professions Education', *BMC Medical Education* 19, no. 1 (2019).

37. Cooper, 'Design Research', 13.

38. Richard Buchanan, 'Wicked Problems in Design Thinking', *Design Issues* 8, no. 2 (1992); Richard Buchanan, 'Design Research and the New Learning', *Design Issues* 17, no. 4 (2001).

39. Buchanan, 'Design Research and the New Learning', 10.

40. For a report of how good identity design and Eric Gill's 1914 'Wonderground map' of the London Underground helped distract Londoners from the realities of the Great War and enticed them to use the Tube, see Emma Jane Kirby, 'The Map That Saved the London Underground',https://www.bbc.com/news/magazine-25551751

41. Christos Giachritsis, *Generating Simulations to Enable Testing of Alternative Routes to Improve Wayfinding in Evacuation of over-Ground and Underground Terminals* (Teddington: BMT Group Ltd, 2014), 2.

42. Jonathan N. Wand, Kenneth W. Shotts, Jasjeet S. Sekhon, Walter R. Mebane, Michael C. Herron, and Henry E. Brady, 'The Butterfly Did It: The Aberrant Vote for Buchanan in Palm Beach County, Florida', *American Political Science Review* 95, no. 4 (2001): 793.

43. Tomas Lopez, 'Poor Ballot Design Hurts New York's Minor Parties . . . Again', https://www.brennancenter.org/our-work/analysis-opinion/poor-ballot-design-hurts-new-yorks-minor-partiesagain.

44. Lawrence Norden, David Kimball, Whitney Quesenbery, and Margaret Chen, *Better Ballots* (New York: Brennan Center for Justice at New York University School of Law, 2008).

45. Anne Quito, 'Branding the World's Newest Country', Works That Work,https://worksthatwork.com/4/branding-south-sudan.

46. Kara-Jane Lombard, ed., *Skateboarding: Subcultures, Sites and Shifts* (London: Routledge, 2016), 7.

47. Buchanan, 'Wicked Problems in Design Thinking', 10.

48. Paola Antonelli, *Humble Masterpieces: Everyday Marvels of Design* (New York: Regan Books, 2005).

49. Naoto Fukasawa and Jasper Morrison, *Super Normal: Sensations of the Ordinary* (Baden: Lars Müller Publishers, 2006), 99.

50. Ibid., 100.

51. Ibid., 5.

52. Daniel Miller, *The Comfort of Things* (Cambridge: Polity Press, 2008), 287.

53. Mihaly Csikszentmihaly and Eugene Rochberg-Halton, *The Meaning of Things: Domestic Symbols and the Self* (Cambridge: Cambridge University Press, 1981), 87.

54. Krippendorff, *The Semantic Turn*, xiv.

55. Press and Cooper, *The Design Experience*, 32.

56. Ibid.

57. Amartya Sen, 'Does Business Ethics Make Economic Sense?' *Business Ethics Quarterly* 3, no. 1 (1993): 50.

58. Jane Fulton Suri, 'The Experience of Evolution: Developments in Design Practice', *The Design Journal* 6, no. 2 (2003): 39.

59. Jenny Preece, *Human-Computer Interaction* (Harlow: Pearson Education, 1994), 19–23.

60. A. Cox, P. Oladimeji, and H. Thimbleby, 'Number Entry Interfaces and Their Effects on Errors and Number Perception' (paper presented at the Proceedings of the IFIP Conference on Human-Computer Interaction—Interact 2011, Lisbon, 2011).

61. Donald Norman's first book is a classic on the subject of how poorly designed artefacts lead us on to make mistakes. Donald Norman, *The Design of Everyday Things*, revised and expanded edition (New York: Basic Books, 2013).

62. Marie Hartmann, Kaja Misvær Kistorp, and Emilie Strømmen Olsen, 'Using Prototyping and Co-Creation to Create Ownership and Close Collaboration: Reducing the Waiting Time

for Breast Cancer Patients', in *This Is Service Design Doing*, edited by Marc Stickdorn, Adam Lawrence, Markus Hormess and Jakob Schneider (Sebastopol: O'Reilly Media, 2018), 252–55.

63. Department of Health of the United Kingdom, *Design Bugs out—Product Evaluation* (Runcorn: Department of Health, 2011).

64. Anna Meroni and Daniela Sangiorgi, *Design for Services* (Surrey: Gower Publishing Limited, 2011).

65. Design Council, *Design for Public Good* (London: Design Council, 2013), 6.

66. Frog Design, *The Business Value of Design* (San Francisco: Frog Design, 2017); Jeneanne Rae, 'The Power & Value of Design Continues to Grow across the S&P 500', *DMI Journal* 27, no. 4 (2015).

67. Benedict Sheppard, Garen Kouyoumjian, Hugo Sarrazin, and Fabricio Dore, 'The Business Value of Design', *McKinsey Quarterly*, no. 4 (2018).

68. Benedict Sheppard, John Edson, and Garen Kouyoumjian, 'More Than a Feeling: Ten Design Practices to Deliver Business Value',https://www.mckinsey.com/business-functions/mckinsey-design/our-insights/more-than-a-feeling-ten-design-practices-to-deliver-business-value.

69. Buchanan, 'Design Research and the New Learning', 12.

70. Richard Buchanan, 'Human Dignity and Human Rights: Thoughts on the Principles of Human-Centered Design', *Design Issues* 17, no. 3 (2001): 38.

71. CABE, *The Value of Good Design* (London: Commission for Architecture and the Built Environment, 2002); CABE Space, *The Value of Public Space: How High Quality Parks and Public Spaces Create Economic, Social and Environmental Value* (London: Commission for Architecture and the Built Environment, 2014).

72. Steven Lang and Julia Rothenberg, 'Neoliberal Urbanism, Public Space, and the Greening of the Growth Machine: New York City's High Line Park', *Environment and Planning A: Economy and Space* 49, no. 8 (2016); Kevin Loughran, 'Parks for Profit: The High Line, Growth Machines, and the Uneven Development of Urban Public Spaces', *City & Community* 13, no. 1 (2014).

73. According to the official description, 'a superblock (in physical terms) is composed of a set of basic roads forming a polygon or inner area (called *intervía*) that contains within it several blocks of the current urban fabric. This new urban cell has both an interior and exterior component. The interior (*intervía*) is closed to through vehicles and open to residents, primarily. The exterior forms the basic road network on the periphery, and is approximately 400 metres wide for use by motorized vehicles'. Agencia de Ecología Urbana de Barcelona, 'Superblocks', http://www.bcnecologia.net/en/conceptual-model/superblocks.

74. Ibid.

75. Russell L. Ackoff, *Re-Creating the Corporation: A Design of Organizations for the 21st Century* (New York: Oxford University Press, 1999), 5–8.

76. For example, André Schaminée, *Designing With and Within Public Organizations: Building Bridges between Public Sector Innovators and Designers* (Amsterdam: Bis Publishers, 2018). See also the further reading section at the end of this volume.

77. I thank César Astudillo for this formulation.

78. Dianne Moy and Chris Ryan, 'Using Scenarios to Explore System Change: Veil, Local Food Depot', in *Design for Services*, edited by Anna Meroni and Daniela Sangiorgi (Surrey: Gower Publishing Limited, 2011), 165.

79. Daniela Sangiorgi and Kakee Scott, 'Conducting Design Research in and for a Complex World', in *The Routledge Companion to Design Research*, edited by Paul A. Rodgers and Joyce Yee (London: Routledge, 2015), 116.

80. Ezio Manzini, 'Social Innovation and Design—Enabling, Replicating and Synergizing', in *Changing Paradigms: Designing for a Sustainable Future*, edited by Peter Stebbing and Ursula Tischner (Aalto: Aalto University School of Arts, Design and Architecture).

81. For example, Manzini, *Design, When Everybody Designs*. See also further reading section at the end of this volume.

82. Kees Dorst, *Frame Innovation: Create New Thinking by Design* (Cambridge: The MIT Press, 2015).

Chapter Four

Necessary Objections and a Call to Action

In the previous chapter, we considered several arguments in favour of viewing design as a profession. We ended with the conclusion that design is capable of providing an important service that is significantly beneficial and strategic to society; because of that, we provisionally called it a profession. Although, at first glance, the arguments in favour of the public service element of design might seem to be substantial enough, they need to be tested further with counterarguments. This is what we will do in this chapter.

It might be useful here to make a brief methodological aside and say a bit about objections and why we will be considering them. In philosophy, a common way to test an argument is to consider possible arguments that can be posed against our position. It works like a kind of debate: arguments in favour of a claim are made and then arguments against it are raised in the way of *objections*. The objections aim to show that the claim is mistaken. By considering objections, we test the argument to see whether the objections can be overcome, and, if so, to what extent. When an argument resists an objection, the argument is strengthened, which makes it more compelling. Sometimes, it may be necessary to improve the argument to deal with an objection. Because of that, the stronger the objections we can think of, the more our case would be strengthened if our arguments (in the original or in an improved form) survive the challenges.

In this chapter, thus, we will challenge the legitimacy of design's claim to professional status. We will do so by raising three plausible objections against considering design a profession; these objections are based on three topics: manipulation, consumerism, and unintended consequences. The goal of the discussion is not to exhaust the topics in themselves, as each of them merits a whole volume in its own right[1].

After the objections have been presented and discussed, I will offer responses to them. To end the chapter and round off the first part of the book, I will make the case for design professional ethics and formulate a proposal for a course of action.

FIRST OBJECTION: MANIPULATION

The objection goes like this: design contributes to manipulation, which goes against the criteria of serving the public, and hence design should not be considered a profession.

Let's explore the topic to substantiate the objection. Manipulation is a form of influence that seeks out to affect somebody else's actions without their knowledge or valid consent. The manipulator aims to achieve their interest by leveraging the other person's emotions and judgements, but it is 'neither coercion nor rational persuasion'[2].

The so-called *dark patterns* are a good example of manipulative designs. They are effective when judged against business objectives such as increasing revenue, while at the same time do not seem to have a primary orientation to the community interest. Dark patterns are design tropes that exploit psychological principles mainly discovered in the field of behavioural economics and experimental psychology in order to trick (potential) customers of a digital service into engaging in determinate behaviours that are not necessarily beneficial for them[3]. They could also be seen simply as 'tricks used in websites and apps that make you buy or sign up for things that you did not mean to'[4]. In chapter 1, we reviewed Nynke Tromp's influence framework and saw how design can be mobilised to influence and steer people into desirable behaviours; do you remember the fly in the urinal? Psychologists have shown that there is an impressive list of 'cognitive biases', which are ways in which our mind plays tricks on us, to be blunt. Many of these biases could also be used exploitatively. One of these is the so-called Scarcity Effect, whereby the perceived scarcity of an item affects the perception of value said item has for a potential buyer increasing the item's desirability[5]. 'BUY NOW AND DON'T MISS OUT, ONLY ONE ITEM LEFT' or 'ONLY TWO SEATS AVAILABLE AT THIS PRICE, BOOK NOW!' are messages that take advantage of the scarcity effect that we frequently encounter when we shop online or try to book a flight. When our perception of scarcity is not grounded in reality but has been purposively altered by, for instance, an online retailer that always includes scarcity appeals next to their products even when there is plenty of stock, we can say that we have been manipulated. Biases are instrumentalised in this way to manipulate (potential) customers and trick them into performing actions that are advantageous to a company but detrimental to them in terms of autonomy or consent.

Sometimes the deception is more sophisticated. For example, a prospective online buyer of a flight ticket might expect that the flight's fare will tend to increase as the date of departure nears. More savvy customers may also know that price also depends on how full a plane is, and many could reasonably expect a price for a flight not to be fixed in the way product at the grocer's is; this pricing method is called 'dynamic pricing' and it has become popular in recent years. Still, some customers might ignore that airlines track them by placing 'cookies'[6]* in their computer, or might know that but be unaware of their effects. What perhaps many prospective customers do not expect is that a dynamic pricing system might raise prices when it detects that the customer returns to the website to make a purchase. The system algorithmically assumes a returning customer is likely to be undeterred by an upcharge as they seem more interested and committed. This is more than a wild speculation; there have been reports of airlines using consumers' personal online data to set flight prices, which in May 2018 led American senator Chuck Schumer to call for the U.S. Federal Trade Commission to investigate the airline industry[7].

To take another related example, consider the options that people are sometimes presented with when paying for goods or withdrawing money in a foreign currency. In the card option payments or in the interactions with the ATM cash dispensers, travellers are often offered the option to pay for the amount in their own home currency. This is called 'dynamic currency conversion'. Research carried out by the European Consumer Organisation has shown that choosing this option was detrimental for the customer's interests in almost all cases that were reviewed. Consumers were unable to make an informed decision because of the various nudging tactics embedded in the design of the interface and the menu structures, such as colours, the size of buttons, or flashing warnings that nudged the consumer to choose the option that left them financially worse off[8].

Similarly, a report from the Norwegian Consumer Council states that Facebook and Google 'have privacy intrusive defaults, where users who want the privacy friendly option have to go through a significantly longer process'[9]. Other findings include misleading wording, giving users an illusion of control, or hiding away privacy-friendly choices.

Some might partially or totally defend these practices by arguing that it is the user's responsibility to avoid being tricked by the dark patterns. One obvious way of doing this, they argue, would be to carefully read the service's cookie policy or terms and conditions instead of simply accepting them without reading. But these policies are extremely lengthy, verbose, and full of legal jargon, which makes them extremely difficult to understand even for legal experts[10]. The very notion of informed consent becomes void when people cannot understand what they are consenting to. So this does not seem to be a reasonable defence of dark patterns.

Others might argue that prospective customers tacitly agree to being engaged in a persuasive strategy: if there is consent, autonomy is not at stake, and there is no manipulation at all. Similar arguments have been put forward to defend advertising[11]. Arguably, people are generally sceptical of advertising. Most people are aware of the fact that if they buy a car that is being advertised alongside white horses, they will not be driving alongside white horses as shown in the ad. Defenders of advertising also show evidence that indicates that consumer choices are not heavily influenced by it. For the defenders, this liberates advertising from the charges of manipulation. But can this reasoning be aptly extrapolated to dark patterns? If one accepts the reports, it seems that deceitful interactive techniques *are* effective at manipulating users; dark patterns often *do* trick consumers into doing things they do not intend. Moreover, by removing cognitive 'friction' and deceiving them, dark patterns rob users of opportunities for reflection and prevent them from becoming aware that they are being tricked. The arguments that support a defence of advertising may not work in favour of dark patterns.

A strong case could thus be made against dark patterns based on how they can undermine the user's (or customer's) autonomy and dignity. It could be argued that it is wrong to trick or pressure people into irrationally doing things they do not want to do, especially when it goes against their own interests. Several commentators have recently discussed how the persuasive mechanisms embedded in new technologies like social media, video platforms, or messaging apps negatively affect our ability to concentrate, meaningfully engage with others, and pursue our deepest goals[12]. These 'alienating' mechanisms aim not necessarily at selling something to the user but rather at maximising 'user engagement' with the website or app. The time spent on a page or app is one of the most important key performance indicators product managers use to determine if the design of a digital service is successful, as frequently this magnitude effectively predicts future lucrative interactions such as subscriptions or purchases. It is no wonder then that design is mobilised to capture and maintain the user's attention and engagement. A rather successful design psychology book tellingly called *Hooked* aims at teaching product managers and designers how to build 'habit-forming products' through subtly encouraging customer behaviour to engage in 'hook cycles' so that they keep returning. Nir Eyal, its author, overtly defends the use of manipulation for the good, while recognising the ethical implications of his approach and encouraging 'designers of habit-forming technology to assess the morality behind how they manipulate users [and to] consider the implications of the products they create'[13].

Much has been written about the ethics of nudging; although its moral status is contested, the matter need not be settled here[14*]. Neither is it my goal for this section to provide a definitive answer on the ethics of manipulation. What is central to our discussion is the recognition that design can be instrumental in operationalising nudging and other persuasion mechanisms in order to manipulate people. This acknowledgement could lead to the critical argument that *because design plays a key role in manipulation, which can undermine people's autonomy and dignity, it should not be considered a profession, as this behaviour does not meet the criteria of serving the public.* Even defenders of manipulation such as Eyal concede that a product is exploitative when the designer of a product does not believe it will somehow improve users' lives.

But we must ask, is this a sufficient charge against the professional status of design? Ought we to invalidate design's claim to being a profession because designers use their design skills and knowledge to engage in manipulative practices that are detrimental to other's wellbeing? I will address these questions next.

REPLY TO THE OBJECTION OF MANIPULATION

This reply will not be a rebuttal of the content of the objection; my aim is not to negate the specific negative arguments included in it. In other words, I will thus not attempt to defend manipulation. My goal is rather to provide an alternative interpretation of what manipulation could entail for design and provide a counterargument to support design's professional status.

For the critics, manipulation poses a problem because design can be instrumental in operationalising persuasion mechanisms used to manipulate people, which can undermine their autonomy and dignity. This markedly contrasts with the many examples we reviewed in chapter 3 that show how design beneficially contributes to society and what the nature of that strategic contribution is. However, throughout the previous discussion, we see how design can fail to contribute to individual or collective wellbeing or can even be detrimental to it. Yet is this realisation sufficient to invalidate our provisional conclusion that design can be considered a profession?

Although it is true that the criticisms convincingly showed that there are relevant cases in which design does not serve society, it does not necessarily follow that design should not be considered a profession just because it fails to serve society in some cases. That is not what the public service element asks of professions. When we defined professions, we argued that for an occupation to make a plausible claim to be a profession, it must have a primary orientation toward community. The criticisms rest on the questionable assumption that to be a profession the public must be served at all times. It seems to be an implausible claim to say that all professions must *unfailingly* serve society. No profession would survive such a standard.

The requirement is to have an important public service element, and to see if a profession meets the criterion we need to look at how the professional practice is exercised *characteristically*. Yet some may question this view by saying that it is not clear what 'characteristic' entails for design. They may assert that a profession must serve society at least regularly, and it is not clear that design at a minimum fulfils this requirement.

This is a valid point. In an established profession, we could just assess how the profession behaves in the majority of cases across many years—or perhaps decades—and determine what counts as characteristic based on statistical frequency. Most doctors do serve the public, and although some doctors do harm their patients, the critics would argue these are negligible cases, defending the professional status of medicine. In a new profession like design, we could explore a different approach and consider that its associated standards of excellence are still under development. It does not have the long history of nursing, law, or architecture, not even of engineering, its cousin profession. Because of this, extending some leniency may be appropriate.

Nurturing the Profession

During these initial times, the profession needs to be nurtured so that standards for professional behaviour may arise and be internalised by practitioners. What is good is learned in practice; one becomes a professional by being exposed and by experiencing the same types of situations as those endured by professionals. Ideally, this internalisation of standards enables a professional to sense satisfaction or regret at the outcomes of their action, a sense that must somehow be shared or endorsed by other, but not necessarily all, practitioners in the same profession.

This means that a different measure than frequency might be needed to assess the legitimacy of design's claim to professional status. Instead of understanding what counts as characteristic in statistical terms, we could better look at it from a qualitative perspective during these formative times. The many cases of design serving society we reviewed in the previous chapter evidence that serving society is in a very relevant sense integral to design practice: *design does serve society and its impact is highly beneficial*. What I propose then is to see these cases of excellent design as *paradigmatic* instances of professional design, that is, as design functioning at its best. It is important to note that these cases are not mere potentialities but actual instances of real design; surely, it would be unreasonable to assign professional status to an occupation based on potentialities. Understood in this way, the design profession can be seen to meet the criteria for counting as a profession.

At this point, both critics and defenders of accepting design as a profession may wonder what can be said about the other cases in which design harms the public. For instance, how do we deal with dark patterns and exploitative designs? Admittedly, society expects a profession to demand of its practitioners to avoid engaging their professional skills for harmful ends. So if design is a profession it should ask of its practitioners to refrain from designing for coercive and manipulative practices as they undermine a person's autonomy and dignity. This seems not to be a problem at least in theory, *precisely* because professionalism carries normative power with regards to the public service element. Professionals could learn and internalise this during their training and during their formative years starting out as practitioners. This means that the very fact of being a professional can guide practitioners in a very specific direction: *one cannot be a good professional without a primary and consistent orientation towards community.* According to our understanding of professions, this is a conceptual truth; it cannot be any other way.

Though conceding that in theory things may work this way, some may insist that in the real world some designers *do* engage in harmful practices regardless of the necessary professional requirement of serving the public. But must this have consequences for the professional status of design? Perhaps we can rephrase the question more broadly: does society expect that the professional status of an occupation must be contingent on its practitioners *never* engaging in harmful practices? This does not seem to be the case; I argued previously that it would be an implausible claim to say that for a profession to maintain its status it must unfailingly serve society. To take a case in point, when evidence surfaces of data fraud in medical clinical trials[15] or engineers[16] are involved in cheating regulatory agencies, the general reaction, apart from public health and safety concerns and calls for criminal prosecution, is one of ethical condemnation of the fraudulent practice but not of the profession per se. The fraudsters are considered 'rotten apples' precisely because they corrupt the profession by going against its service ideal as they put commercial or institutional interests before commitment to society.

Granted, when the scale of the corruption is structural or widespread, the whole profession is questioned and all its professionals suspected of being rotten apples. This happened with the scandals that affected the accounting profession during the last two decades, especially the 2000s, with prominent cases such as Enron[17] or Siemens[18]. When this happens, we think that the profession has been corrupted. In the wake of the scandals, the accounting profession managed to survive as it learned to recognise the absolute primacy of the public interest. Arguably, stronger regulations and new oversight mechanisms implemented in most countries ignited and contributed to that process of soul-searching[19].

In summary, although this reply does not refute the problems that are highlighted by the objections, I do believe my arguments plausibly dispel the challenge they represent for the professional status of design. At the same time, although the professional status of design might survive the objection, design's role in manipulation still poses a serious ethical challenge for the profession.

SECOND OBJECTION: CONSUMERISM

A second matter on which we could ground an attack on the claim that design possesses a sufficiently strong public service element is the role design plays in consumerism. We shall first briefly explore the notion and then its connection to design.

One way of understanding consumerism is simply as a notion that describes the phenomenon of consumption, in a neutral way. In this section, to take the strongest possible interpretation of the objection, consumerism will not be a neutral term. Consumerism here refers to habitual 'consumption that is not intended to address needs (unnecessary consumption) and consumption that addresses needs but in unnecessarily superfluous forms'[20]. At the same time, consumerism is commonly associated with grave social, economic, and environmental problems.

From Needs to Inequality

This understanding of consumerism rests on the frequent, and also contested, distinction between 'needs' and 'wants' (or pseudo needs), which require further exploration. The notion of 'need' should not be taken too narrowly as to include only 'basic' needs such as food and shelter. Following on Aristotle's understanding of needs, Parsons defines them as 'those things required for us to achieve things of value (things that are good, or worthwhile)'[21]. A need is thus something that a person has to have *in order* to live a good life. This evidently means first that the deprivation of basic needs is considered intolerable by virtue of the harm that it would cause to human beings and the impossibility it would create for a good life to happen. At the same time, the view of need we just put forward allows for other needs than basic needs to be important because the good life is about more than merely subsisting. Philosopher Martha Nussbaum posits that although all animals nourish themselves, 'human nourishing is not like animal nourishing, [it is] planned and organized by practical reason and, second, done with and to others'. Eating, for instance, is more than about calorie intake (which is what one needs to survive); it is about sharing food with others, about caring for others, about enjoying the tastes and appreciating the colours and the smells, and it is also about making meaning by sustaining rituals and communal celebrations (which is what one might need to live a good a life).

On the other hand, a want is something we just desire, which is not strategic to living a good life. We can tell that something is a want and not a need because we would not judge a deprivation of a want as seriously detrimental to the good life, let alone as intolerable, as with the deprivation of basic needs[22]*. We could illustrate this with an example: we say that someone *wants* to visit the magnificent Victoria Falls at the border between Zambia and Zimbabwe, but another *needs* health care or to have friends. And this can be said because our intuition is that, while visiting Victoria Falls can indeed contribute to happiness, it is not strategic to a good human life in the way health care or having friends is.

Consumerism arises thus when a society makes a habit of satisfying too many wants through consumption. In our definition, it can also arise when real needs are satisfied in excess. In this case, when needs are satisfied in 'superfluous forms', the very necessity that legitimates them is somehow voided. Needs, then, become suspiciously similar to wants. Examples are easy to imagine: having a regular medical check-up might be a need, but 'over-the-counter' genetic testing ordered without a medical referral might not be; or having a single motorcycle to move around versus having four.

Consumerism is hardly an invention of the design profession. It made an appearance in the 1950s during what is often referred to as 'Cold War Keynesianism', when the American government started promoting full employment, maximised production, and purchasing power to drive the economy[23]. The connection between design and consumer culture is not new either. For instance, when the American National Exhibition opened in Moscow in 1959 during a brief period of relative calm between the American and Russian superpowers, it aimed at showing consumption as an inherent part of American life. The exhibition was carefully designed as a propagandistic event targeting both the Russians and the 'Home Front' by exhibiting a view of the good life epitomised by the liberating effects of unfettered consumption and material welfare. This was to be conveyed by a display of a large quantity of consumer durables like automobiles, colour television sets, microwave ovens, vacuum cleaners, refrigerators, or air conditioners among countless other domestic appliances and even whole kitchens and prefabricated houses, besides non-durable goods such as ice cream or fizzy beverages[24]. Consumerism was not an invention of design, but professional designers were there at its inception. Many of the most prominent American architects and designers actively participated in the design and development of the exhibition, among them Charles and Ray Eames, George Nelson, and Richard Buckminster Fuller.

Albeit in different forms, the trend of consumerism was set forth in the developed world during the 1960s and through the 1980s. Nowadays, consumerism has become a global phenomenon as developing countries have both a growing economy and a growing consumer population. The size of the global middle class increased from 1.8 billion in 2009 to about 3.5 billion people in 2017 (one half of the world's population) and is expected to grow to some four billion by 2021 and reach 5.3 billion by 2030. Most of this growth will be in Asia[25]. At the same time, although fewer people live in extreme poverty than in the past, almost half of the world's population still struggles to meet basic needs[26]. Although the criteria that define the poverty line can be contested (living on less than USD3.20 a day in lower-middle-income countries, and USD5.50 a day in upper-middle-income countries), it is evident that consumerism is no longer restricted to affluent countries.

Consumerism was presented in its first decade as civic virtue, first in the West and later in the whole world. Who would not want faster, bigger, more luxurious goods? Who would not want more choice and increased individual freedom? Who would not want to spend less time on house chores? Until recently, increased consumption was equated with human happiness and consumer choice treated as a right. But there is more than increasing evidence to suggest that consumerism is not a viable recipe for happiness. Materialistic pursuits can undermine our wellbeing, make us feel more anxious, and put us at a greater risk of depression; having, or even merely aspiring to have, more material possessions creates strain and stress, and burdens us with heightened psychological insecurities and weakened ties that connect us with friends, family, and community[27]. Anthropologist Elizabeth Chin gives an account of how consumerism negatively affects African American children from lower-income neighbourhoods and what it is like being poor in a wealthy society. In her account, consumption is a medium through which social inequalities are formed[28]. Societal inequality (that is, the gap between rich and poor), and not only lack of wealth, negatively affects wellbeing. Extensive research has shown that inequality erodes trust between members of a community, reduces life expectancy, increases violence, and causes physical and mental disease[29].

There is a general consensus that by placing a great strain on environmental resources, communities, and ecosystems, consumerism causes lasting effects for all species on the planet, human and non-human. Consumerism, and its associated 'throwaway culture', is widely seen as one of the root causes of environmental degradation[30]. Environmental concerns are also discussed from moral perspectives grounded on global ethics and intergenerational justice; that is, how excessive consumption in affluent nations happens at the expense of people in less industrialised nations and of future generations[31]*.

The case for arguing that design plays a key role in consumerism is getting stronger.

A brief deviation: some of you might have seen or heard of Jacques Carelman's *Catalogue of Impossible Objects* (Catalogue d'Objets Introuvables), an artistic project from the late 1960s structured around a collection of absurd objects created as a parody of the catalogue of a French mail order company. Carelman's objects are fascinating because they immediately convey how unnecessary, impossible, and absurd they are. My favourite is the 'bag for carrying a cat', which is precisely that: a cat-shaped bag to carry a cat. Another of Carelman's objects, undoubtedly its most famous, is the 'coffeepot for masochists', which is a coffeepot with a spout that is placed above the handle. Pouring coffee would scald the user![32]* The catalogue, in the words of the artist, is 'a criticism of our consumer society, to ridicule the necessity of the inhabitants of the big and rich western cities to buy things and shortly after to get rid of them again and so continue consuming'[33]. Ample evidence indicates now that Carelman's trenchant criticism was prescient.

Planned Obsolescence

We have seen in the introduction that design actively participates in the product and service innovation lifecycle, from the introduction of a product or service to the market to its termination; design is present throughout. One of design's salient roles is to enhance the desirability of the product or service by adapting it to the user's needs, expectations, and context of use, with an advantage that more often than not is centred on symbolic factors (by striking chords with people to stir their emotions) and on an increase in convenience (for instance, a reduction of physical and cognitive effort).

Because design is involved during all stages of the product lifecycle, there are many concrete features of consumerism for which design is instrumental. We shall concentrate on one of them, namely on the design of a product's 'planned obsolescence', which is the intentional and calculated shortening of a product's lifespan. This reduction in duration can be achieved, for instance, by manufacturing internal parts that are not sturdy enough to resist prolonged average use or by contributing to a 'perceived' obsolescence by releasing and promoting new models that make the older ones seem obsolete in the perception of the consumer. Planned obsolescence is thus driven by market forces, cost reductions, or the appeal of fashion, and it is one of the main obstacles for sustainable consumption as it contributes to the quantity of household waste generated within industrialised nations[34]. As a general heuristic that simplifies a complex reality, sustainability scholar Tim Cooper suggests that 'longer-lasting products are a prerequisite for sustainable consumption'[35]. By designing an intentional built-in obsolescence, design contributes to creating waste and to increasing global warming through the greenhouse gases created in the production and distribution process for the new products.

Purposively making a device difficult to repair once it has stopped functioning correctly is often necessary for planned obsolescence. This happens either by precluding users from performing minimal maintenance tasks themselves such as replacing a battery without somehow damaging the product (as happens with watches or smartphones) or by preventing the repair itself by, for instance, making it almost impossible to disassemble the product (this happens, for example, with Apple's EarPods earphones). Obsolescence can also be achieved by discontinuing the supply of spare parts or by making them so expensive that repair becomes unattractive for the user. Prompted by complaints from consumers, the European Commission took measures in October 2019 that include requirements for reparability of appliances such as refrigerators or dishwashers[36]. According to reports in the United States, around twenty states are said to have 'right to repair' legislation in progress[37].

Many have criticised designers, but few have done so as poignantly as Victor Papanek in a passage that is worth quoting in length:

> Never before in history have grown men sat down and seriously designed electric hairbrushes, rhinestone-covered shoe horns, and mink carpeting for bathrooms, and then drawn up elaborate plans to make and sell these gadgets to millions of people. . . . By creating whole new species of permanent garbage to clutter up the landscape, and by choosing materials and processes that pollute the air we breathe, designers have become a dangerous breed. And the skills needed in these activities are carefully taught to young people[38].

The necessary role design plays in consumerism poses serious doubts about the commitment of the design profession in relation to its public service element; this substantiates a critical argument questioning the legitimacy of its claim to professional status. Indeed, in this section we explored several ways in which design strategically and instrumentally enables consumerism, which has tremendous negative consequences for society. Among these, the climate emergency and the environmental crisis at large are for some the overarching ethical issues of the day. They represent an existential threat for humans and non-human animals, for plants, and for the whole natural world itself, and, even, for future generations[39].

This too seems to be a serious challenge for the professional status of design. In the next section, we will consider a reply to this objection.

REPLY TO THE OBJECTION OF CONSUMERISM

For the critics, given that design plays an important role in consumerism, a phenomenon that has been linked to psychological, social, and environmental harms, serious doubts arise about the quality of design's commitment to the public. Just like I did when I dealt with manipulation, I will not attempt to refute this specific negative argument. Again, I will seek to provide an alternative interpretation of what this argument could mean for design.

It is my contention that the defence I presented previously when dealing with manipulation can be aptly mobilised to defend the professional status of design also from consumerism. That is, that a new profession merits extra leniency. Some of you might agree with me, whereas others may not be fully persuaded and be inclined to find that the challenge consumerism poses for design goes to its very core. Because of that a different approach to the reply may be necessary.

Design Is Not an Island

The consumerism objection is indeed strong, especially because designers are *closely* involved in the development of consumer culture. Again, I will not try to refute the attack, but I will attempt to disarm its possible implications for the professional status of design by providing a different interpretation for the challenge.

We could start by considering that design is not the only profession that is enabling consumerism. A great number of (putative) professions are involved in it: lawyers, notaries, financial planners, financial analysts, engineers, accountants, psychologists, economists, business consultants, managers, human resource experts, technical writers, anthropologists, sociologists, etc. Of course, not all engineers and not all lawyers, and certainly not most anthropologists or sociologists work in sectors related to consumer culture. But ultimately, given the depth and breadth of consumerism in our culture, arguably only very few professions would survive an attack if an accusation of contributing to consumerism would taint an occupation and prevent it from making a claim to professional status.

This seems to be an unproductive course of analysis. We ought to be aware of the wider socio-cultural and structural context of consumerism and come to see it is a broad societal problem that *also* affects the professions. At a practical level, moreover, it would not make much sense to strip a great deal of them of their professional status; all we would get from that is professionalism without professions.

A Nascent Critical Tradition

Perhaps we can probe the matter more constructively. Is a forceful critique of consumerism fully incompatible with defending a professional status for design? It need not be. Accepting the fact that we live in a consumer society and that being a consumer is a central part of the human experience in most societies does not carry the implication that a designer must adopt an acritical or supportive view of consumerism. On the contrary, reflecting on an ethics of consumption is one of the great challenges at the present time for design and for the other professions involved in consumer culture. In fact, the general community orientation of professionalism could offer designers a direction for this reflection. There is already a certain consistent corpus of critical reflection produced within design on the perils of consumerism and on other related social issues[40].

It might be illustrative to cite a few canonical passages to give a sense of the critique. In 1964, the British graphic designer Ken Garland published his seminal *First Things First* manifesto, at a time when the British economy was booming, and mass consumption was entering homes in wealthier nations:

> We do not advocate the abolition of high pressure consumer advertising: this is not feasible. . . . But we are proposing a reversal of priorities in favor of the more useful and more lasting forms of communication. We hope that our society will tire of gimmick merchants, status salesmen and hidden persuaders, and that the prior call on our skills will be for worthwhile purposes[41].

More or less at the same time that Garland published his critique, Victor Papanek started to write *Design for Real World*, which was published in English in 1971. The first sentences of its preface are one of the most often quoted passages in design literature:

> There are professions more harmful than industrial design, but only a very few of them. And possibly only one profession is phonier. Advertising design, in persuading people to buy things they don't need, with money they don't have, in order to impress others who don't care, is probably the phoniest field in existence today. Industrial design, by concocting the tawdry idiocies hawked by advertisers, comes a close second[42].

The last sentence of the preface has been cited fewer times, but is no less important and urgent:

> As socially and morally involved designers, we must address ourselves to the needs of a world with its back to the wall, while the hands on the clock point perpetually to one minute before twelve[43].

If we fast-forward fifty years to more recent writings, we find that the critical tradition not only goes on, but also gained a more strategic vantage point. Hereby there is a growing recognition that design can meaningfully contribute to tackling complex, systemic problems with an approach that is 'place-based and regional, yet global in its awareness'. Design scholar Terry Irwin writes in 2015:

> Fundamental change at every level of our society, and new approaches to problem solving are needed to address twenty-first-century 'wicked problems' such as climate change, loss of biodiversity, depletion of natural resources, and

the widening gap between rich and poor. Transition Design is a proposition for a new area of design practice, study, and research that advocates design-led societal transition toward more sustainable futures. This reconception of entire lifestyles will involve reimagining infrastructures including energy resources, the economy and food, healthcare, and education[44].

These passages show that a nascent but robust critical tradition exists within design. This is not the place to include more writings and manifestos or to enumerate countless design initiatives and movements with a focus on social or environmental issues. It must suffice to create the awareness about the existence of a broad movement within design that contests the 'notion of continuous production and consumption and its inherent, unsustainable, economic growth'[45]*.

These critical approaches also affect designers working on corporate projects who are changing their perspective and are more focussed on designing 'products that consumers will want to keep for a prolonged period'[46]. The issue of planned obsolescence and consumerism in general, however, is obviously not a question for designers only. It is a complex interplay of different actors beside designers, producers, consumers, and policy makers.

Design educator Katherine McCoy lamented in 1993 that 'we have trained a profession that feels political or social concerns are either extraneous to our work or inappropriate'[47]. My feeling on the issue is that this has somewhat changed in the last two decades, as the training of designers has incorporated many social and environmental themes into the curriculum, but alas, perhaps not on a sufficiently wide scale. In design we do not find the gamut of educational literature and textbooks we find in other new professions such as nursing or social work. Given the urgency of the matter the pace of change needs to accelerate and permeate the profession; obviously, this book is also a contribution in that direction. A profession and its professional apparatus—universities, associations, conferences and events, journals, and other forms of knowledge sharing—are a suitable platform to promote a change that can influence practitioners, as well as the whole ecosystem of educators, students, companies, policy makers, and civil society at large.

The challenge of consumerism is indeed a radically serious problem, but design is generating its own antibodies to resist the assault. This internal resistance can also be seen as a clear manifestation of the two main elements of professionalism in action: an intellectual reflection on the profession that is needed to generate the critical perspectives and the commitment to the wellbeing of society that guides that reflection.

If my arguments so far have been persuasive, although both objections (the role design plays in manipulation and consumerism) do pose serious challenges to design, they do not offer sufficient reason to fundamentally question its public service element and reject the legitimacy of design's professional status. If anything, the upshot of the objections I have raised points us in the direction of the necessity to sustain and perhaps increase reflective inquiry into the ethical challenges of the design profession.

THIRD OBJECTION: UNINTENDED CONSEQUENCES

In this section, we shall introduce the concept of unintended consequences (or effects) and examine how they pose a serious problem to design professionalism. The aim is not anymore to directly challenge the professional status of design, but to question the extent to which design has control over its own service to society. This objection assumes that the commitment and orientation toward society *are* sufficiently present in design to make it count as a profession, but argues that this commitment is tainted because its effects escape the control of the designers.

Unintended consequences are the unplanned outcomes of the implementation of an action. In technology studies, the notion is invoked to mean that 'things happened outside the scope of the original intent because no one could have known in advance how a technology would be used'[48*]. Unintended consequences can be negative, positive, or neutral, but the actual valuation finally depends on the observer's perspective. Some unintended consequences are called 'perverse' because they produce exactly the opposite of the intended result.

To illustrate with an example: according to some reports, the introduction of the energy-saving lightbulb may have actually enticed people to use them in places previously left unlit, contributing thus to *increasing* energy consumption or cancelling out the energy saving[49]. Similarly, using energy-efficient domestic appliances can save people money on energy, which can be used to buy other things that they were not previously buying, which can still lead to a similar or higher energy consumption. This is called the 'Jevons paradox', named after William Stanley Jevons, who in 1865 observed that an increased efficiency of coal use led to the production of more goods per unit of coal, which in turn lead to *higher* coal consumption[50].

When we speak of unintended consequences we assume thereby a high degree of uncertainty at the time of deciding regarding the eventual outcome. When the consequences of our action can be foreseen with greater certainty, it is then preferable to speak of 'side effects' to refer to the consequences that were not directly intended but somehow foreseen[51*]. We do that with aspirin, for instance—we say an upset stomach is a possible side effect because we know it could happen, yet we would not call that an unintended consequence.

We could consider the case of lead-based paint to explore this issue. Lead-based paint offers many objective benefits: accelerating drying, resisting corrosion, and increasing durability among them. Lead, however, is now widely known to be a cumulative toxicant that affects multiple body systems. In the case of paint, one can assume that manufacturers of paint do not directly intend to cause harm to people. Does this mean that lead poisoning can be considered an unintended consequence of lead-based paint? Hardly anymore. In fact, lead poisoning has been reported since Roman times[52], and efforts to ban lead paint date back to the early 1920s, but it is still widely available throughout the developing world[53]. Perhaps lead poisoning could have been an unintended consequence a century ago, but given the available evidence it is no longer legitimate to continue calling it that *even* if poisoning people is not the intention of the producers.

Thanks to available information, trade-off analyses can be made that indicate that the health risks of lead-based paint for humans and non-human organisms are simply too high, and the potential negative effects of lead-based paint greatly outweigh its obvious benefits. Given that exposure to lead is known to be always harmful, lead-based paint becomes inadmissible and is considered to be a major public health concern. On occasions, some potential harms may be tolerated, as happens with medicines or transportation vehicles. Mostly, the degree to which these risks are permissible depends on the potential good they could bring about and the severity of the harms they may cause.

Epistemic Uncertainty and the Difficulty of Calculating Trade-offs

This objection relates to the notion of 'epistemic uncertainty' we discussed in chapter 1; unintended effects are a manifestation thereof. It is because designers operate under epistemic uncertainty that unintended effects occur. Moreover, because design deals with intractable, 'wicked' problems, its effects can extend indefinitely in the future. Because of this, calculating its trade-offs can be intrinsically more difficult than calculating those of lead-based paint or even of a particular medicine, for which highly regulated, albeit imperfect, clinical trials are put in place where researchers, regulators, medical associations, and society at large learn about the safety and potential efficacy of a medicinal drug. Although some very influential voices from the artificial intelligence community are calling for similar tests for algorithms given the permanent side effects they cause on society, such tests are nowhere near being a plausible scenario for design as a whole[54].

Consider now the example of cars. It will surprise no one to say that cars create greenhouse gases and air pollution. Alas, as of now, even electric ones do so, despite being considerably greener. They still need dirty non-renewable energy for both their production and to power the batteries in most countries[55]. They also injure and kill people. What is more, they are a con-tributing factor for the appearance of suburban sprawl, which has profound negative social and environmental effects. Many scholars have linked car culture to environmental and urban degradation[56]. Furthermore, as Jane Ja-cobs famously argued, cars are a contributing factor toward making cities less habitable as car-centric urban policies destroy the social fabric of com-munities by isolating individuals as a result of privileging car mobility[57].

At the same time, though agreeing about the harm cars cause, it seems reasonable to argue that it is not obvious who could have prevented the situation and who should be blamed for it now. Some sceptics could, more-over, even question if anybody should be blamed at all; if only because the effects of car emissions on the environment were not discovered until the early 1950s, years after the mass adoption of cars[58]*, which makes this a clear-cut case of unintended consequences. This issue is also related to what is known as the 'problem of many hands', whereby 'due to the complexity of the situation and the number of actors involved, it is impossible—or at least, very difficult—to hold someone reasonably responsible'[59].

A clarification is in order here. By unintended consequence I do not mean what Wade L. Robison calls 'error-provocative designs', which are designs that are going to provoke errors no matter how intelligent, motivated, or well-trained the user is[60]. Remember the 'butterfly ballot'? That is one exam-ple. And those doors that do not communicate whether they need to be pushed, pulled, or slid are another. My favourite is the USB-A port. You might have seen the internet meme: 'TWO WAYS TO PLUG IT IN—TAKES THREE TRIES'[61].

No, we are referring to actual effects that are outside the designer's origi-nal intentions for their design but go beyond the individual and affect the social realm, even at a small scale. Consider, for example, a new type of unexpected traffic risk posed by the first generation of electric and hybrid cars. Instead of the common rumbling sound of a gas engine, these cars only produce an almost inaudible hum when moving at very low speeds. This makes it difficult for pedestrians to be aware of their proximity and creates a dangerous situation for all pedestrians but especially for vulnerable persons who rely on a combination of acoustic and visual warning signals[62]. Fortu-nately, this issue is not difficult to fix with the fitting of a so-called acoustic vehicle alerting system, which is already mandatory in EU countries and in the United States starting in September 2020[63].

Similarly, design gets mobilised to deal with problems that it unintended-ly helped create; that is, to 'solve' unintended effects. Designers, for in-stance, create effective anti-pollution masks for urban cyclists, but the need for these masks arises only because polluting cars exist in the first place. Sigmund Freud made a similar point almost a century ago: 'If there had been no railway to conquer distance, my child would never have left town and I should need no telephone to hear his voice'[64].

This discussion of unintended consequences opens the door for a serious challenge against the public service element of design. The problem design faces is that something may appear at first beneficial to society (or for the sake of argument, harmless) when potential consequences are assessed on the basis of available information, but turn out to cause massive harm when implemented, as we have seen in the examples reviewed.

Unpredictable Patterns

We can complicate the point even further by noting that the consumption, adoption, and use of products is not a passive act but a dynamic process through which people engage with products in ways other than those initially intended or foreseen by designers. This is an issue that has been explored by many sociologists, historians and philosophers of technology, and anthropologists who emphasised the significance of the social element in the functioning of artefacts[65].

An important contribution is the notion of 'multistability', developed by the philosopher of technology Don Ihde to refer to the 'different trajectories of use' any product can have[66]. All technologies can possibly exceed the intent of the designer, as they could display many other possibilities than those originally intended as the designed functions. 'No technology is "one thing"', Ihde argues[67]. One can use a hammer to hit a nail, which is presumably its intended use, but a hammer can also become a piece of art when used by an artist as expressive material. Evidently, it could also be used as a murder weapon or to crack nuts. These possibilities are not fully determined by the properties of the product itself, but by users and uses embedded in specific cultural and political contexts. Multistability makes thus unintended effects an intrinsic part of design.

Fortunately, not all unintended effects are necessarily negative. There are a myriad of examples of *multistable* designs whose consequences appear positive or neutral at first sight: technical slings designed for lifting cargo containers or other heavy materials can be used for fitness exercises, such as so-called suspension training in which users work against their own body weight. Lighters can be used to light a cigarette or a candle but also to open a beer bottle or held in the air at a music concert to honour a musician or to request an encore. The flashlight functionality available in cellphones quickly became used by rock concertgoers who began to wave their phone flashlights over their heads in the most emotional part of the gig as a replacement for the classic, but by then dwindling, cigarette lighters. Of course, we are usually more concerned with the negative consequences. For instance, LGTB dating apps that are initially designed as community-building or communication tools might end up being used for entrapment and abuse[68]. However, as one should keep in mind that design's effects can extend indefinitely into the future, this evaluation in terms of positive or negative is always temporary and, in most cases, relative to the observer's position. Multistability acquires a dramatic character when the weapon industry adopts innovations originally developed for 'civilian' use such as mixed reality smart glasses or motion-sensing input devices, which were initially intended for entertainment[69].

Although there are different views regarding the degree of interwoven-ness between users, artefacts, and the context in which they are used[70*], there is general consensus that the way a product is used is determined by more factors than merely the intention of the designer and the product itself. Especially for designs with symbolic meanings, its social and cultural signifi-cance is contingent on the context in which that product is embedded. To refer to this indeterminacy and a fallacious sense of control, Ihde speaks of the 'designer fallacy', for 'the notion that a designer can design into a tech-nology, its purposes and uses'[71]. In line with Ihde, philosopher of technology Peter-Paul Verbeek warns us to stay on guard because 'there is no unequiv-ocal relationship between the activities of designers and the mediating role of the technologies they are designing'[72].

Unintended consequences are the factual manifestation of the 'epistemic insufficiency' under which designers operate; because of this, they, as De-Martino said of economists, might have 'influence without control'[73]. The third objection could be then summarised as follows: *design's contribution to society has a volatile and unpredictable nature as it depends on many factors that escape the control of the designer. Moreover, as the purview of design becomes broader and strategic, its impact on society is ever more significant. Because of this, well-intending designers can cause serious harm as they try to do good.*

UNINTENDED CONSEQUENCES REVISITED: TAMING THE UNCERTAINTY

The discussion of unintended consequences highlights several elements that have important implications for design. Although the objection does not necessarily seek to challenge its professional status, it does seriously ques-tion the extent to which designers control their outcomes, which raises con-cerns regarding the harms they can cause as they try to solve other problems.

Yet is it always true that design's contribution is so volatile? And is it always the case that designers cannot exercise control? In light of the intrin-sic epistemic complexities and uncertainties of technological multistability, can we do anything other than throwing up our hands in the air and hope for the best?

Anticipating Effects

We surely can do more than that. During a project, designers can work to anticipate potential negative effects of their design decisions. That is one way to *tame* uncertainty. Admittedly, fully foretelling the use and effects of design is impossible, but that is simply because fully foretelling the future in a non-deterministic world like ours is an epistemic impossibility. This realisation should not make us despair as it does not entail having to abandon critical reflection and speculation on the possible future scenarios that are envisaged and delineated.

In fact, and just to illustrate this possibility, product manufacturers already do this to protect themselves from liability claims, especially 'strict' liability. Although it varies per country, under strict liability a manufacturer could be held liable for injuries sustained by users in certain *unintended* uses, if these should have been *reasonably foreseen* by the manufacturer. A quick example may clarify the issue: a toy containing small parts designed for children aged between three and six years may also may also have some characteristics that are appealing to children under three years of age. In this case the manufacturer needs to foresee this possible unintended use *before* releasing the toy to the public and include a warning about the small parts presenting a choking hazard. For a similar reason, collectibles such as dolls or cars intended for grown-ups may be labelled with a 'this is not a toy' label. The manufacturer foresees that children may want to play with them, which would be an unintended use, and decides to include that label to protect themselves from liability. To reiterate, I include this vignette about liability only to illustrate the real possibility of performing a mental exercise to imagine future plausible scenarios. I do not want to create the impression that avoiding liability is central to our discussion. Still less do I want to encourage designers to make decisions out of fear of litigation.

There are certainly other design-centric ways that designers could use to explore multistability other than being concerned about strict liability. Several scholars propose different courses of action to deal with the uncertain trajectories of design. Don Ihde outlines a series of 'prognostic pragmatics' that could serve minimal heuristic purposes and guide foresight and reflexion around the possible multiple trajectories of a design, albeit in general terms and without determining particulars[74]. Considering that designed artefacts mediate human experiences and practices in the world, designers would do well to harness their imagination skills to do a 'mediation analysis', as recommended by Verbeek[75]. Mediation analysis is guided by practical and philosophical principles, and it aims at exploring and anticipating possible uses of an artefact. It can reduce uncertainty and better enable us to make an informed prediction on the possible ethical effects it could bring about, if

only because unintended consequences are often—but not only—due to failures of 'imagination and anticipation'[76].

Anticipation and Mental Simulation

Imagining, evaluating, and simulating possible future scenarios are key aspects of the design activity. Researchers call this process 'mental simulation'
to refer to the generation of hypotheses about possible future scenarios in
order to explore alternative courses of action or speculate about how a situation can evolve[77]. This process plays a key role in understanding design
situations and in decision making. Designers systematically use simulation
for making predictions about the future use of a design ('people will plug in
the device and then start it by pressing the power button'), for instance; for
evaluating formal or aesthetic issues ('their brand is red, that colour will
clash with this photograph, it won't look nice'); but also for assessing the
likelihood of a particular behaviour ('that power button won't work: people
will look the grooves on its sides and will try to turn it like a knob instead of
pushing it like a button'). Through mental simulation designers highlight the
adequacy or inadequacy of a particular design alternative. Thanks to a process of mental simulation, they generate and assess design solutions and sub-
solutions both intuitively and analytically. The process covers everything
from the simulation of partial or very specific parts of a solution to the
simulation of integrated and complete solutions. These simulations are often
not purely mental. Designers also use tools and techniques (drawing, for
instance) to aid the simulation process. Donald Schön describes this process:

> The designer asks himself, in effect, 'What if I did this?' where 'this' is a
> move whose consequences and implications he traces in the virtual world of a
> drawing or model. Making a design move in a situation can serve, at once, to
> test a hypothesis, explore phenomena, and affirm or negate the move[78].

Schön offers us a fairly eloquent description of the mechanics of the
process that designers follow to explore the consequences and implications
of a design move. His description makes it easy to envision how these speculative what-if moves might take place in the designer's mind, and how the
designer could mobilise principles and knowledge to make decisions. This
allows the designer to generate solutions that often seem to work as intended.
In chapter 3 we reviewed numerous examples; there is no need to introduce
new ones to make the point anew. Multistability allows for other uses, but
that does not make all design solutions volatile.

Guiding Principles and Foresight Strategies

Naturally, simulation and evaluation do not happen in a vacuum. There is a vast body of knowledge that designers can use when generating and evaluating their designs. They can use technical standards, guidelines, laws, and norms. For example, designers can also use design principles related to specific disciplines such as the Jakob Nielsen's ten heuristics for user interface design or Google's 'Material Design' principles. Scientific findings regarding perception, models of human behaviour, or ergonomics can also be used to inform or assess their designs. Aesthetic aspects are also judged against known and tacit aesthetic ideals, such as *minimalism* or *symmetry* (present, for instance, in influential Swiss graphic design). Referring to this, Bryan Lawson writes: 'designers usually develop quite strong sets of views about the way design in their field should be practised'[79]. Sometimes, designers develop their own principles and include ethically laden issues. One such example is Dieter Rams' ten principles for good design, which emphasise designing products that are long-lasting and honest[80]. In sum, there is a multiplicity of conceptual and theoretical tools to assist designers in their technical judgement.

Simulation in general, and particularly evaluation, is not only about deciding features of use, aesthetic aspects, or functionality in the strict sense. It is also about detecting risks and harms by envisioning possible unintended effects as well as intended ones. So the volatility and the unpredictability of design cannot be eradicated but they can be domesticated through a commitment to exercising foresight. Arguably, guidance might come from different sources such as colleagues, upbringing, social norms, philosophy, professional organisations, or personal morality[81].

Besides declarative knowledge, design has generated ethics-centred frameworks and methods that designers can use in their daily practice. To take one example, there is a diversity of methods and approaches that can be grouped under the 'Design for Value' umbrella, which seek to 'develop technology in accordance with the moral values of users and society at large'[82]. Value-sensitive design is perhaps the most popular approach, but there are others[83].

A successful simulation that detects all possible unintended effects seems to be impossible in principle, but guaranteed results cannot be expected from any other profession either. Perfection is not what professionalism is about. It would be unrealistic to expect a designer to be able to predict the future as if they were Laplace's demon, the mythical intellect imagined by the French scholar Pierre Simon Laplace in the early 1810s, which is capable of knowing the future.

Admittedly, even if superhuman powers are not to be expected of designers, these exercises in foresight are bounded by several intertwined factors that affect them directly: the multistable nature of products, the epistemic insufficiency in which designers operate, as well as the quality and extent to which they exercise their imagination and anticipation capacities. Arguably, there are many other external constraints that may be relevant too: the institutional culture where the designer operates, expected time given for a project, vicinity of deadlines, budgets, or team composition.

To conclude, the unintended consequences of design do pose a serious challenge for the profession. Moreover, the many examples from this chapter illustrate that serious ethical entailments arise from unanticipated effects, but also from the role of design in manipulation and consumerism. Notwithstanding this, if my arguments are cogent, the objections do not pose an unresolvable challenge. They only highlight the necessity of tackling the complex ethical issues that design generates.

Design contribution to society is not fully volatile and unpredictable: design does function according to plan more often than not, even if multistability allows for different uses than the ones originally intended. Moreover, designers seem to be well-equipped to tame at least some of the uncertainty and exercise a reasonable degree of foresight, inasmuch as that can be done in an uncertain world.

WHERE ETHICS AND DESIGN MEET

The examples we have considered throughout these first four chapters could be categorised into two main categories or junctures where ethics and design meet[84]. The first is connected with the design decisions that designers make *during* design activity and could be termed the *ethics of designing.* The second juncture could be called the *ethics of the designed,* and it is related to assessing and theorising about the ethical significance of the outcomes of design activity *once* they have entered the realm of artefacts that exist in the world. Naturally, one could signal other minor points of connection and also alternative ways of conceptualising the intersection between design and ethics[85]*.

The first juncture, the *ethics of designing,* concerns the ethical challenges and demands designers encounter *during* the practice of design; that is, as designers seek to envision what does not yet exist. To illustrate, imagine your task is to design a registration form for a digital service. Your client or your boss asks you to include a field so that people enter their gender. You wonder whether this field needs to be included at all. Then you consult with your boss or the client and, let us assume for the sake of argument, that they give you a very good reason for including the field. You then wonder if it needs to be a mandatory or an optional field. And many other questions arise. Would you use a pull-down menu with closed options or an open field so people can enter how they identify as in their own words? Would you include binary or non-binary options? Would you allow people to select multiple genders? Would you ask for preferred pronouns instead of their gender?

In sum, the ethics of designing is mainly about generating and assessing possible courses of action during design activity; one could say that it is roughly concerned with figuring out what to do and what to avoid doing related to the different future scenarios that every design alternative opens up. At the same time, the ethics of designing also involves the post facto examination of the actions, decisions, and intentions that may have played a role in design decision making. This dual nature of the ethics of designing mirrors the two types of professional reflection signalled by Donald Schön: in-action and on-action reflection[86].

The second juncture, the *ethics of the designed*, involves a reflection about and around the effects of an existing design on the world as a whole, including both human and non-human stakeholders, and even on the future. It could also involve scrutinizing the potential effects of *hypothetical* arte-facts: for instance, artificial intelligence doctors, cars on Mars, and end-of-life-care robots. The locus of analysis is thus on an *outcome* of design, whether real or merely a possible concept. Furthermore, the analysis is not only on the outcomes themselves, but also on the new possibilities to access different lifeworlds that those designs open up.

The conclusion that serious ethical entailments arise from professional design activity should not be surprising: design is fraught with ethical questions that require deeper ethical reflection. But it seems deeply inadmissible to leave it at that and eschew the ethical burdens. As a famous philosopher once affirmed, taking our conclusion seriously means acting upon it. What is more, *the legitimacy of the professional status of design is contingent on dealing seriously with these and similar challenges; the continuity of the leeway an emergent profession can expect is predicated on the degree of commitment that professionals exhibit towards professional ideals*.

But what could we do, then, if we want to take our conclusion seriously? What could we do to act upon it? From here, there are at least two courses of action; we will discuss them in the following.

First Course of Action: Deprofessionalisation

The first course of action would be to revisit the radical critique, the attack against professionalism we reviewed in chapter 2. Using Ivan Illich's termi-nology, we could argue that design causes 'iatrogenesis', a notion Illich uses to refer to sickness caused by doctors and different levels of 'pathogenic medicine, . . . damage that doctors inflict with the intent of curing'[87]. This notion seems apt to be applied to design, which could be seen to be iatrogenic because of the harms designers cause while intending to do good. This, in turn, could lead to a renewed call for the deprofessionalisation of design. After all, the design profession seems to be failing the very people it is purported to help. Our discussion of manipulation, consumerism, and unintended consequences could provide the basis for developing a claim in this direction.

However, this is not the course of action I would suggest taking. There is, I believe, very little to be gained by deprofessionalising design. The problem with this course of action is that design's ethical afflictions will not just go away with its deprofessionalisation. The need for design services will not go away because goods and services would still have to be designed; someone else than professional designers would provide them.

While persuasively claiming that professional design *is nothing but* an enabler of a socio-environmental catastrophe at a planetary scale[88*], the advocates of deprofessionalisation may at worst contribute to fostering the very behaviours they seek to curtail. After all, non-professional designers would still do the work, but would be primarily concerned with technical rationality and not with the ethical considerations that are *expected* of professionals. The critical reflection would occur *outside* the profession, the realm where it is most needed, and be ineffectual. Business as usual would simply carry on, and the moral project of serving people in the 'accomplishment of their individual and collective purposes' that design professionalism represents would be endangered.

Second Course of Action: An Inquiry into Design Professional Ethics

There is a different and more promising course of action to deal with the challenge posed by the critics: an enquiry into design in the form of design professional ethics. The challenges designers encounter in the exercise of their profession call for promoting and developing a reflective enquiry into design professional ethics as a plausible and necessary first step toward actively practicing their commitment toward society. As I just pointed out, maintaining professional status is contingent on adequately dealing with these challenges; because of this, the inquiry will focus on the most basic level of the design profession: its very grounds. This second course of action deserves much further discussion, which we continue in the next chapter.

To wrap up, in this chapter we have reviewed and discussed three serious challenges targeting the quality of design's public service element, and with it the ethical core of design professionalism. The challenges prove to be critical in themselves. At the same time, however pressing, they do not by themselves constitute a mortal threat to the professional status of design as long as designers remain structurally committed to realising the profession's public service objectives. An inquiry into professional ethics, which is a necessary first step in this direction, will be delineated further in the next chapter.

NOTES

1. See the further reading section at the end of the book.

2. Robert Noggle, 'The Ethics of Manipulation', Stanford Universityhttps://plato.stanford.edu/archives/sum2018/entries/ethics-manipulation/.

3. I thank strategic designer César Astudillo for this formulation.

4. Harry Brignull, 'Dark Patterns',https://darkpatterns.org/.

5. Martin Eisend, 'Explaining the Impact of Scarcity Appeals in Advertising: The Mediating Role of Perceptions of Susceptibility', *Journal of Advertising* 37, no. 3 (2008).

6. A 'cookie' is a small text file that a website saves on the user's computer or mobile device every time a site is visited. Among other things, it enables the website to remember the user's actions, browsing history, display preferences, or log in status over a period of time.

7. David Butler, 'Consumers Union Praises Senator's Call for Ftc Investigation of Airline "Dynamic Pricing"', news release, 2018.

8. BEUC The European Consumer Organisation, *Dynamic Currency Conversion: When Paying Abroad Costs You More Than It Should* (Brussels: The European Consumer Organisation, 2017).

9. Forbrukerrådet, *Deceived by Design* (Oslo: Forbrukerrådet, 2018).

10. Kevin Litman-Navarro, 'We Read 150 Privacy Policies. They Were an Incomprehensible Disaster',https://www.nytimes.com/interactive/2019/06/12/opinion/facebook-google-privacy-policies.html.

11. Michael J. Phillips, *Ethics and Manipulation in Advertising* (Westport: Quorum, 1997).

12. Adam Alter, *Irresistible: Why Are You Addicted to Technology and How to Set Yourself Free* (London: Vintage, 2017); Cal Newport, *Digital Minimalism: On Living Better with Less Technology* (London: Penguin Business, 2019).

13. Nir Eyal, *Hooked: How to Build Habit-Forming Products* (New York: Portfolio Penguin, 2014), 176–78.

14. Many ethical objections have been raised against nudges, as well as strong arguments in their defence. For a taxonomy and an assessment of the objections, see Carl Sunstein, 'The Ethics of Nudging', *Yale Journal on Regulation* 32, no. 2 (2015). For a contrarian position, see Mark D. White, *The Manipulation of Choice: Ethics and Libertarian Paternalism* (New York: Palgrave Macmillan, 2013).

15. See, for example, Stephen L. George and Marc Buyse, 'Fraud in Clinical Trials', *Clinical Investigation* 5, no. 2 (2015); Katherine Eban, *Bottle of Lies: The inside Story of the Generic Drug Boom* (New York: Ecco Press, 2019); Editorial Board, 'Strengthening the Credibility of Clinical Research', *The Lancet* 375, no. 9722 (2010).

16. For example, Luc Bovens, 'The Ethics of Dieselgate', *Midwest Studies In Philosophy* 40, no. 1 (2016). In the 'Dieselgate' scandal 11 million VW diesel cars were equipped with a 'defeat device' that would engage in full emissions control during regulatory testing to improve results but not on real roads.

17. Ronald R. Sims and Johannes Brinkmann, 'Enron Ethics (Or: Culture Matters More Than Codes)', *Journal of Business Ethics* 45, no. 3 (2003).

18. Hartmut Berghoff, '"Organised Irresponsibility"? The Siemens Corruption Scandal of the 1990s and 2000s,' *Business History* 60, no. 3 (2018).

19. Cecil W. Jackson, *Detecting Accounting Fraud: Analysis and Ethics* (Essex: Pearson Education, 2015).

20. Conrad Lodziak, *The Myth of Consumerism* (London: Pluto Press, 2002), 2.

21. Glenn Parsons, *The Philosophy of Design* (Cambridge: Polity, 2016), 138.

22. This point is informed by the discussion provided by Lodziak, *The Myth of Consumerism*, 1–7. I am, however, more lenient than Lodziak in my understanding of needs.

23. Christophe Bonneuil and Jean-Baptiste Fressoz, *The Shock of the Anthropocene: The Earth, History and Us* (London: Verso, 2017), 163.

24. Greg Castillo, *Cold War and the Home Front: The Soft Power of Midcentury Design* (Minneapolis: The University of Minnesota Press, 2010).

25. European Commission, 'Poverty, Middle Class and Purchasing Power',https://ec.europa.eu/knowledge4policy/foresight/topic/growing-consumerism/poverty-middle-class-purchasing-power_en.

26. World Bank, *Poverty and Shared Prosperity 2018: Piecing Together the Poverty Puzzle* (Washington, DC: World Bank, 2018).

27. Tim Kasser, *The High Price of Materialism* (Cambridge: The MIT Press, 2002).

28. Elizabeth Chin, *Purchasing Power: Black Kids and American Consumer Culture* (Minneapolis: University of Minnesota Press, 2001).

29. Richard G. Wilkinson and Kate Pickett, *The Spirit Level: Why Equality Is Better for Everyone* (London: Penguin Books, 2010).

30. See, for example, J. R. McNeill and Peter Engelke, *The Great Acceleration* (Cambridge: The Belknap Press of Harvard University Press, 2014). See also the further reading section at the end of the book.

31. For an introduction to global environmental and climate ethics from the perspective of global ethics, see Heather Widdows, *Global Ethics: An Introduction* (Durham: Acumen, 2011), 228–49.

32. The coffeepot became an iconic image in the design world as it was prominently featured in the cover of Norman's canonical book. Donald Norman, *The Design of Everyday Things*, revised and expanded edition (New York: Basic Books, 2013). The coffeepot for masochists is usually used to illustrate unusable or uncomfortable products; this interpretation is not wrong, but there is a more sophisticated reading behind the object's obvious absurdity and ridiculousness.

33. Bilbao International, 'Exposición: "Objetos Imposibles" De Jacques Carelman', http://www.bilbaointernational.com/en/exhibition-impossible-objects-by-jacques-carelman/.

34. Tim Cooper, 'Planned Obsolescence', in *Encyclopedia of Consumer Culture*, edited by Dale Southerton (Thousand Oaks: SAGE Publications, 2011).

35. Tim Cooper, 'Slower Consumption: Reflections on Product Life Spans and the "Throwaway Society"', *Journal of Industrial Ecology* 9, no. 1-2 (2005): 55.

36. European Commission, 'New Rules Make Household Appliances More Sustainable', news release, 2019.

37. Roger Harrabin, 'EU Brings in "Right to Repair" Rules for Appliances', BBC,https://www.bbc.com/news/business-49884827.

38. Victor Papanek, *Design for the Real World*, second revised edition (Chicago: Academy Chicago Publishers, 1984), ix.

39. Robin Attfield, *Environmental Ethics: An Overview for the Twenty-First Century*, second edition (Cambridge: Polity Press, 2014); James Garvey, *The Ethics of Climate Change: Right and Wrong in a Warming World* (London: Continuum, 2008).

40. See, for example, Elizabeth Resnick, *The Social Design Reader* (London: Bloomsbury, 2019).

41. Ken Garland, 'First Things First',http://www.kengarland.co.uk/KG-published-writing/first-things-first/.

42. Papanek, *Design for the Real World*, ix.

43. Ibid., xiv.

44. Terry Irwin, 'Transition Design: A Proposal for a New Area of Design Practice, Study, and Research', *Design and Culture* 7, no. 2 (2015): 229.

45. A description of many of these movements can be found in Alastair Fuad-Luke, 'Adjusting Our Metabolism: Slowness and Nourishing Rituals of Delay in Anticipation of a Post-Consumer Age', in *Longer Lasting Products: Alternatives to the Throwaway Society*, edited by Tim Cooper (Surrey: Gower, 2010), 133. For a more radical perspective, the interested reader might want to consult Arturo Escobar, *Designs for the Pluriverse: Radical Interdependence, Autonomy, and the Making of Worlds* (Durham: Duke University Press, 2017), chapter 5. See also Sasha Costanza-Chock, *Design Justice: Community-Led Practices to Build the Worlds We*

Need (Cambridge: The MIT Press, 2020), for an exploration of how community-led design can be operationalised against structural inequalities.

46. Cooper, 'Planned Obsolescence', 1096.

47. Katherine McCoy, 'Good Citizenship: Design as a Social and Political Force', in *The Social Design Reader*, edited by Elizabeth Resnick (London: Bloomsbury, 2019), 138.

48. Sheila Jasanoff, *The Ethics of Invention: Technology and the Human Future* (New York: W.W. Norton & Company, 2016), 24. For other definitions and comprehensive theoretical treatment of the concept, see Adriana Mica, 'Unintended Consequences: History of the Concept', in *International Encyclopedia of the Social & Behavioral Sciences*, edited by James D Wright (Amsterdam: Elsevier, 2015).

49. Peter-Paul Verbeek, *Moralizing Technology: Understanding and Designing the Morality of Things* (Chicago: The University of Chicago Press, 2011), 58.

50. Richard York, 'Ecological Paradoxes: William Stanley Jevons and the Paperless Office', *Human Ecology Review* 13, no. 2 (2006).

51. I tried to keep this discussion as non-technical as possible and, in order to do so, I took some terminological liberties. The whole discussion, obviously, is related to the doctrine of the double effect. In this doctrine 'intention' must be understood as a *direct* connection between foreseen outcome and intention. Contrarily, a person is said not directly to intend the foreseen outcome, if the result, however regrettable, is not the end they *directly* seek nor means to their ends. This doctrine accounts for the permissibility of an action that causes a serious harm as an unintended effect of achieving some good end. Particularly important is the distinction the doctrine makes 'between what we do (equated with direct intention) and what we allow (thought of as obliquely intended)'. Philippa Foot, 'The Problem of Abortion and the Doctrine of Double Effect', *Oxford Review* 5 (1967).

52. D. A. Gidlow, 'Lead Toxicity', *Occupational Medicine* 54, no. 2 (2004): 76.

53. Perry Gottesfeld, 'The West's Toxic Hypocrisy over Lead Paint',https://www.newscientist.com/article/mg21829190-200-the-wests-toxic-hypocrisy-over-lead-paint/.

54. Olaf J. Groth, Mark J. Nitzberg, and Stuart J. Russell, 'AI Algorithms Need FDA-Style Drug Trials',https://www.wired.com/story/ai-algorithms-need-drug-trials/.

55. Jake Whitehead, Robin Smit, and Simon Washington, 'Where Are We Heading with Electric Vehicles?' *Air Quality and Climate Change* 52, no. 3 (2018).

56. Bonneuil and Fressoz, *The Shock of the Anthropocene*; and McNeill and Engelke, *The Great Acceleration*.

57. Jane Jacobs, *The Death and Life of Great American Cities* (New York: Random House, 1961).

58. For a fascinating account of how Caltech researcher Arie Haagen-Smit first linked air pollution to cars in the early 1950s, see Chip Jacobs and William J. Kelly, *Smogtown: The Lung-Burning History of Pollution in Los Angeles* (Woodstock: The Overlook Press, 2008), 69–99.

59. Ibo Van de Poel, 'The Problem of Many Hands', in *Moral Responsibility and the Problem of Many Hands*, edited by Ibo Van de Poel, Lambèr Royakkers, and Sjoerd D. Zwart (New York: Routledge, 2015), 50.

60. Wade L. Robison, *Ethics within Engineering: An Introduction* (London: Bloomsbury, 2017), 22–28.

61. The definitive book on unusable, confusing objects is Norman, *The Design of Everyday Things*.

62. Stephan Brand, Maximilian Petri, Philipp Haas, Christian Krettek, and Carl Haasper, 'Hybrid and Electric Low-Noise Cars Cause an Increase in Traffic Accidents Involving Vulnerable Road Users in Urban Areas', *International Journal of Injury Control and Safety Promotion* 20, no. 4 (2013).

63. European Commission, 'Commission Welcomes Parliament Vote on Decreasing Vehicle Noise', news release, 2014,https://ec.europa.eu/commission/presscorner/detail/en/IP_14_363; U.S. National Highway Traffic Safety Administration, *Federal Motor Vehicle Safety Standard No. 141, Minimum Sound Requirements for Hybrid and Electric Vehicles* (Washington, DC: Federal Register, 2019).

64. Sigmund Freud, *Civilization and Its Discontents*, reprint edition (New York: W. W. Norton & Company, 2010), 61.

65. For example, Wiebe E. Bijker, Thomas Parke Hughes, Trevor Pinch, and Deborah G. Douglas, eds., *The Social Construction of Technological Systems: New Directions in the Sociology and History of Technology*, anniversary edition (Cambridge: The MIT Press, 2012). See also the further reading section at the end of the book.

66. Don Ihde, 'Technology and Prognostic Predicaments', *AI & Soc* 13, no. 1-2 (1999).

67. Ibid., 47.

68. Heba Kanso, 'Amid Egypt's Anti-Gay Crackdown, Gay Dating Apps Send Tips to Stop Entrapment', Reuters, https://reut.rs/2itGG1d.

69. See, for example, Vincent Boulanin and Maaike Verbruggen, *Mapping the Development of Autonomy in Weapon Systems* (Solna: Stocholm International Peace Research Institute, 2017).

70. Social constructionists hold that society and technology are fully interwoven. Moderate realists reject this view, while allowing for a moderate degree of social embeddedness and the recognition that artefacts can carry symbolic meanings and have normative powers. I do not focus much attention on these debates because, at least as I see it, they are not central to my arguments. Moreover, most contemporary participants in the debate agree that technological artefacts are not mere neutral instruments that users simply use and different degrees of interplay between users and technologies take place. See also the further reading section at the end of the book.

71. Don Ihde, 'The Designer Fallacy and Technological Imagination', in *Philosophy and Design: From Engineering to Architecture*, edited by Pieter E. Vermaas, Peter Kroes, Andrew Light, and Steven A Moore (Dordrecht: Springer Netherlands, 2008), 51.

72. Verbeek, *Moralizing Technology*, 97.

73. George DeMartino, 'Epistemic Aspects of Economic Practice and the Need for Professional Economic Ethics', *Review of Social Economy* 71, no. 2 (2013): 175.

74. Ihde, 'Technology and Prognostic Predicaments', 50–51.

75. Verbeek, *Moralizing Technology*, 117.

76. Jasanoff, *The Ethics of Invention*, 24.

77. Gary Klein, *Sources of Power* (Cambridge: The MIT Press, 1998), chapter 5; Daniel Kahneman and Amos Tversky, 'The Simulation Heuristic', in *Judgment under Uncertainty: Heuristics and Biases*, edited by Daniel Kahneman, Paul Slovic, and Amos Tversky (New York: Cambridge University Press, 1982).

78. Donald Schön, 'Problems, Frames and Perspectives on Designing', *Design Studies* 5, no. 3 (1984): 132.

79. Bryan Lawson, *How Designers Think: The Design Process Demystified*, fourth edition (Oxford: Architectural Press, 2006), 160.

80. Dieter Rams, *Less but Better / Weniger, Aber Besser* (Berlin: Die Gestalten Verlag, 2014).

81. Some of these come from Richard Buchanan, 'Design Ethics', in *Encyclopedia of Science, Technology, and Ethics*, edited by Carl Mitcham (Detroit: Macmillan Reference, 2005), 507.

82. Jeroen Van den Hoven, Pieter E. Vermaas, and Ibo Van de Poel, *Handbook of Ethics, Values, and Technological Design* (Dordrecht: Springer, 2015), 1.

83. Ibo Van de Poel and Peter Kroes, 'Can Technology Embody Values?' in *The Moral Status of Technical Artefacts*, edited by Peter Kroes and Peter-Paul Verbeek (Dordrecht: Springer, 2014); Batya Friedman and David Hendry, *Value Sensitive Design: Shaping Technology with Moral Imagination* (Cambridge: The MIT Press, 2019).

84. Robison, *Ethics within Engineering*, 123. Robison refers to the 'argument from design' and the 'argument from effects'.

85. Parsons, for instance, focusses on three main ways in which design and ethics are connected: 1) ethical dilemmas that emerge during creation, 2) questioning the choice of what to create, and 3) how design can come to influence existing ethical notions. Parsons, *The Philosophy of Design*, 130. He also refers to other 'less obvious' points of connection (p. 169).

86. Donald Schön, *The Reflective Practitioner: How Professionals Think in Action* (New York: York: Basic Books, 1983).

87. Ivan Illich, *Medical Nemesis: The Expropriation of Health* (New York: Pantheon Books, 1976), 32.

88. Though not literally framed as a critique against the design profession but against the 'design industry', a similar argument is made in Joanna Boehnert, *Design, Ecology, Politics: Towards the Ecocene* (London: Bloosmbury, 2018), 37–40.

Part II

An Inquiry into
Design Professional Ethics

Chapter Five

Charting an Inquiry into Design Professional Ethics

We ended the previous chapter with the conclusion that the moral challenges designers face in the exercise of their profession call for promoting and developing a reflective enquiry into design professional ethics. Performing this inquiry can also be seen as indicative of the level of commitment professionals have toward their profession. In this chapter, our main objective is to flesh out that program. For that, we will define how this inquiry specifically concerns professional designers and explore its context and its relation with the larger inquiry of design ethics and philosophy of design at large. A direction will be stipulated, as well as its main goals and aims; possible difficulties and limitations will also be presented.

CHARTING THE GROUND FOR THE INQUIRY

There are frequent calls for integrating ethics into the design process; we reviewed some of them in the previous chapter. To cite another recent example, design scholar Jeffrey K. H. Chan insisted from the pages of the influential journal *Design Issues* on 'the paramount need for ethics in design today'[1]. But despite the general consensus that ethical reflection ought to be incorporated into design practices, ethics has not been integrated on a wide-scale basis[2]. I do not believe this is particularly due to a lack of design methods with a focus on ethics or to shortcomings in the existing practical approaches. In the previous chapters, I mentioned methods and approaches such as value-sensitive design, transition design, and design for social innovation, and other methods will be discussed in the later chapters. Nor is there a lack of ethical guidelines, which in recent years have increasingly focussed on the design of autonomous intelligent systems[3]*. Along these lines, my contention is that developing new methods or ethical design guidelines will not help change the situation much.

I must hasten to add that neither am I advocating for codes of ethics nor for other declarative rules such as strict guidelines of dos and don'ts. Formulating a code of ethics (or calling to apply an existing one) in the hope that professionalism will follow seems to be the wrong way around[4]. Codes may be useful as a guide, but in order to apply a code, a professional will still have to deal with the complexity of open-ended ethical considerations. A code of ethics could possibly make sense as one outcome of an inquiry into professional ethics, but it would not be a suitable starting point.

Why not? Because there is a problem that lies at the most basic level of the profession, at its very foundations. A general awareness seems to be missing that would enable a designer to roughly know how to proceed when they are involved in ethically laden situations. This awareness, which is part of so-called *ethical expertise*, alerts them as to what ethical factors need to be considered in a situation, and how they can be assessed. By 'ethical expertise' I mean 'the possession of ethical and moral knowledge, and the ability to use them to solve an ethical conundrum in a proficient way'[5].

Just like codes, guidelines are of little use if a designer is unable to recognise which relevant ethical values might be at stake or simply that a situation calls for ethical reflection at all. Open-ended methodological frameworks, on the other hand, could enable a designer to detect these situations that call for ethical attention, but as I just commented, they are not widely used. And we are back to the beginning.

Richard Buchanan may help us start pinpointing what is going on and help us move forward: he asserts that designers 'are better able to discuss the principles of the various methods that are employed in design thinking than the first principles of design, the principles on which [their] work is ultimately grounded and justified'[6].

The word 'able' in this assertion admits different interpretations. It could mean that they *do not know how to reason* about the right ends of their profession; it could mean that they are less able because they *do not know the first principles*; it could also mean that they *do not have the opportunity* to discuss said principles. Of course, the interpretations are closely related—one could be lacking know-how because of not having the opportunity to develop it. And perhaps this is what Buchanan has in mind.

Designers become quite proficient at reasoning about technical means precisely because they have access to countless opportunities to expand their know-how besides the actual work: publications, conferences, courses, education, etc. There is no reason to think designers suffer from some kind of cognitive limitation that is specific to them and prevents them from reflecting on the principles that ultimately ground and justify their profession. We could safely assume, then, that creating the opportunities would lead to an increased capacity to engage in this reflection around principles and justifications.

Buchanan's words, regardless of the interpretation we choose, signal a serious actual shortcoming with an important implication. Being able to reason about what justifies and grounds one's practice is essential to professionalism, if only because those 'first principles' pragmatically guide the professional in promoting, ensuring, or safeguarding some aspect of others' wellbeing.

If my previous analysis was right, the implication of all this is that the first step in any exploration into design professional ethics must be centred on the fundamental ethical grounding of the profession. It is not an easy feat, but if we want to take the conclusion seriously that ethical entailments arise from professional design activity, we need to act upon this realisation. I am of the view that what should occupy the profession as a whole now is charting its main matters of concern, its ends, its challenges at large, its unresolvable disputes, and its conflicts of value. Professional ethics can help us in this exploration; this inquiry can contribute to building a general awareness that is missing.

This approach has manifestly nothing to do with ethical oaths, codes of conduct, and ethical guidelines. Furthermore, it does not seek to develop a practical method or framework for ethical designing. At the same time, it also adds something novel to the broad field of design ethics in one important respect: it focuses on professional ethics, which differs from the current approach to ethical analyses of design and technology.

The Relation between Design Ethics and Design Professional Ethics

Design ethics in the comprehensive sense is a widely discussed subject approached by scholars and researchers coming from different fields and different epistemological traditions; the scholarly literature is vast[7*]. In this volume, design ethics has a very specific focus, which Richard Buchanan describes as following: 'design ethics concerns moral behavior and responsible choices in the practice of design'[8].

Although Buchanan does not mention the notion of 'professional ethics', it seems clear to me that his discussion is about professional designers[9*]. Buchanan's definition fits rather well with our understanding of professional ethics as it was outlined in chapter 2 (the study of the decisions, reasoning, and actions of professionals). But there is a caveat with the term 'design ethics': it is not specific enough to unambiguously and solely refer to the ethics of the professional activity of design; that is, to design professional ethics.

The term 'design ethics' just seems to need too much context to be understood as referring to the 'ethics of designing' (the ethics of the practice and the activity), let alone being understood as the 'ethics of *professional* designing'. 'Design ethics' could also be taken to include or refer to the 'ethics of the designed' (the ethical dimension of artefacts, their consequences, their very existence, etc.). Design ethics in this sense would be an exercise on 'applied ethics', centred on analysing and discussing designed material and immaterial artefacts to extract ethical insights that could serve to formulate or reformulate theories. Some examples of this approach are Ibo Van de Poel and Lambèr Royakkers' discussion of highway safety, Peter-Paul Verbeek's 'post-phenomenological' analysis of obstetric ultrasound devices, or Sheila Jasanoff's examination of data vulnerabilities on the internet [10]. Surely, the two understandings of design ethics are related, and even intertwined, if only because the ethics of the designed can *decisively* inform the ethics of designing.

Because of all this, and to avoid ambiguities, I propose to reserve the term 'design ethics' as a comprehensive term to refer to the study of the all the ethical dimensions of design, and to use the term 'design professional ethics' to refer to *the study of moral behaviour and responsible choices in the practice of professional design*. Design professional ethics, then, can be seen as an approach that looks at the ethics of designing from the vantage point of the ethical commitments designers have by virtue of being professionals.

'Design professional ethics' could thus be considered to be a subfield within 'design ethics', which, in turn, could be a subfield of the 'philosophy of design' [11]*. Design professional ethics is thus the reflection on and assessment of the professional moral commitments of designers, and their acting and reasoning qua professionals. Meanwhile, design professional ethics has also a practical purpose as it seeks to guide designers in their dealing with the ethical tangles that arise in practice.

Figure 5.1 offers a visual representation of the elements we have been discussing so far and the relations between them. It is also important to note, first, that only the main elements that have been discussed are included here. This representation is thus not exhaustive as both the philosophy of design and design ethics include other spheres of inquiry [12]*. Second, that the 'ethics of designing' is an area of inquiry that much exceeds 'professional ethics'; and third, that both the 'ethics of designing' and the 'ethics of the designed' are not separate disciplines, but rather loci of inquiry, which not only overlap, but also influence and inform one another.

Although, as I mentioned before, the scholarly literature on design ethics is vast; adopting the 'profession as a lens' [13] for studying design activity is, as far as I see, still a rather uncommon perspective. Although an inquiry into professional ethics is a frequent approach in other professions such as journalism, engineering, medicine, teaching, or nursing [14], the quantity of sources

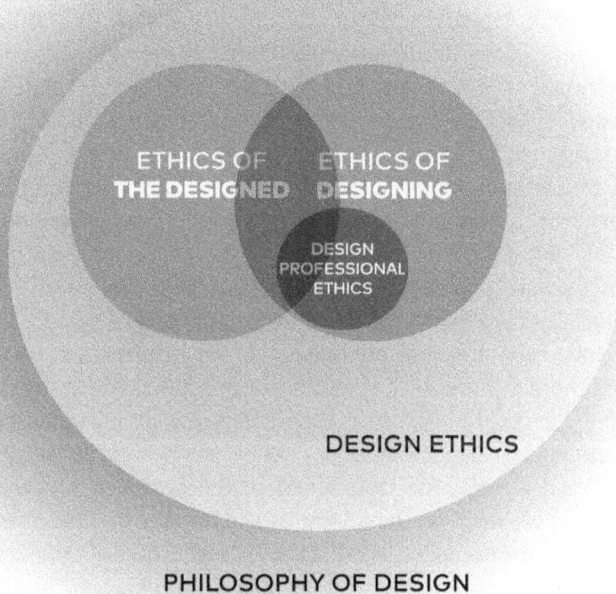

ETHICS OF
THE DESIGNED

ETHICS OF
DESIGNING

DESIGN
PROFESSIONAL
ETHICS

DESIGN ETHICS

PHILOSOPHY OF DESIGN

Figure 5.1. The Context of Design Professional Ethics within Design Ethics

that consider design activity from the vantage point of professions and professionalism is comparatively small.

Two pioneering publications in English that discuss the ethics of designing in the sense intended here are Alain Findeli's *Ethics, Aesthetics, and Design* and Carl Mitcham's *Ethics into Design*[15]. Naturally, there are other, more recent works that can be cited here that do discuss design from a professional ethics perspective; that is, reflecting on what designers do in their capacity of professionals. These volumes, however, while sharing the same object of interest, take a narrower approach to issue, focussing on duty and obligation. For instance, Jean Russ' *Sustainability and Design Ethics* is at large concerned with some broad ethical issues, especially values and flourishing. However, Russ narrows down the perspective when discussing professional ethics, mostly touching upon issues of due diligence, obligations, and codes of ethics. In the discussion of flourishing, the ethical dimension of professionalism is primarily conflated with notions of duty[16].

Similarly, Adrian Shaughnessy's *How to Be a Graphic Designer without Losing Your Soul* deals with broad professional ethics issues such as cultivating one's integrity[17]or the opportunities a designer has 'to be a designer with preoccupations other than being a force for consumerism or commercial propaganda'[18]. There is also a discussion of codes of ethics and professional organisations. However, the notion of 'profession' is not taken as a moral concept, nor is 'professional ethics' explicitly mentioned. Along these lines, 'professional skills' are touched upon in a sense that is related to professional competence: how to deal with clients, how to run a studio, or how to protect your design work from 'irate clients'. But Shaughnessy definitely delves into topics of professional ethics: for instance, he calls on designers to give ethical concerns the same weight they give to stylistic or commercial considerations. He also asks some rhetorical questions that are important to designers as professionals *and* as persons ('How ethically clean are we?'), but, alas, he conflates this again with rules and an 'ethical code'[19].

Intended for different publics and with different levels of theoretical reflection, both volumes are valuable contributions. However, they deal, either implicitly or explicitly, with a *narrow* view of design professional ethics (obligations and codes of ethics). Alas, when they do discuss subjects like professional virtues, the flourishing of the designer, and the professional's commitment to society, they do not seem to consider these subjects as falling within the purview of professional ethics and do not connect a broad ethics with professionalism in the way we intend to do here.

Two Levels of Specificity

In what specific way is design professional ethics different than 'regular' ethics? Design professional ethics has two specific levels of ethical reflection that are relevant for designers but are not at all relevant for non-designers. The first level of specificity has to do with the active participation of designers in the matters being ethically considered; the second has to do with the professional status of designers. I will delineate these issues of specificity in more detail in the following by resorting to the example of the online form and the field for gender that was used earlier.

The first level: it is *designers*—as opposed to other people not directly involved in design projects—who face ethical quandaries as they move through the different stages of the design of the form; put differently, it is people in their *role as* designers that face these particular choices. It is designers who are the moral agents involved in those situations, not laypeople.

Some might aptly point out that non-designers can also be directly involved in a design project. This is admittedly true even when only considering the design project in a strict sense, and leaving aside the manufacture, distribution, or promotion of the designed artefact or service. During the design project clients, managers, producers, and prospective users and other stakeholders might be involved. The question remains whether design professional ethics normatively pertains to them; to me the litmus test is whether these 'other' individuals could be counted as designers or not: as we saw in chapter 1, everyone designs, but not everyone is a designer. It is reasonable to think that a good psychologist participating in design projects would want to act according to psychologists-professional ethics; similarly, lawyers would arguably find more adequate normative guidance in legal professional ethics, a programmer in computing professional ethics, and so forth. As for users who participate in a project as informants or co-creators, because they are non-professionals, anything other than 'regular' (everyday) ethics seems to be unnecessary, and implausible.

Naturally, people who do not participate in a design project, either lay-people in different roles (parent, friend, student, etc.) or in different professions, may also be ethically concerned with the dignity of persons, gender rights, or the need for inclusivity that are relevant to the online form case. However, they do not face the *same* ethical questions one faces as a designer, if only because they are not tasked with designing a form. People outside a design project might apparently deal with the same topic, but they do so from a different vantage point and with a radically different purpose than designers (and others directly participating in it). Having to make a design choice makes the ethical challenge specific to designers.

The second level of specificity is related to the professional status of designers. In part I of this book, I have defended a view of professionalism that is grounded not only in expertise and skill but first and foremost in having a primary orientation toward community and the common good. This dimension of professionalism is what makes deeply contextual ethical reasoning not only desirable but necessary. What is more, design situations rarely involve esoteric life and death dilemmas based on the trolley problem[20], but primarily making more mundane ethical design decisions as to whether it is ethically acceptable to ask somebody to select their gender in a registration form. (Some people might, for instance, object to it on privacy grounds, whereas others who identify as non-binary might find it extremely hard if not impossible to provide an answer.)

In medical ethics, these types of normatively important and pervasive ethical issues are termed 'everyday ethics', as opposed to 'dramatic ethics', which are based on extreme and less frequent medical cases, which will not be encountered by most doctors in practice[21]. Cloning or separation of conjoined twins are examples of the 'dramatic' type, and dealing with potential parental sexual abuse or having too little time per patient due to the limitations of overstretched and underfunded health services are more likely to be 'everyday' issues.

In design, the choice of material for a chair, the images used in the promotion of a product, or whether to use facial recognition instead of a PIN code for identification when using a service are examples of 'everyday' ethical issues; whether it is ethically acceptable to design a gym in a permanent settlement on Mars for billionaires who want to escape Earth would be a 'dramatic' one.

The upshot of all this is that design professional ethics is relevant to professional designers on two counts: by virtue of being designers and by virtue of being professionals. Being a designer defines the scope of action, whereas being a professional defines a way of acting and reasoning about actions. Although the normative entailments of being a professional can globally guide the designer in how to deal with both types of dilemmas, it is the 'everyday' type that is more relevant to our inquiry.

A DIRECTION FOR THE INQUIRY

If we want this inquiry into design professional ethics to be centred on raising and fostering an awareness of the fundamental ethical grounding of the profession, the words of Victor Margolin could set an appropriate course for the inquiry to take. This course fits neatly with what we have been discussing so far and explicitly integrates the public service element of design. His words can also be taken as a prudent but optimistic exhortation, which is worth quoting in full:

> The future we are facing deeply implicates designers who work across many different professional fields. They are, in effect, the agents whose skills produce the milieu of products and services in which we live. To the degree that this milieu does not enhance and affirm human potential and well-being, we must hold designers at least partially accountable. We need to foreground the question of how to create an ethics of designing that can suggest humanly satisfying directions for future work[22].

I endorse the spirit of Margolin's words wholeheartedly, but I disagree with him in that we need to 'create' an ethics of designing. My view is that what we need to do this time is not to create, but to uncover—to reveal—something in design that is covered over. This something, it must be assumed, has to be there. Although it is not there in the whole of design, in each and every instance, as the objections presented in the previous chapter illustrate, it must be assumed to be there at least in its paradigmatic instances. The assumption is that what needs to be uncovered and revealed is hidden in plain sight: in the exemplary cases that show what design can be when practised at its best.

What ethical aspects do these instances have in common? For whose primary benefit are design methods and techniques mobilised? And, more importantly, to pursue what greater ends? What we need to reveal, then, is the overarching purpose of design that we pursue. Philosopher James Allen speaks of 'the issue or outcome of what we do, and do precisely because it has this as its issue'[23]. In other words, what we need to uncover is an awareness of the ends of the practice—its *telos*, in Greek—the 'for the sake of which we act'. For what do we ultimately design?

Goals, Purposes, and Goods

We have been speaking of 'goals', 'purposes', and 'goods', and a clarification of these terms is in order because they will prove relevant for our discussion and are conceptually close to each other, which could well provoke some confusions. These terms could be divided in two main groups: the *goals* of an activity, on the one side, and *purposes* and *goods*, on the other. Goals are the literal things one needs to do or obtain in order to achieve the purpose (the goal of playing chess is to checkmate the other king). The terms 'purpose' and 'goods' (and also 'ends' and '*telos*') tend to be used interchangeably, but although we can take purposes and goods to be two sides of the same coin (together they form the 'for the sake of which we act'), there is, as far I see it, a subtle nuance between purpose and good. The *purpose* is the reason why something is done (the purpose of chess is, for instance, to have fun, to train one's mind, or to make friends); *goods* denote the outcome of achieving the purpose (for instance, the ability to be patient and thoughtful or the friendships that are made).

When used as a noun, the philosophical understanding of the term 'good' denotes thus the 'good end' or 'good outcome' of an action or a process, which is an *intrinsically* valuable state. Goods are highly desired by most people because they have value in themselves. Their value could be *non-instrumental* (they are an end in themselves or they have intrinsic value, for instance, 'friendship') or *instrumental* (they have value in themselves but *also* serve to achieve another good: 'health', for instance, is valuable in itself because it is good to be healthy but it has also instrumental value in that it enables one to pursue *other* purposes in life). This conception of goods as being intrinsically valuable contrasts markedly with how the term is understood in economics, where goods are instrumental material *means* to satisfy wants and needs. Money, for example, is a good in economic terms, but philosophically it has no intrinsic value; it only serves to buy things that may (or may not) have intrinsic value, such as a ticket to see a concert (if the musicians are good, the concert can provide 'joy', which is an intrinsic good).

Importantly, sometimes an activity, the goal, and the purpose are fully intertwined. Consider, as an example, contemplating a landscape during a hike in the mountains; here the activity, the goal, and the purpose are the same: contemplating the landscape. Conversely, when scanning the landscape (activity) with the intention of spotting somebody (goal) with whom we went hiking with to go back home together (purpose), goal, purpose, and activity are separate elements. If our friend were to show up in front of us unexpectedly, the scanning of the landscape would cease as its purpose has been achieved by other means. Moreover, the good that is achieved in the first case (an aesthetic experience, for instance) is different than the one that could be gained by finding our friend (the convenience of riding back together and a nice conversation about the hiking, for instance).

The Importance of Purposes and Goods

It is in these ultimate ends where we will find reasons for acting; our individual motives only become *good* reasons when they serve the goals and purpose of the profession, as philosopher Adela Cortina postulates[24]. It is because reasons and purposes are thus connected that gaining greater understanding of the purpose of design is so important.

There is, furthermore, no use in attempting to provide these reasons directly, through a top-down approach, and because of that this inquiry is conducted. Formulating codes of ethics, we saw previously, is not a suitable way to achieve ethical expertise nor to gain understanding of the goods of the design. Along the same lines, Carl Mitcham, following Aristotle, posits that one 'cannot articulate and reflect on what one does not already have. Ethics cannot come from on high, as it were, to articulate guidelines for action'[25]. A more promising way, then, is to get to this awareness about the principles, and the overarching purpose of design by adopting a bottom-up approach. By recognising and examining what we consider excellent *designing* we could learn much about the ends of design. We do not need to create an ethics of designing as Margolin suggested, we just need to *uncover it*.

The task is straightforward: we have to find the special touch that is specific to design and that makes it different from the other professions. The task at hand is to present a coherent account of the particular way in which the design profession serves individuals and society attain strategic goods.

If we want to perform an inquiry into design professional ethics, we need to find this particularity, which cannot be solely about the methods or techniques of design, but first and foremost about the particular goods that are achieved through design. After all, design methods and techniques become 'goods' in an ethical sense only when marshalled for meaningful professional purposes. Put another way, we need to be able to plausibly explain in what particular way the design profession promotes the general professional goal of serving others. Toward what ends do designers need to align their individual motives so that they become reasons? It would not be enough to say that designers design symbols, products, interactions, and systems. These elements form an excellent categorisation apparatus, but cannot be the ultimate ends, whereas valuable in themselves these are means toward an end.

Despite the difficulties signalled by Buchanan, designers ought to engage in this inquiry *precisely* because they are unable to discuss the first principles on which their profession is 'ultimately grounded and justified'. Also, the exploration needs to be reflective; reflection is, after all, one of the most salient characteristics of professional performance. In order to do their work, professionals need to inevitably reflect as they face constant novel and unexpected situations and challenges that exceed their original training and the expertise acquired thus far. It is through reflection that professionals can understand their practice and develop awareness about the ethical consequences of their performance. Through situated action and reflection designers can develop the ethical expertise they are missing, much in the same manner as they develop their design skills.

Reflection and Autonomy

Designers must do this work themselves; *reflection cannot be externalised.* Some authors have suggested that if the ethical challenges are too much for designers, they should be relieved of the burden and have an ethics specialist take care of that[26]. If these authors are right, this would be disastrous for design as a profession: being able to function autonomously—without constant tutelage—is paramount for professionalism. It can reasonably be expected of a professional that they are able to deal with the challenges that arise when the profession is practiced characteristically, just like nurses, teachers, social workers, doctors, or lawyers are able to deal with the ethical issues they encounter in their typical professional activity. This does not entail that they ought to be able to deal autonomously with *all* the ethical issues they encounter; indeed, being able to recognise limitations and ask for external advice, be it from peers or from members of other professions, whenever necessary, is also a sign of responsible professional behaviour. Many professionals already do this when they ask for the opinion of ethics committees (for example, hospital ethics committees) or ethics specialist when they encounter issues that are too complex.

Relatedly, in some cases, external controls may be necessary to help ensure that design outcomes meet the ethical thresholds that society stipulates through regulation; this would be similar to the way that products undergo regular safety and quality controls before launch, and drugs undergo clinical trials. In chapter 4, we saw that influential writers from the artificial intelligence community are advocating in that direction. No designer is infallible, and given design's ethical import, such controls could be a positive development. Moreover, they would not threaten to undermine professional autonomy as they would be conducted in hindsight; that is, after the designer made their best decisions, but before the product reaches the user. Controls like this, however, would not eliminate the need for design professional ethics.

THE NEED FOR BROADNESS

Ethics was presented in the introduction as being much more than about discerning between right and wrong or being concerned with duties and obligations, which we referred to as ethics in the *narrow* sense. Alasdair MacIntyre describes this narrow view of ethics as 'the tendency to think atomistically about human action and to analyse complex actions and transactions in terms of simple components. Hence the recurrence in more than one context of the notion of "a basic action"' [27]*. To reiterate, in this volume, we adopt an encompassing view of ethics that goes beyond basic actions to also cover discussions about the nature of the good life, about how we should live, and about how we want to be as persons. This is often referred to as ethics in the *broad* sense. At this point, we need to examine this distinction in more detail to justify the appropriateness of the broad approach.

In the professional ethics literature, the narrow sense is a frequent approach; authors often deal with the subject by covering issues such as confidentiality, privacy, conflict of interest, relations between professionals and clients, or ethical principles such as beneficence, non-maleficence, or justice. This is also recurrent in standards of professional practice [28]*. Narrow ethical questions are variations of 'How ought I to act in this case?' or 'What ought I to choose (now)?' or 'What is the right thing to do (in this situation)? Naturally, these are generic templates; in a concrete ethical deliberation, we would phrase this type of question differently. For instance, if we go back to the example of the designer and the form, they might ask themselves something like 'Is it OK to include a field for gender?' or 'Should I include a field for gender? I read somewhere that the most inclusive option is not to include it at all, but marketing needs the data'. Perhaps the ethical saliency of the issue is worded not as a question but as an uncertainty deserving further exploration: 'I've got to include a field for gender, but it doesn't feel right, I don't know'.

It is worth restating that the precedence of the broad approach over the narrow does not in the least imply that the narrower sense is trivial or unimportant. On the contrary, practical questions about what to do or how to act in concrete cases dominate our everyday perception of ethics and are often the initial prompt for deeper ethical deliberations. However, I am of the view that to do professional ethics *exclusively* in the narrow sense might be a misdirected and misguided endeavour, especially when we are dealing with difficult issues (or wicked problems) that cannot be directly addressed in terms of 'right or wrong' or of duties and obligations. Arguably, given design's complexity, a great deal of the ethical issues faced by designers will not be solved through the narrow approach of right and wrong.

Philosopher of technology Shannon Vallor argues that the moral dilemmas presented by today's technologies cannot be considered in the same way applied ethics traditionally has dealt with ethical dilemmas such as the ethics of capital punishment, torture, or eating meat. She concedes that it is reasonable to ask moral question like 'Is torture right or wrong?' and that applying various moral principles might be conducive to concrete answers. But she also shows how this approach might not make much sense when applied to technologies that deeply change the way we live and do so in unpredictable ways. She argues that there is something 'plainly ill-formed' about questions such as 'Is Twitter right or wrong?' Or: 'Are social robots right or wrong?' Focusing on acts does not help much either: 'Is tweeting wrong?' or 'Is it wrong to develop a social robot?' are not useful questions[29].

The reason for the unsuitability of the *is-X-right-or-wrong approach*, Vallor argues, has to do with the fact that modern technologies present 'open developmental possibilities for human culture as a whole, rather than fixed options from which to choose'[30]. This emphasises the fact that design not only causes unintended effects, but it can also change human culture in profound ways. For instance, a domestic appliance such as a vacuum cleaner does not only remove dust; it fosters, at the same time, the emergence of new standards of cleanliness according to which people will determine whether a house is clean. Similarly, the availability of instant messaging apps changes how people make appointments to go out with friends or to visit family. Because of all this, our inquiry into professional ethics needs to be broad, as a narrow approach would not be really helpful in guiding a big part of design activity that is more concerned with designing broad courses of action than evaluating and choosing among a closed set of alternatives.

The Ethics of Wicked Problems

Although Vallor focuses primarily on 'emergent' modern technologies, her realisation about 'open developmental possibilities' could in principle be relevant for all types of complex and wicked problems that design aims to tackle. After all, the complexity and 'wickedness' of design problems is what prevents the existence of fixed options from which to choose. It has long been shown and discussed that dealing with wicked problems requires a different approach than 'tame' problems do. The 'wickedness' in wicked problems encompasses their ethical dimension too; because of this a systemic approach (that is, a broad approach), instead of a narrow one, is necessary.

Evidently, when dealing with a wicked problem, one *could* approach it with a 'tame' problem mindset. For example, one could remain within the narrow sense of ethics and try to answer specific issues at stake and ponder whether vacuum cleaners are right or wrong or are desirable or ethically indefensible. Because ethics in the narrow sense does not prescribe a concrete direction for an answer, ethical deliberation will need to be carried out, and answers may vary depending on the analyses. Most people, I presume, would be inclined to find reasons for arguing in favour of vacuum cleaners as they alleviate drudgery and save time, whereas some fewer others might argue that they are not acceptable as there is no real need to vacuum dust, given that one can effectively dust mop a floor, which is more sustainable. Others might accept vacuum cleaners but reject equipping them with artificial intelligence on privacy grounds, and so on. The issue could be settled by evaluating which argument weighs more heavily. The actual outcome of the debate is almost irrelevant to our discussion; the point here is that it would be wholly misdirected to assess a vacuum cleaner as if it were a stand-alone device and simply by discerning if sustainability is more or less important than cleanliness or efficiency.

A more sensible course of action would be to realise that to appropriately answer the ethical issues that affect particular situations, it might be necessary to inscribe the narrow questions into the broader sense of ethics. A vacuum cleaner is part of a much larger problem involving profound issues related to gender, domesticity, comfort, power, and justice, among other facets of the situation. In this broader sense, we find questions that do not directly refer to practical issues, as to whether to choose one alternative above the other, but are more concerned with the old Socratic question of how one should live, or the more modern phrasing of what it means to live a good life. Imagine we are as designers tasked with designing a new type of household appliance; reflecting on what it means to live a good life means reflecting on what desirable family dynamics are, in terms of, for instance, emotional bonds, gender roles, sustainability, wellbeing, privacy, and, why not, convenience. Possibly, we may come to the realisation that many different family dynamics can be desirable and worth enabling[31*].

We could further illustrate all this with another example: *smart* cradles and rockers for babies. These smart devices are equipped with sensors and some degree of automation. They feature soothing vibrations and movements, 'womb-like conditions', and gentle sounds that seek to simulate a human embrace. Among other things, they respond automatically to the baby's cry by activating heartbeat sounds and a 'calming womb motion', according to their manufacturers.[32] It would be quite injudicious to ask whether smart rockers are right or wrong. Asking whether it is right or wrong to participate in the design of such devices would also show a lack of judgement. Of course, we could ask narrow questions like: 'is it okay to include heartbeat sounds or to simulate a human touch?' But this would too narrow as a first step. We need to consider the device and its ethical dimension as a whole and in its context of use. A designer seeking to engage in ethical deliberation around the design of these devices needs to go beyond the narrow sense and consider addressing broad ethical questions like, for instance, what might these devices *do* to a parent or carer *as a person*, how might the baby's life as *a whole* be affected by them, or how might the relationships between children and parents or carers change because of them.

Manufacturers and reviews from users claim that these devices do in fact help babies sleep better, cry less, and relieve colic and gases[33]; imagine now that empirical research shows that those claims are sound. Be that as it may, this would not tell us much about what we should think about *the way* in which it alleviates crying. Alleviating crying is desirable, but so is human contact and paternal and maternal comfort. Do we think something important is lost *as well* as gained? Both using and not using the device seem to lead to 'authentic and substantial goods'; it is not a choice between right or wrong but between rival goods. So, to be meaningful, the design and evaluation of such devices has to be closely related to conceptions of what good parenting is. This, in turn, is closely connected to what a good life is and what living *well* means.

The reflection on design ethics need not stop there. Many other questions may arise. How might a device like this influence how people relate to their babies' discomfort and, when the babies grow older, to their children's discomfort? Do they wish for a smart bed for adolescents that would alleviate their teenage grievances? How might a smart cradle influence our notions of care? How does a smart rocker harm or improve parenting skills? And in what direction? To enable and foster what purposes? When confronted with the sustained crying of a baby, besides a deep sense of empathy, many parents develop capacities for self-control and endurance. How are these capacities affected by the use of these devices?

Some could argue that most of these questions could be answered empirically, but of course, that is not the central point. Science might inform us that babies sleep more, have less pain, that parents become detached, or, on the contrary, that they sleep better and are more caring than without the device. By contrast, moral reflection asks, in this case, whether it is ever good or in what measure it is appropriate to use a smart cradle to achieve that goal; this is a value-laden evaluation, and it has to do with what we think is good. So empirical research might tell us that these devices 'work' (that is, the baby does not cry) but will not tell us whether smart devices for babies are good. And we are precisely interested in this type of answer that can come from philosophy, not from empirical research. Evidently, these ethical reflections and evaluations concern, perhaps even primarily, the ethics of the designed, and fall within the purview of other disciplines in philosophy, such as philosophy of technology or philosophy of health and happiness. In any case, the answers to these questions could decisively inform design professional ethics.

LIMITATIONS, DIFFICULTIES, AND PERSPECTIVES

Our inquiry into professional ethics will not offer immediate 'yes or no' or 'right or wrong' answers to our particular ethical questions or dilemmas, but it will help highlight what is at stake in them, thus guiding and informing designers toward the goods they pursue in their professional activity.

Professional ethics needs to be understood as an intellectual struggle for deeper awareness and as a sustained reflection around the purpose of the practice of professional design and the complex decision making processes associated with it. Ethical deliberation might be triggered by practical, narrow questions, but if we want to take design professional ethics seriously, these questions need to be tackled taking a broad view of ethics, which is necessarily loaded with discussions of ends and conceptions of the good.

Professional ethics in a broad sense raises questions that can be worded in a more personal form too: 'What kind of professional do I want to be?' 'What is really important to me as a professional?' 'What types of symbols, products, interactions and systems do I want to design?' 'What for?' 'To promote what ends?' 'With whom?' 'And for whose benefit?' Philosopher Bernard Williams posits that ethical deliberation so understood 'presses a demand for reflection on one's life *as a whole*, from every aspect all the way down'[34].

However, although we can ask these questions in the first person, to be ethical in the broad sense they need to relate to the question of how *one* should live. This means that the question is thus not *only* about me (and you, actually), but about 'the good life' in general. The point here is that to discern whether to choose option A over option B when we are designing, we need to be aware or have some sense of some ultimate conception of the shared professional good we are pursuing which is related to the good life. This awareness is what makes it possible for our individual motives to become *reasons*, thus exceeding the particular contexts of our own meandering lives and, especially, our own particular preferences and desires.

Understanding professional ethics in this way presents important difficulties for the designer. MacIntyre famously distinguished between two different kinds of obstacles for any contemporary attempt to envisage human life as a whole: one philosophical and one social. The first is related to the dichotomy between the narrow and the broad senses of ethics we discussed earlier. The second obstacle has to do with the way 'each human life [is partitioned] into a variety of segments, each with its own norms and modes of behavior. So work is divided from leisure, private life from public, the corporate from the personal'[35]. To refer to this phenomenon, MacIntyre coined the term 'compartmentalisation'[36].

A negative consequence of compartmentalisation is a *fragmented self*, which arises when the different domains that are present in someone's life in the form of roles (professional, friend, parent, employee, citizen, and so on) are divided into strictly isolated compartments. This division means that the ethical deliberations one performs generate irresolvable tensions between one's different social roles. The problem occurs when these tensions go beyond a dynamic, and even healthy, tension that can be embraced and perhaps be meaningfully harmonised. In practice, compartmentalisation means doing something as a professional that one finds ethically inadmissible as a friend or as a citizen. A fragmented self is seriously diminished as a moral agent because it lacks integrity, and it is incapable of transcending the limitations imposed by its own different roles. So perhaps the question, in the end, is not 'what kind of professional do I want to be?' but rather 'what kind of person do I want to be?' Naturally, this issue deserves deeper attention, and we will return to it chapter 9.

To round up, we started this section arguing for the need to gain a deeper understanding about the particular ways in which the design profession serves the public. We will find this somewhere between two extremes. On the one extreme, we find the general public service element that is intrinsic to professions. On the other, we find the paradigmatic cases of design, that is, instances of design at its best. It is between these two extremes that we find the space for design professional ethics. A promising way to start would be to reflect on the design profession by performing first an anthropology of the nature of the practice it accommodates. This will enable us to find that particular *designerly* way toward the good.

But before we attempt to approach this task, we need to pause with the discussion of design for a chapter and zoom out again to expand on the theoretical philosophical foundation upon which our professional ethics will be grounded. This is what we will do in chapter 6. This foundation will complement the mindset we delineated in this chapter and give us additional philosophical vocabulary that will accompany us during the last third of the book.

NOTES

1. Jeffrey K. H. Chan, 'Design Ethics: Reflecting on the Ethical Dimensions of Technology, Sustainability, and Responsibility in the Anthropocene', *Design Studies* 54 (2018): 196.

2. Aimee Van Wynsberghe and Scott Robbins, 'Ethicist as Designer: A Pragmatic Approach to Ethics in the Lab', *Science and Engineering Ethics* 20, no. 4 (2014): 948–49.

3. For example, IEEE Global Initiative on Ethics of Autonomous and Intelligent Systems, *Ethically Aligned Design: A Vision for Prioritizing Human Well-Being with Autonomous and Intelligent Systems*, Version 2 (Piscataway: IEEE, 2018). Primarily aimed at technologists doing design, this document is set to advance the public discussion around the ethical and social implications of autonomous and intelligent systems.

4. This is a point eloquently made by David Carr in his treatment of the ethics of the teaching profession, see *Professionalism and Ethics in Teaching* (London: Routledge, 2000), 44.

5. David Casacuberta and Ariel Guersenzvaig, 'Using Dreyfus' Legacy to Understand Justice in Algorithm-Based Processes', *AI & Society* 34, no. 2 (2019): 315.

6. Richard Buchanan, 'Human Dignity and Human Rights: Thoughts on the Principles of Human-Centered Design', *Design Issues* 17, no. 3 (2001): 36–37.

7. See the further reading section at the end of the book.

8. Richard Buchanan, 'Design Ethics', in *Encyclopedia of Science, Technology, and Ethics*, edited by Carl Mitcham (Detroit: Macmillan Reference, 2005), 504.

9. He mentions the word professional nine times, explicitly discusses the 'practice of design', and later to the 'activity of designing'.

10. Ibo Van de Poel and Lambèr Royakkers, *Ethics, Technology, and Engineering : An Introduction* (Malden: Wiley-Blackwell, 2011), 147–55; Peter-Paul Verbeek, *Moralizing Technology: Understanding and Designing the Morality of Things* (Chicago: The University of Chicago Press, 2011), 23–28; Sheila Jasanoff, *The Ethics of Invention: Technology and the Human Future* (New York: W.W. Norton & Company, 2016), 156–61.

11. Others may insist that 'design professional ethics' should be a subfield of 'professional ethics', which is a field of study in its own right within philosophy. Be that as it may, this particular point is not central to my argument.

12. Also, other junctures than the represented between design and ethics may be determined as discussed in note 84 in the previous chapter.

13. Michael Davis, 'Profession as a Lens for Studying Technology', in *The Ethics of Technology: Methods and Approaches*, edited by Sven Ove Hansson (London: Rowman & Littlefield International, 2017), 83–96.

14. For professional ethics in journalism, see, for example, Sandra L. Borden, *Journalism as Practice: Macintyre, Virtue Ethics and the Press* (New York: Routledge, 2010). For engineering, see, for example, Michael Davis, *Thinking Like an Engineer: Studies in the Ethics of a Profession* (Oxford: Oxford University Press, 1998). For medicine, see, for example, Edmund D. Pellegrino and David C. Thomasma, *The Virtues in Medical Practice* (New York: Oxford University Press, 1993). For teaching, see, for example, Chris Higgins, *The Good Life of Teaching: An Ethics of Professional Practice* (West Sussex: Wiley-Blackwell, 2011). For nursing, see, for example, Alan E. Armstrong, *Nursing Ethics: A Virtue-Based Approach* (London: Palgrave Macmillan, 2007).

15. Alain Findeli, 'Ethics, Aesthetics, and Design', *Design Issues* 10, no. 2 (1994); Carl Mitcham, 'Ethics into Design', in *Discovering Design: Explorations in Design Studies*, edited by Richard Buchanan and Victor Margolin (Chicago: The University of Chicago Press, 1995). Findeli's paper is an adaptation of material that was previously published in French in the early 1990s (see his paper for references).

16. Jean Russ, *Sustainability and Design Ethics*, second edition (Boca Raton: CRC Press, 2019).

17. Adrian Shaughnessy, *How to Be a Graphic Designer without Losing Your Soul*, new expanded edition (New York: Princeton Architectural Press, 2010), 23–26.

18. Ibid., 8.

19. Ibid., 108–10.

20. Be that as it may, my colleague David Casacuberta and I do not think it makes much real sense either to believe that discussing trolley problem can help develop artificial intelligence, Casacuberta and Guersenzvaig, 'Using Dreyfus' Legacy to Understand Justice in Algorithm-Based Processes'.

21. Natalie Zizzo, Emily Bell, and Eric Racine, 'What Is Everyday Ethics? A Review and a Proposal for an Integrative Concept', *Journal of Clinical Ethics* 27, no. 2 (2016).

22. Victor Margolin, 'Design, the Future and the Human Spirit', *Design Issues* 23, no. 3 (2007): 15.

23. James Allen, 'Why There Are Ends of Both Goods and Evils in Ancient Ethical Theory', in *Strategies of Argument Essays in Ancient Ethics, Epistemology, and Logic*, edited by Mi-Kyoung Lee (Oxford: Oxford University Press), 244.

24. Adela Cortina, *Hasta Un Pueblo De Demonios: Ética Pública Y Sociedad* (Madrid: Taurus, 1998), 151–52.

25. Mitcham, 'Ethics into Design', 183.

26. Glenn Parsons, *The Philosophy of Design* (Cambridge: Polity, 2016), 151.

27. Alasdair MacIntyre, *After Virtue: A Study in Moral Theory* (Notre Dame: University of Notre Dame Press, 2007), 204. MacIntyre does not explicitly refer to a narrow sense of ethics, but he clearly has this in mind.

28. See, for example, AIGA, 'AIGA Standards of Professional Practice',https://www.aiga. org/standards-professional-practice. As an illustration: '7.5: A professional designer shall not knowingly make use of goods or services offered by manufacturers, suppliers or contractors that are accompanied by an obligation that is substantively detrimental to the best interests of his or her client, society or the environment'.

29. Shannon Vallor, *Technology and the Virtues: A Philosophical Guide to a Future Worth Wanting* (New York: Oxford University Press, 2016), 27–28.

30. Ibid.

31. Acknowledging that there are multiple desirable types of family dynamics should not be conflated with moral relativism: *not all* family dynamics are worth enabling or consolidating. The issue of pluralism versus moral relativism deserves further discussion but exceeds the scope of this volume.

32. Babocush (https://www.babocush.com/) and Snoo Smart Sleeper (https://happiestbaby. co.uk/products/snoo-smart-bassinet) are two examples of these devices.

33. This are actual claims from the sellers or manufacturers of the products. See previous note.

34. Bernard Williams, *Ethics and the Limits of Philosophy* (Abingdon: Routledge, 2006), 5. Italics in the original.

35. MacIntyre, *After Virtue*, 204.

36. Alasdair MacIntyre, 'Social Structures and Their Threats to Moral Agency', *Philosophy* 74, no. 289 (1999): 322.

Chapter Six

A Philosophical Foundation for Our Inquiry

In this chapter, I will introduce the philosophical tradition that will serve as the foundation for the following steps in our inquiry into design professional ethics: *virtue ethics*. This might not be surprising to some of you, given our manifestly broad approach to ethics, the asserted prevalence of the 'good' over the 'right', and the frequency of expressions like 'living well' or 'living a good life'. Besides introducing virtue ethics, in the following pages, we will review two important alternative ethical theories that could have provided a grounding for our discussion; namely, Kantianism and utilitarianism. A brief appraisal of these theories will highlight the ways virtue ethics seems to be better able to provide an adequate general foundation for our discussion than the alternative theories. We will end the chapter by engaging with the ideas of one of the main contemporary philosophers within virtue ethics, Alasdair MacIntyre, on whose analysis of practices as the bedrock of morality we will rely.

A PRIMER TO VIRTUE ETHICS

In the remaining chapters of this book, we will be dealing with many concepts that are associated with the virtue ethics tradition such as *flourishing*, *goods*, *practices*, or *practical wisdom*. These notions will serve for understanding the deeper meaning of design professional work; to contextualise them, instead of merely offering stand-alone definitions of them, in this section we will consider the main tenets of virtue ethics and its emphasis on the person's dispositions and motives. It is the very notion of 'virtue' (singular) and its associated terms that interest us in this section; although we will enumerate some 'virtues' (plural), we are less interested in specifying an exhaustive list of relevant virtues like we find in other books.

The long history of virtue ethics can be traced back in the East to the teachings of the Chinese philosophers Confucius (551–479 BCE) and Mencius (371–289 BCE), and to Plato (428/427–348/347 BCE) and Aristotle (384–322 BCE) in the West. It was the prevalent view in Western society until the Age of Enlightenment, when principle-based ethics began to dominate during the early nineteenth century. Virtue ethics re-emerged in the late 1950s catalysing a reaction to the deficiencies of principle-based ethics[1]*.

Virtue ethics is concerned not with isolated acts but with the nature of the good life and with living well. While there is a variety of streams in virtue ethics, all share a main focus on the individual moral agent, and the traits and dispositions that make them virtuous (hence virtue ethics). Also, in making ethical evaluations, they all use terms that describe people's character or qualities (just/unjust, tolerant/intolerant, tactful/tactless, etc.), rather than judgements about the rightness or wrongness of particular external acts.

Flourishing and the Virtues

According to Aristotle, living well is moving toward the 'good life'. The good life is associated with the Greek word *eudaimonia*, which can be translated as 'human flourishing', 'happiness', or 'wellbeing'. Eudaimonia should not be understood to refer to a temporary state of pleasure, but to a life as *a whole*: 'the state of being well and doing well in being well'[2]. Philosopher Roger Crisp describes it as 'whatever makes a human life good for the person living it'[3]. A basic aspect of Aristotelian thinking is that every skill, every action, and every rational choice should aim at some good[4]. Previously, we mentioned the notion of *telos*, the final end or good 'for the sake of which we act'; in an Aristotelian fashion, we can thus say that the ultimate end of a human life is *flourishing*.

For Aristotle, flourishing as a human being is equivalent to living well and acting well as a whole; it is characteristic of a virtuous person to be able to deliberate 'about what is good and beneficial for himself, not in particular respects, such as what conduces to health or strength, but about what conduces to living well as a whole'[5]. So the virtues are those qualities that enable a person to live well, but they are not just means to that end; they are, at the same time, *central* to that person and are a part of their flourishing. What constitutes the good for a person, MacIntyre summarises, 'is a complete human life lived at its best, and the exercise of the virtues is a necessary and central part of such a life, not a mere preparatory exercise to secure such a life'[6].

Having flourishing as the ultimate overarching end of a human life might seem too narrow a purpose or even a straitjacketing one for some, but it need not be. A pluralistic understanding of what flourishing is can accommodate a multitude of avenues for living well, which allows everyone to define how one understands their own life purposes in the different spheres of one's life[7]. There is here a second, equally important point that can be made: at least *some* conception of the good is necessary to start pursuing it. MacIntyre maintains that 'without some at least partly determinate conception of the final *telos* [that is, our final purpose] there could not be any beginning [to a pursuit of the good]'[8]. This initial understanding, which we get from our primary and secondary socialisation into a particular culture, from our up-bringing, from role models, and so forth, is what enables us to deepen our knowledge about our own purpose and our conception of the type of life we want to pursue.

The Virtues Examined

For MacIntyre the virtues are those dispositions that enable us to overcome obstacles and endow us with knowledge of ourselves and of the good[9]. Because persevering, caring, showing courage and integrity, practising practical reason, or achieving intellectual functionings *are* goods in themselves, attaining them *is* living well. Virtue ethics focuses on the *excellence* not only of the result but also of 'the way it was brought about and the kind of reasoning that led to it'[10]. MacIntyre provides us with an aphorism-like dictum that nicely rounds up this idea: 'the good life for man is the life spent in seeking for the good life for man'[11].

Perhaps it is time to entertain an example to illustrate all this: imagine a student of design; becoming a designer is a learning process that requires years not only of study but also of apprenticeship[12]. To become a 'good' designer, technical knowledge and skill are necessary but not sufficient; the process demands, among other qualities, perseverance, diligence, courage, practical reason, capacities for self-reflection, intellectual capacities, integrity, care, or solidarity. These virtues are, in a way, instrumental to becoming a designer, but at the same time, they exceed their instrumental role and become *constitutive* of that person allowing them to flourish as a human being; as MacIntyre argued previously, they are not a 'mere preparatory exercise' for life. Granted, to thrive as a designer, also at least some 'luck' will be necessary, as the person aiming to become one will be part of an external social world which is plagued by contingent and structural circumstances that may be beneficial or detrimental to that goal.

There is more to be said about the notion of virtue. Philosopher Julia Annas understands 'virtue' as a 'feature of a person, a tendency for the person to be a certain way . . . a disposition which expresses itself in acting, reasoning, and feeling in certain ways'[13]. The virtues are thus not only about *being* in certain ways; being virtuous predisposes one *to act* in certain virtuous way. It is the action that makes one virtuous: if one is honest, brave, or creative, then one acts honestly, bravely, or creatively. Virtue ethics is also about more than rationality or analytical calculative thinking: Shannon Vallor remarks that 'the virtuous person not only tends to *think* and *act* rightly, but also to *feel* and *want* rightly'[14].

Along these lines, ethicist Miguel Alzola points out that a virtue has four elements: an *intellectual*, an *emotional*, a *motivational*, and a *behavioural* component[15]. This can be illustrated with an example. Consider Alex, a graphic designer who is asked to create an advertisement for a gambling house. Suppose also that they have designed ads for state-run lotteries in the past, centred on having fun and on contributing money to good causes while also having a chance to get lucky and win some money or, even, a huge jackpot prize. Suppose, too, that Alex consciously reflected on the controversial status of advertising for lottery gambling, and that they truly concluded that the people who run state-run lotteries, while wanting to sell more tickets, are not out to create gambling addicts. In sum, Alex thinks that not *all* advertising is wrong, nor is buying a lottery ticket; thus they feel comfortable about designing these ads, as long as responsible play is encouraged. But Alex has recently been approached by a small online bookmaker, Paul. This potential client wants our designer to target demographic groups other than their normal clientele: people in lower-earning brackets, ex-gamblers, and young people. Paul asks Alex to represent gambling not just as fun, but as a way out of poverty, and to use sex and glamour to promote the view that online gambling is the 'new fun-tier'[16]. The strategy includes reminding ex-gamblers of how good it felt to have the dream of hitting a jackpot.

Alex immediately realises that these ads are different in kind than the ones they made for state-run lotteries in the past, and they start making sense of in what way they are different. Alex *feels* that their integrity is being compromised, as they experience a sense of fragmentation provoked by an irreconcilable conflict between the different roles they have (professional designer, parent, concerned citizen). They intellectually *understand* that this new proposed representation of gambling is deceitful and can contribute to so-called problem gambling. At the same time, Alex understands that as a designer they have commitments toward clients and that designing ads is part of their job. Alex also feels personally concerned for people who might think that gambling is a way out of poverty and is deeply worried that their teenage child might encounter similar ads soon, but they are also worried that Paul might see things in a different way.

These insights have enough *motivational* power for Alex to commit to act upon them; they recognise that the easier thing to do would be to just design the ads; despite this, they commit themselves to challenge the brief for the design of the ads because 'it doesn't feel right'. The kind of person Alex wants to be would not contribute to putting others' livelihoods and health in jeopardy.

The *behavioural* element of a virtue allows Alex to realise the commitment into concrete actions: they think a first reasonable step would be to talk to Paul and argue for a change of approach toward responsible goals for the ad campaign, but Alex is sceptical about this. They expect to encounter a slamming door as a reply; in this case Alex's integrity will be subjected to a stronger test and they will have to make an even more difficult decision in order to maintain it; the most dramatic of which might be to reject the project. Naturally, Alex, being virtuous, might find other ways of dealing with the issue that might be satisfactory as long as they are able to sustain the sense of wholeness that characterises personal integrity.

Character and the Virtues

Annas adds that virtues are *persistent*, *reliable*, and *characteristic*. A virtue is persistent because the honest person is able to act honestly despite challenges and difficulties; it is reliable because the bravery of the person can be expected to be there when necessary; and it is characteristic because it is a deep feature of the person[17]. A virtue also develops every time we encounter new situations in which we need to respond selectively. This is important because what forms our characters in certain ways is our upbringing and education *together* with our realised capacities for virtuous acting and reflecting. Aristotle insisted on the importance of practice and habituation for the virtues: 'becoming just requires doing just actions first, and becoming temperate, temperate actions'[18]; so how one lives one's life determines one's character. In order to develop the virtues, one needs to be exposed to and experience situations that require virtuous action and reflection. Also, one must also be able to sense satisfaction or regret at the outcomes of one's action, sensations that must be somehow shared or endorsed by other, but not necessarily all, virtuous persons. This evidently requires situated reflection: *a virtue is not a mere routine habit*. Every situation will require a different, contextual type of response, which mere habituation would not be able to produce.

To illustrate this, consider now the example of a service designer working in a project with the aim of designing a medical imaging environment that would include equipment such as x-ray radiography machines, magnetic resonance imaging scanners, or ultrasound devices. A designer or a design firm might even build their expertise around designing for health care; conceivably, a designer in this area might face similar projects several times in their career. Of course, knowledge and skill can be translated from one project to another, and the designer might develop capacities for dealing with routine matters intuitively. Nevertheless, every design project *will* have particular challenges and contexts that will require new reconfigurations and new responses that cannot be routinised. Each design project will have its own starting point and its own ending point; the designer and others running the project will need to determine every time what a suitable approach is given the specifics of the projects. The particularities of every design project will affect not only the methodological, but also the ethical challenges a designer might face. These challenges might have to do with having to deal, for instance, with different people, different cultural and geographical contexts, or different needs. For instance, designers from the award-winning design consultancy Fuelfor Healthcare Innovation ask themselves: 'how do we make sure that people—patients as well as staff—are not left feeling disoriented, uncomfortable, and stressed?'[19] Our imaginary designer will have to ask themselves a similar question, and every time an answer will have to be given anew. Questions like these are fraught with ethical dilemmas, and every decision a designer makes during the project can be seen and taken as an opportunity to exercise the virtues and to reflect upon the outcomes; this will yield a sense of satisfaction or regret over one's action that can enable the designer to further develop their capacities for moral action.

This example leads us onto the topic of *practical wisdom*, a fundamental virtue that allows us to know what to do in any given situation. Because of this, practical wisdom is understood to be an *intellectual* virtue, developed through experience and associated with knowledge (in contrast with *moral* virtues such as courage or temperance, which are associated with goodness). Practical wisdom becomes relevant when a situation demands an adequate contextual understanding of it; someone is said to have practical wisdom when they possess sufficient knowledge to effectively do something and have the right reasons for acting. Practical wisdom provides a person with the ability to think about 'what can be otherwise'[20]; that is, deliberating over and choosing among possible alternatives, but also about devising new alternatives. Aristotle notes that this deliberation in the case of a virtuous person, tends to aim 'in according with his calculation, at the best of the goods for a human being that are achievable in action'[21]. In conclusion, practical wisdom is about acting, not only about deliberating; its end is discerning what should or should not be done[22] and seeking to act upon this discernment. Practical wisdom will be revisited in chapter 8.

Beyond Individualism

Another important point we must emphasise is that virtue ethics operates beyond the individual level. In his *Politics*, Aristotle famously defined human beings as 'political animals'[23]; this is taken to mean that we humans are naturally sociable, and it is only by living together with others that we can achieve fulfilment. We have language capacities that serve to discuss what is 'advantageous and what is the reverse', and possess 'a perception of good and evil, of the just and the unjust'[24]. It is important to note that thanks to language, what is advantageous and what is the reverse are constantly formulated and reformulated through a myriad of discussions among members of a community: with spouses, parents, or children, at work, in pubs, at places of worship, at neighbourhood meetings, and so on. The virtues, thus, are what equip us for living together and for belonging to the various communities of which we are a part; in virtue ethics, human life cannot be fully pursued without others. This contrasts markedly with *homo economicus*, the prevalent model of human behaviour in neoclassical economics, which posits that calculated self-interest is the primary reason for action. *Homo economicus* is a relentlessly self-maximising, perfectly rational, calculating individual (allowing for different degrees of inclusion of one's immediate family but only *because* they help promote one's interests). The idea of exclusively pursuing the individual interest is based on a notably wrong-headed theory of human nature[25]*.

Because of the implausibility of living as isolated individuals, there needs to be an intertwining 'of the *telos* pursued by individuals and by their community in its shared sense of *telos*. Probably the best way of describing this is that the good for individuals and the *common good* must be interrelated'[26]. In philosophy, the common good is generally understood as the public and shared interests of a community, whereby members stand in a social relationship with one another; 'it requires members not only to act in certain ways, but also to give one another's interests a certain status in their practical reasoning'[27]. It is 'common' because it is pursued by a community and not merely by isolated individuals; what is more, it is often the case that is difficult to tell who, whether immediately or in the longer run, would benefit from it and to what extent[28].

Even in contemporary societies, not only in Ancient Greece, there is an evident tension between the individual and social concerns; the promotion of the social or communal goods might compete with individual interests and put key principles such as pluralism and personal autonomy under pressure. In certain circumstances, individual interests might have to be renounced in favour of the common good; food rationing during wartime, queuing and waiting at airport security checks, or having fewer free parking places in cities are some examples of the common good trumping individual interests.

There is a caveat, though. Under illiberal and dictatorial governments, the tensions can result in the nefarious prevalence of the common good over individual concerns; such evil prevalence is explicitly exemplified in the Nazi slogan 'The good of the community before the good of the individual' (in German, 'Gemeinnutz Geht Vor Eigennutz')[29]. Because of this, the issue of the subordination of the social to the individual is a major issue of discussion in liberal-democratic societies; compulsory vaccination versus recommended vaccination is an example of different societies having different views on weighing the common good versus individual rights. For the same reason, many contemporary authors have emphasised the importance of pursuing the common good while at the same time finding an acceptable middle ground between individual autonomy and individual interests on the one side and social responsibilities and concerns on the other[30]*.

So far, we have not put any emphasis on the different varieties of virtue ethics; we have thus far focused on the *Eudamonist* version, which is structured around human flourishing. There are, however, other accounts of virtue ethics; although they all agree on important issues as we saw previously, they differ on 'what we should do in particular contexts and how we should live our lives as a whole'; these other accounts will be left undiscussed[31].

Criticisms of Virtue Ethics

Virtue ethics is not without criticism, and several authors have formulated challenges to the theory[32]. I will consider two important ones in the following.

The first objection is often referred to as the 'situationist' challenge, which is based on findings from social psychology coming from notorious social psychology experiments such as the Stanford Prison Experiment or the Milgram Obedience Experiment that attempted to investigate obedience to authority and the effects of perceived power[33]. The gist of the criticism is that in several experimental situations many participants behave as a response to the situation they are in, and not moved by character traits or dispositions, which 'shows that there are no such things as character traits and thereby no such things as virtues for virtue ethics to be about'[34].

Several responses have been provided to the situationist challenge: the first is to note that situational influences are fully compatible with virtue ethics; the second is to look to the participants who *did* behave virtuously and showed exemplary moral resistance to situational pressure, exactly as predicted by the virtue ethical account instead of looking at the ones who did not[35]. A third counterargument is to point out that these findings are based on one-off artificially crafted experimental situations and do not consider the long term; also, what they reveal is not a lack of character but conflict between traits: for instance, compassion versus deference to authority[36]. Although not succeeding at rebutting virtue ethics, the situationist analyses are important because they show that the 'cognitive landscape' that surrounds decisions *does* influence moral reasoning (I will say more about the notion of cognitive landscape later, when discussing the shortcomings of principle-based theories).

The second objection against virtue ethics concerns the lack of guiding capability of virtue ethics to act as a guide for moral behaviour. The theory does not prescribe a single principle of action as do Kantianism or utilitarianism; virtue ethics does not have clear-cut guiding principles that could, at least in theory, be applied more or less algorithmically to assess alternatives. A response to this objection can be two-fold. The first response is to transform virtues and vices (that is the opposite of virtues) into a normative guide for action. Rosalind Hursthouse suggests that 'not only does each virtue generate a prescription—do what is honest, charitable, generous—but each vice a prohibition—do not do what is dishonest, uncharitable, mean'[37*].

Similarly, ethicists Justin Oakley and Dean Cocking propose a 'regulative ideal' as a criterion of right action. A regulative ideal is an internalised normative disposition that is based on a conception of the good. Specifically discussing professional ethics from the perspective of virtue ethics, they argue that the aim of a profession must not merely be serving others, but that they must help people 'attain certain goods that play a crucial strategic role in our living a flourishing life for a human being'[38]. This conception of the good can guide agents when transformed into a regulative ideal, which consist of standards of *correctness* (for assessment along the lines of right and wrong) and *excellence* (embodied in the virtues and values) which govern actions and motivations 'beyond the merely correct or incorrect'[39]. Along these lines, an internalised regulative ideal provides a guide for action and a benchmark against which actions can be assessed, so that a person is 'able to adjust their motivation and conduct so that it conforms—or at least does not conflict—with that standard'[40].

Regulative ideals are important to our inquiry, and we will rely on the notion in part III. The idea is straightforward: a good professional has internalised a conception of the proper ends of the profession as a regulative ideal and acts according to those ends when practising the profession. Afterwards, the professional can evaluate those acts against that ideal to determine whether the profession was practised well. This indicates that the objection that virtue ethics cannot guide behaviour can be rejected.

The second answer to the objection of the putative low guidance capacity of virtue ethics is to say that explicit principles are not at all necessary because the virtuous person would know what to do intuitively. General principles might be necessary for novices and intermediate moral decision makers, but, with some exceptions, experts do not need declarative ethical expertise in the form of principles. Developing a virtuous character is developing tacit or non-declarative ethical expertise; in most cases, a virtuous person would have sufficient ethical expertise and would simply *know* what to do in a given situation, or else, it would have enough conceptual tools to elaborate a selective answer without needing principles when tacit knowledge fails, for example in radically new situations[41].

So although virtue ethics seems to resist the challenges, a complex view emerges indicating that virtue is more than a simple conception centred on mere character alone. Virtue ethics includes deliberations, emotions, motivations, and behaviour as well as a clear sense of the ends that are pursued, possibly in the form of regulative ideals that can guide action. Is this complex view a solid and adequate foundation for our inquiry? Before we can answer that question, two plausible alternative approaches will be briefly reviewed.

ALTERNATIVE APPROACHES: PRINCIPLE-BASED ETHICS

Virtue ethics contrasts markedly with the other main traditions of ethical inquiry, which can be said to be principle-based ethics[42*]. Principle-based ethics base their ethical analysis on certain rational evidence (in the form of general principles) that can serve to discern whether an action is ethically acceptable. Different theories apply general principles in different ways. They can be taken either as a starting point for deriving rules or norms that give rise to *duties* or obligations to act in a particular way, or as a measure to assess the *consequences* of particular acts in order to determine their right-ness or wrongness. These different uses determine different subgroups within principle-based theories: the two most important are *deontological* theories based on duties and *consequentialist* theories based on the consequences of a given action.

In the folloinwg, we will review the main tenets of Kantianism and utili-tarianism, respectively the most widely discussed and influential deontologi-cal and consequentialist theories. Although a critical inspection is beyond the scope of this volume, a brief discussion is necessary to properly comprehend the adequacy and usefulness of virtue ethics as a general foundation for our discussion.

Kantianism

Kantianism takes one's motives and intentions as the central element in a moral evaluation. Because of our capacity for rational deliberation, we can guide our actions by reason and resist instinct. Because, for Kantians, our *will* is the only thing we can truly control, we are capable of governing ourselves by principle and are accountable for what we will. Furthermore, we act morally if we act according to the right reasons, even if the consequences of our acts generate suffering. What counts, then, is what moved us to act in the way we did; the consequences of our acts are fully irrelevant. For Imma-nuel Kant (1724–1804), the founder of this school of thought, we ought to act *from* duty, not merely in compliance with it. A famous example that illus-trates the primacy of duty above anything else is that for a Kantian it would be wrong to lie in order to save a person from a murderer.

Only an action done for 'a good will' is a right action. But, then, how can we tell if an action is right? To provide a guide to acting rightly, Kant formulated his Categorical Imperative, which is the basic principle of Kantian ethics. The categorical imperative is a principle that every rational person must accept, and which is the basis of all other moral rules. It is 'categorical' because it is *true in all situations*. Kant famously expressed it as: 'act only in accordance with that maxim through which you can at the same time will that it become a universal law'[43]. In present-day language, what Kant is saying is that one ought to act in such a way that one would want for it to become something that everyone else should do.

So, before acting, one could ask oneself, 'would I want everybody at all times to do what I am about to do?' If their *rational* answer is yes, then the act is ethically acceptable, otherwise it is not. So, for example, imagine you are visiting the Petrified Forest National Park in Arizona. You might want to take a specimen of petrified wood home with you. Then you can ponder whether it is ethically admissible to do so; to know that with certainty, you could ask yourself whether you would want *everybody* to take home specimens of petrified wood after every visit. Put differently, would you want there to be a universal rule that says 'It's okay to take home specimens of petrified wood'? You probably would not, would you? Because we answer negatively, it is certainly wrong to take home a specimen of petrified wood.

Kant formulated two other versions of the categorical imperative, so there is more to Kantianism than what we have covered here. But because what we have omitted in this brief summary is not central to our discussion, it seems convenient to wrap up at this point and summarise by saying that for Kantianism: 1) rightness can be found in the nature of the action itself, and on acting from duty; and 2) ethical rules must be able to become universal rules without exceptions.

Utilitarianism

Utilitarianism, a school of thought whose main proponents were Jeremy Bentham (1748–1832) and John Sturt Mill (1806–1873), proposes the complete opposite of Kantianism. For utilitarians, all that counts are the *consequences* of one's acts for the affected people and the world as a whole, not one's motivations. Simply stated, utilitarianism prescribes that one should choose the action that maximises good consequences.

Utilitarianism is well known by the utilitarian principle that posits that an action is right when it achieves the greatest happiness of the greatest number; It is best summarised in Mill's oft-quoted 'Greatest Happiness Principle': 'actions are right in proportion as they tend to promote happiness, wrong as they tend to produce the reverse of happiness'[44]. Under this principle, the more happiness an act produces, the better that act. Unlike Kantianism, which can be quite intricate, utilitarianism is fairly straightforward: one ought only to take actions that maximize net happiness by weighing the consequences (that is, expected happiness against expected pain, or harms versus benefits to those affected by the decision). All that is necessary, thus, is to define the relevant alternatives, and to calculate the most net happiness. And it is a simple calculation: happiness minus pain. At least in theory.

But in practice it is not that simple: how are we to calculate the net happiness of *every* single act? That is impossible. A more plausible alternative to that type of utilitarianism (called 'act utilitarianism') consists in using rules ('rule utilitarianism') formulated upon an estimate of the *aggregate* general consequences of particular types of acts. In this way, instead of assessing the wrongness or rightness of a particular act, rule utilitarians try to find a rule that applies to the situation at hand. Unlike Kantianism, the rule arises not from a principle, but from an experience-based generalisation. So, for instance, in rule utilitarianism, murder or slavery are considered wrong because they *in general* tend to produce bad consequences (more pain than happiness) for those involved. A logical entailment of this is that murder or slavery are not inherently wrong; they are wrong because they generally bring about bad net consequences. A less extreme example is speeding; it is wrong insofar as it generally leads to traffic accidents, injuries, deaths, and material losses, which reduce happiness in society.

How can we know what utilitarian rules we ought to follow in order to decide? For example, some rule utilitarians posit that by being part of our individual consciences, others require these rules to be part of the public knowledge or be built into public institutions[45]. Going back to our example of the petrified wood, a rule utilitarian, thus, would also say that it is wrong to take a specimen of petrified wood home with you; their justification, however, would be different to that of a Kantian. A utilitarian might say that the happiness of the many people taking petrified wood chips home with them would eventually weigh less than the sorrow provoked by the present and future generations of visitors who would no longer be able to visit the park because it has been depleted of petrified wood. Because of that, the rule is enshrined in a law (and in our consciences).

Some utilitarian writers, most notably Peter Singer, argue that the interests of all sentient beings and not only those of humans should be given equal consideration because they can experience pain and pleasure[46]. The utilitarian calculation, therefore, needs to compute non-human animal pleasure and pain as well. This brief outline only covers the basic utilitarian tenets, and there is much more to utilitarianism than this, but for the purposes of my argument, a deeper treatment is not necessary.

Why Kantianism and Utilitarianism Are Not a Good Fit for Design

Needless to say, I do not intend to settle the long ongoing discussion between virtue ethics versus Kantianism versus utilitarianism as the substantive general foundation of morality[47]. There are many problems with both Kantianism and utilitarianism, but we will cover only a few relevant shortcomings of these theories. The goal here is to plausibly show why virtue ethics offers, as a whole, a more promising avenue for grounding design professional ethics than a principle-based ethics.

Both theories have difficulties providing an explanation for moral motivation. Why should one do a particular act or follow a particular principle? These theories do not sufficiently *move* people toward action; acting requires more than a rationalistic justification[48]. We already touched upon the mistaken notion of *Homo economicus* and the myth it represents; along these lines, although humans *are* rational beings, they are not only that—they have feelings and emotions that sway them in particular ways toward action. Kantians tell us what we ought to do, but do not go further in fostering our will power to act in that direction. Both Kantians and utilitarians generally spurn emotions and feelings and favour pure rationality, but findings from neuroscience show the crucial importance of emotions in decision making[49]*. If we want a theory that provides more meaningful motives for acting morally than reason alone, it seems justified to pursue a theory that goes beyond what principle-based ones can offer. Virtue ethics could be such a theory, but before committing to it, let us test the potential suitability of principle-based theories from the perspective of design.

Because design problems are highly dynamic and context-based, it does not seem reasonable to think that we can find out what the right moral course of action is by simply converting what one has in mind into a 'universal law' as Kant suggested. We can easily accept that it is desirable for all people to be kind to strangers; but can we say the same about the smart cradles we considered in chapter 5? Can we say that *all* parents should use them at *all* times? Pressured to make a choice in these terms, we might say that we would want *all* parents to use such a cradle under *some* hypothetical circumstances, while under different circumstances we would want no baby to fall asleep cradled by an autonomous machine[50]. Moreover, envisioning possible design scenarios in order to test whether the preferred one could be reasonably transformed into a universalisable maxim would be *extremely* complex, if only because there are simply too many factors to be considered. To apply Kant's categorical imperative to a design decision is just too hard to fathom due to the sheer number of design scenarios that open up with every different design idea.

Similarly, although a very strong case can be made from a utilitarian perspective that we ought to help those in need 'unless, to do so, we had to sacrifice something morally significant'[51], utilitarianism becomes daunting under the epistemic insufficiency that characterises design. How are we to calculate? Based on what relevant description are we to define the elements that are being weighed against each other? And considering the multistability of artefacts, and the unintended effects, how are we supposed to factor this into the calculation of net happiness?

Principle-based theories are excessively formal and focus exclusively on calculative analysis and argumentation, and although they lay claim to rationality, they can lead to mutually incompatible answers. The famous 'trolley problem', originally posed by philosopher Philippa Foot, illustrates this conflict between Kantianism and utilitarianism: imagine the 'driver of a runaway tram which he can only steer from one narrow track on to another; five men are working on one track and one man on the other; anyone on the track he enters is bound to be killed'[52]. What should the driver do? In a simplified interpretation, utilitarian reasoning would indicate that the driver should go to the track where we find one person, whereas a Kantian would say that the driver should do nothing as killing is wrong. By letting the tram follow the tracks, the driver would not be killing anyone, but only letting them die. Paradoxically, this phenomenon might be conducive to 'ethical relativism', where what is morally correct depends on personal preference and the arbitrary choice between one theory and the other. Considering this important shortcoming of these theories Shannon Vallor summarises a frequent and compelling argument in favour of virtue ethics: 'moral principles simply codify, in very general and defeasible ways, patterns of reasoning typically exhibited by virtuous persons'[53].

Good behaviour does not arise from following moral rules; it is the other way around: good rules *reflect* the virtuous patterns of reasoning and behaviour. Nevertheless, virtue ethics does not eschew ethical principles; guidance from rules and principles could very well be relevant *as well* as 'contextual, embodied, relational, and emotional', considerations that are ignored by rationalistic philosophers[54]. Practical wisdom is the virtue that enables us to navigate the epistemic insufficiencies, whereas our own purposes of life and conceptions of the good life provide a general directionality for our decisions.

In the case of professional design, the direction can come from two sources: first, from the overarching ends of professionalism (the public service element), and second from the specific ends of the design profession (the specific substantial goods that the profession seeks to provide), which supply reasons for acting in one way or another. Of course, our individual motives as professional designers become *reasons* when they are aligned with a common understanding of the good. This alignment is what enables us to achieve agreement, albeit partial, over important matters. Moore summarises the dynamics of practical virtue:

> the virtuous individual, drawing on all of her practical experience and practical wisdom, seeks to make a judgement which, in that particular situation, is consistent with her overall judgement as to what is truly good for her as a

person, as well as for others and for the community, and acts appropriately on that judgement[55].

A Rejoinder

A satisfactory design professional ethics is a professional ethics that is able to provide guidance to professionals that deal with wicked problems, which, to reiterate, are problems that intrinsically lack a definitive formulation and, more importantly, immediate and ultimate tests for solutions. The task of applying the categorical imperative (or calculating net happiness) to the myriad of design alternatives that are generated in a project, from concept to implementation stage, in order to choose the 'right one' is simply too massively complex. Having to do so without a definite problem formulation and high certainty over the outcomes would be inhumanly colossal, if not outright impossible. Virtue ethics is not fully unconcerned with calculations, but its focus is on the realm of moral responsiveness to the particular demands of the situations. Because of this, its emphasis lies on the habitual disposition to do the right thing, and to do those things *because* they are good or contribute to the good.

Granted, the principle-based approaches could possibly work for designers working on artefacts whose effects can be well anticipated thanks to professional experience and a body of technical knowledge. These sources could inform designers about how the designs could work and how they could be received. A designer working on a project for a regular chair, a ballpoint pen, the layout of a classic newspaper, or markers for toddlers will find a lot of usable information that will enable them to make *good* decisions.

To illustrate, designers know (or should know) that little children will put markers in their mouths, so the ink needs to be non-toxic; they know that toddlers have not achieved full manual dexterity, which calls upon them to design thick, big markers that are easy to grab; they should also know that toddlers will use the markers to hammer with. From an ethical perspective, it is absolutely unjustifiable to design a marker for toddlers with toxic ink or that is too small and not sturdy enough; doing so would compromise the toddler's safety and spoil their fun. There are many rules and regulations that can help designers avoid committing these errors: some materials are forbidden or highly regulated, even the forms of some designs are restricted because they could become dangerous (for instance, children could climb horizontal rails in balustrading). To sum up, there is a vast amount of technical knowledge that enables a designer to perform a utilitarian analysis when designing a chair, a ballpoint pen, a marker for toddlers, and so on.

However, while a utilitarian analysis could be of use when assessing a discrete feature of a product like those, it would be unwise to attempt to calculate the potential net happiness of a whole project like a smart cradle or the Superblocks. In projects like this, we cannot tell with full certainty how such designs will affect society. Considering also that their effects can take a long time to manifest, if we do perform a calculation, it would only provide us with an unwarranted sense of certainty.

There are more grounds for not taking principle-based theories as the backbone of a design professional ethics. If we want to develop a profession-al ethics that addresses real-world ethical reasoning, we need to consider how people think, reason, and act in real professional settings, which are complex and dynamic. Decision making scientists inform us that decisions never oc-cur in a vacuum and involve much more than will (favoured by Kantians) or calculative rationality (favoured by utilitarians). Decisions are made against a backdrop called 'cognitive landscape' that includes many factors which influence the decision itself: the framing of the decision (the terms in which choices are formulated), features of the situation (time pressure, for in-stance), the emotional state of the decision maker, the goals (well or ill-defined or conflicting), the stakes at play (how much the decision maker cares about the outcome), the tools that are available to make the decision, the reasons why the decision maker is deciding, and the extent to which they have experience with similar decisions[56]. Acting well in complex situations requires exercising the virtues, especially practical wisdom for dealing with a conflict between virtues (for example, a desire for thoroughness against wanting to meet a deadline). The virtues are precisely what enable one 'to perceive, experience emotions, deliberate, decide, and act in a proper way'[57].

Virtue ethics can withstand challenging conditions of uncertainty by fo-cussing on motive, on a horizon for the action, and on what is necessary to decide well, rather than relying on fixed principles and prescribing clear-cut answers. Journalism scholars Stephen Klaidman and Tom L. Beauchamp write that 'virtuous traits are especially significant [in environments that are] too pressured to permit prolonged and careful reflection'[58]. By being less focused on deliberation and more on habitual disposition, virtue ethics seems also to be able to function better under time pressure. Furthermore, and importantly, virtue ethics provides moral motivation for acting well by an-swering the question 'Why act ethically?' that principle-based theories leave unanswered. Virtue ethics, contrarily, reconciles our rational and our emo-tional internal spheres in its explanation of how and why we are moved to action, offering a plural account of motivation.

To conclude, virtue ethics appears to be a more promising alternative than Kantianism and utilitarianism as a viable philosophical foundation for design professional ethics. Naturally, a demonstration of the suitability and effec-tiveness of virtue ethics for supporting our inquiry into design professional ethics will be indirect and contingent on the extent to which the findings of this inquiry are cogent and persuasive.

ENTER ALASDAIR MACINTYRE

Alasdair MacIntyre (born 1929) is a philosopher who, starting with the publication of *After Virtue* in 1981 (revised in 1984 and 2007)[59], became a major figure in late-twentieth- and twenty-first-century ethics and political philosophy. His ideas around *practices* and *internal goods* are one of the central conceptual pillars for our inquiry into professional ethics.

This section offers a précis of his conceptual apparatus for the understanding of these notions. MacIntyre's treatment of practices is part of a larger critique of modernity in general and of modern liberal individualism in particular: modernity destroyed the traditional ways in which people dynamically made sense of their lives and substituted them with mere rules and principles (such as the utilitarian principle we covered previously), leaving people without an adequate moral compass. Although I will be drawing on some of his views on modernity, in this book I will not delve in depth into his critique of capitalist modernity, nor will I subject his highly influential account and his political philosophy to further scrutiny[60].

MacIntyre draws on Aristotle's account of human action[61]* to develop a new stream of virtue ethics; because of this, he is often called a neo-Aristotelian philosopher. For Christopher Lutz, one of MacIntyre's main commentators, his version of virtue ethics 'begins with the interests of agents and the shared interests of members of communities. As such, it unites the virtues with practical reasoning and community life'[62]. MacIntyre derives his account from an anthropology of practices. He does not seek to propose timeless truths about the world based on principles and premises, but aims to find, formulate, and reformulate *approximations* to those truths based on particular cases[63].

Practices

For MacIntyre, it is in *practices* where virtues and character develop; it is in a practice where the goods can be achieved, and it is in achieving those goods together with other practitioners where new and modified conceptions of the good can emerge. In an oft-cited and rather intricate passage, MacIntyre defines practices as:

> any coherent and complex form of socially established cooperative human activity through which goods internal to that form of activity are realized in the course of trying to achieve those standards of excellence which are appropriate to, and partially definitive of, that form of activity, with the result that human powers to achieve excellence, and human conceptions of the ends and goods involved, are systematically extended[64].

MacIntyre's definition highlights five central ideas[65]*:

1. *Practices are coherent and complex:* they are not isolated activities, but need to have an organising purpose that is coherent enough to aim at some goal in an organised way. Not everything is a practice: medicine is a practice, but stitching is not; engineering is a practice, but fastening nuts and bolts is not; baking bread is a practice, but sifting flour is not.

2. *Practices are social and cooperative activities:* they are human activities that are embedded in a larger social context and require or gather like-minded others.

3. *Practices have internal goods:* internal goods are goods that are intrinsically associated with the practice. For instance, making an elegant move in chess, caring for patients in nursing, fostering autonomy in teaching, coming up with an effective solution in design. We will discuss internal goods more extensively as we move along.

4. *Practices have standards of excellence:* without standards the internal goods cannot be achieved; they guide the practitioner providing a sense of directionality and purpose. These standards arise in what MacIntyre calls a 'tradition' and are historically determined by the community of practitioners. Surfers, for instance, generally frown upon using a surfboard with a motor, graphic designers reject combining too many different typefaces, architects value the quality of materials, and so forth.

5. *Practices are extended:* one way to understand this is to connect this extension to striving for the best in a technical sense—by reaching the standards of excellence, higher thresholds are defined, which requires higher and higher abilities and knowledge from practitioners. The standards are the steppingstones from where the practice grows, but they are not fixed for ever: 'the standards are not themselves immune from criticism, but nonetheless we cannot be initiated into a practice without accepting the authority of the best standards realized so far'[66]. To illustrate, in the West, since the late nineteenth century, and after three thousand years of history, doctors no longer recommend bloodletting, as the method has been discredited as a treatment. A wider way to understand how practices are extended is to connect it to wider aspects of human life[67]. So they are extended not only technically, but also by new understandings of what 'better' means, and what types of ends can be considered legitimate; the practice is also extended when its practitioners act upon those new understandings. The practitioners themselves are morally 'extended': they become better human beings. For example, Microsoft workers did not want a technology that was developed for entertainment (the augmented reality headset HoloLens) to be used for warfare[68]; these workers did not believe enhancing warfare technology could be seen as a legitimate end for their design

skills. By acting upon this realisation, they showed constancy, integrity, and courage. In this way, practices, and this is why they are so crucial, provide us with opportunities to grow as persons. In chapters 8 and 9, we will come back to these issues, which are central to my account.

We must note two things before ending this section. First, arguably all professions and many occupations (especially skilled trades) sufficiently satisfy the conditions posed by MacIntyre; so, lawyers, electricians, farmers, secretaries, doctors, pilots, dentists, musicians, or actors all belong to a practice. These are social activities that are complex enough and have internal goods and standards of excellence. Naturally, and this might be too obvious to require noting, not all practices are professions or skilled trades: volunteering in a community centre or parenting are practices too. Second, and more specifically, design *is* a practice. I will not argue in detail that it is; the arguments that were provided in chapter 3 to substantiate the claim that design is a profession could be marshalled here to convincingly argue that design is a practice too. I have no doubt that these arguments would provide more than sufficient grounds to substantiate this claim.

An Example of a Practice: Surfing

Before we explore the basic structure of MacIntyre's ideas, it might be convenient to start with an example to revisit the notions of purposes and goods from a different angle; to do so, let us consider the example of the practice of surfing. Just like in the earlier example of contemplating the landscape, catching waves is equally an activity, a goal, and a purpose. Catching and riding waves is the literal goal of surfing because catching waves is what one needs to do to achieve the purposes of surfing.

Surfing, however, goes beyond merely catching and riding waves. Some surfers might have the purpose of exercising, having fun, procuring something to show-off with on the social networks, or they might just want to kill time. Other surfers, however, might have more profound purposes and assert that surfing's ultimate end is about much more than all that. They might declare that surfing is about being out there in the ocean, overcoming the fear when the surf is big, exercising self-control, enduring brutal 'wipe-outs', knowing how to position oneself in the water with nothing else other than one's knowledge, being able to 'read' the waves to know how and when they will break, and, finally, catching waves *in a particular way* (like surfers do).

Naturally, surfing *is* about riding waves and performing manoeuvres, but committed surfers would also say that surf is primarily about being 'one' with the wave, about never 'dropping-in' or 'snaking' on someone (both are usually caused by greed and usually prevent the other surfer from catching *their* wave), it is about claiming a wave for oneself *only* when one is better positioned and *absolutely sure* they will catch the wave (claiming a wave and letting it go unridden is a serious ill in surfing). Others might add that it is also about showing appreciation of the qualities of other surfers and about helping fellow surfers in trouble.

These are some of surfing's purposes, but what about its goods? For some, the goods might be 'likes' and comments on Instagram after they post a picture of a good manoeuvre, having fun, or getting fit. Committed surfers might agree that these *are* goods but would say that through surfing one can achieve more *genuine* goods; goods that are intrinsically connected to the genuine purposes of surfing. They might say that surfing gives them the opportunity to perform manoeuvres that can only be executed on waves (getting 'tubed' inside the barrel of the wave is the quintessential surf manoeuvre), but also to feel 'surfed-out' after a long day of surfing (which is much more than just being physically tired). Surfing allows surfers to develop courage, perseverance, humility to recognise one's own limitations, to foster respectful attitudes and care toward others and the environment, and, why not, experience a sense of awe and wonder for the power of the uncontrollable ocean from within. These goods seem to be genuinely attached to surfing, whereas likes or killing time seem to be goods that are less uniquely connected to it. These enumerations of purposes and goods are only indicative; there are many ways of surfing and many ways to achieve the genuine goods it offers. Surfing legend Gerry Lopez aphoristically made the same point: 'surf is where you find it'[69].

This discussion raises a pertinent question: what enables a surfer, or any practitioner for matter, to tell if a purpose is a true purpose? What can help them discern if a good is genuine? The short answer is: the virtues. The practice of surfing gives the surfer the opportunity of developing the virtues: excellence, perseverance, care, and humility. The virtues, according to MacIntyre, 'enable agents to identify both what goods are at stake in any particular situation and their relative importance in that situation and how that particular agent must act for the sake of the good and the best'[70].

Internal and External Goods

MacIntyre emphasises the difference between two types of goods that can be attained through practices. He distinguishes between *external* and *internal* goods. External goods are the rewards of participating in a practice that could be gained in ways *other* than by participating in that particular practice. Conversely, internal goods are those goods that can only be gained by engaging in that practice. Going back to the example of surfing, one could argue that one can kill time or obtain 'likes' in many other ways than by surfing; because of that, these are external goods. Contrarily, being one with the wave, performing manoeuvres, and caring for fellow surfers in trouble are goods that can *only* be achieved through surfing; because of this, these are internal goods.

Although MacIntyre insists that the internal goods of a practice can *only* be gained through a specific practice, I am not fully convinced that *all* internal goods can only be obtained through a practice that provides them. Certainly, some internal goods are exclusive: being one with the wave is exclusive of surfing, whereas solving a mathematical problem elegantly is exclusive of mathematics. In my understanding, goods need not be exclusive to be internal, though they might be. It seems to me that the capacity for self-control the surfer develops is not altogether that different than the self-control developed by a climber or a deep-sea diver, even if the practices are dissimilar. But that some internal goods are not exclusive does not entail that they are external. Helping a fellow surfer in need is certainly not an external good as it is genuinely attached to surfing, and is an important part of it, albeit perhaps less constitutively than being one with the wave. At the same time, the virtues that come to pass when a surfer helps a fellow surfer in trouble, as far as I can see, are not that different from those that enable a cyclist to stop to help a fellow cyclist.

Now let us turn to exemplify design goods: crucial internal goods of design could be found, for instance, in performing the very act of designing; in balancing creativity with multiple constraints and actually getting something made; in conducting usability testing and seeing that your design works as expected; in working with other designers on a project and exchanging ideas and working as a team; in the personal relationship that arises in these interactions; in the human ingenuity and creativity that emerges while dealing with complex problems; in the joy of moving from an abstract idea to a functioning artefact that contributes to people's well-being; in the patience, tact, and thoughtfulness a designer needs to exhibit when dealing with difficult clients; in successfully handling a conflict with a client about the design process; in being open to what clients have to say about their own domains and learning from them; in the industriousness necessary for bringing a project to a good end; and so on and so forth. These are only possible internal goods, and the list does not intend to exhaust the topic.

External goods are easy to mention: money, fame, social status; unsurprisingly, these are always more or less the same indicators of success. To reiterate, however important these goods could be obtained in other ways: just like a doctor can become famous not necessarily by being a good doctor, but by being a television host, or a designer can obtain fame by being a good public speaker, and not necessarily by being a good designer. Similarly, a designer can become rich by being a good entrepreneur, by making good investments, or by winning the lottery, not necessarily by designing well.

Why is the difference between internal and external goods important? For MacIntyre, every practice involves a shared sense of a purpose: a common understanding of the 'for the sake of which we act'.

What does it mean to understand design in terms of a shared purpose? It means that design is a practice insofar as designers pursue a common over-arching purpose; that is, design is not a practice merely because most members studied design or know how to make prototypes or have an aesthetic sense. Though these elements are important, understanding the practice of design by describing the characteristics of the practitioners and the activity would be incomplete. The practice of design is surely defined by what designers know (their knowledge and skills) and what they do (designing), but *also* by how they judge what they pursue (the standards of excellence) and why they do (the goods of design).

The point here is that when the internal goods of a practice are obtained, the practice is sustained and extended: the practice grows when its practitioners focus on purposes that are intrinsically attached to it, instead of pursuing external goods. The more a purpose is intrinsic to that practice, the more genuine it is. Design as a practice is sustained and extended *only* when designers pursue the *telos* of design (designing excellent products, being thoughtful with clients, respecting users, mentoring young designers, and so on). Putting the *primary* focus in the pursuit of external goods might perhaps benefit the individual designer, but hardly the practice as a whole.

However, the point MacIntyre makes is not that we have to reject external goods; after all, 'no one can despise them altogether without a certain hypocrisy'[71]. External goods *are* goods in a true sense. The basic idea is that achieving these goods must be contingently attached to the pursuit of the internal goods of the practice. External goods (money or fame, but also 'likes' on Instagram) are attached to practices 'by the accidents of social circumstance'[72]. Relatedly, external goods are also important and necessary because they help sustain the practice: a practice cannot be sustained without them. An independent designer, for instance, needs to get paid for their work to ensure their livelihood, but also to sustain their design studio, to pay for materials, infrastructure, and so forth. Being recognised by clients might contribute to getting more requests from existing and new clients. Pursuing external goods is thus undeniably indispensable; yet, from the perspective that we are discussing, the *good* practitioner would not pursue them for their own sake, but for the sake of the internal goods of the practice.

As MacIntyre argues, being able to distinguish between 'genuine from merely apparent' purposes, is key for the quality of our practical reasoning[73], which is the virtue that enables us to know what to do in complex circumstances. It is worth keeping in mind that this knowledge about what to do is for the virtuous person necessarily connected to acting upon the knowledge: 'the conclusions of our practical reasoning *are* our actions'[74]. That is why it is so important to understand which purposes and goods are genuine and worth pursuing.

Another relevant distinction within the realm of internal goods is made by Geoff Moore, who differentiates two types of excellences[75]. The first type is the *good product*; the internal good, in this case, is associated with excellence in the outcome of the activity. In the case of design, we could mention, for example, designs that are honest, beautiful, sustainable, usable, pleasant, comfortable, and so forth. The second type has to do with goods related to the practitioner *as a person*, with their flourishing as individuals. By pursuing the excellences of the product, a designer also develops morally as a person. A designer dealing with a difficult project struggles to reach an adequate solution and while they do so, they also develop, for instance, self-discipline, constancy, thoroughness, or ingenuity. If working together with end-users, they might also develop empathy, solidarity, tolerance, etc. The important thing is that even if they fail to achieve a minimally satisfactory solution, they still *might* be able to develop the personal virtues nonetheless. Whether they do or not is a matter of how they approach and pursue their tasks and with what purposes they do so.

Practices are competitive environments in the pursuit of both external and internal goods. External goods are 'objects of competition in which there must be losers as well as winners'. Designers pitch against one another for work, and if one gets the project that is a victory for the winner and a loss for the loser. It is a zero-sum game. The quest for internal goods may be competitive too, but in a radically different way: it is not the good itself which is the object of competition[76], as practitioners want to bring the practice to the next level and improve on the performance of past and present practitioners.

In contrast to external goods, the excellence in outcomes can benefit the whole practice. The new achievements are 'a good for the whole community who participate in the practice'[77]; so when a designer produces an excellent design, all designers reap the fruits. This communal benefit manifests itself clearly in, for instance, the way designers rely on previous design exemplars when they work on new designs. The use of precedents and referents to inform the generative stages of design is a well-documented phenomenon in design research[78]. Jonathan Ive's design for Apple in the 2000s built on the mathematical precision and clarity of the work that Dieter Rams achieved for Braun in the 1960s. The positive nature of this competition encompasses also new methods or techniques that are first developed for particular projects, and then diffused across the profession: user-centred design methods came into being in software design during the mid-1980s, but reached almost all design disciplines during the 2000s and 2010s.

This phenomenon of building on the past achievements of previous practitioners is obviously not exclusive to design—it is a key feature of all creative practices: one finds it in science and in the arts too. We could counterfactually assert that there would be no wailing clarinet in Gershwin's 'Rhapsody in Blue' without *klezmer* (Jewish music from Eastern Europe)[79]. How would Billie Eilish sound without having listened to Amy Winehouse? And what type of singer would Amy Winehouse have been in a world without Otis Redding? We could imagine that both Eilish and Winehouse would still have become artists, but we would also be pretty sure that they would sound very different, if only because they would have had other influences. What we could *not* at all envision is Eilish or Winehouse becoming great artists *without* being influenced by and engaging with the work of other great artists that preceded them.

The Narrative Order of Our Lives

It is probably already clear from the discussion that internal goods are more important than external ones. Pursuing the internal goods is what brings us closer and enables us to achieve the purpose of the practices in which we engage. In turn, it is by pursuing these goods, enabled by the virtues, that we achieve our own purposes as practitioners and as persons. One is engaged in many practices simultaneously: a profession, a family, perhaps a sport or a hobby, a circle of friends, a neighbourhood community, etc.

The pursuit of the internal goods of the many practices with which one engages becomes a quest to discover and reformulate one's purposes in life; MacIntyre refers to this as one's 'narrative quest'[80]. The quest provides directionality, but not necessarily a destination; as we move along, we learn more about what we seek and about ourselves[81]. The idea of the narrative quest can be further illustrated with a line of a poem written in 1917 by the poet Antonio Machado (1875–1939), 'walker, there is no road, the road is made by walking.' Each of us walks their own path, and only when looking back we are able to see actual the road we made.

In his characteristically convoluted style, MacIntyre offers us a refinement of the notion of virtue connecting it to the pursuit of our quest:

> The virtues therefore are to be understood as those dispositions which will not only sustain practices and enable us to achieve the goods internal to practices, but which will also sustain us in the relevant kind of quest for the good, by enabling us to overcome the harms, dangers, temptations and distractions which we encounter, and which will furnish us with increasing self-knowledge and increasing knowledge of the good[82].

Understood this way, the virtues enable us to strive for excellence in the outcomes of our practices, and concurrently to grow as persons. They help us withstand challenges and overcome the obstacles we encounter as we try to write the story of our life as we live it. MacIntyre summarises the relationship between the larger questions and the discrete decisions we make regarding individual acts: 'I can only answer the question "What am I to do?" if I can answer the prior question "Of what story or stories do I find myself a part?"'[83]

Tradition

The narrative quest has an overall purpose: flourishing as a human being (to reiterate, understood as achieving happiness in the deepest sense, not only as a temporary sense of wellness). Because we are social animals, our flourishing is not isolated but connected to everyone else's; it is not only our flourishing as persons that we are pursuing, but aiming also to contribute to others' flourishing too, albeit tacitly. There is a caveat: these insights might inform us of the purpose, but seem to provide very little knowledge about the way to get there. Flourishing and the virtues appear to be a rather circular itinerary: flourishing is virtuously pursuing the quest for a good human life, and a good human life is one in which we can flourish thanks to the virtues.

We are 'born with a past', MacIntyre writes[84], which means we occupy a place in history, and our lives are embedded in a tradition in which we find ourselves. When we enter into a practice, we 'enter into a relationship not only with its contemporary practitioners, but also with those who have preceded us in the practice, particularly those whose achievements extended the reach of the practice to its present point'[85].

This is important because a tradition offers us different ways to pursue our quest without having to figure out anew what to do every time. A tradition is a moral steppingstone to pursuing our quest: it provides us with conceptions of the good and what it means to live a good life. Paraphrasing the poet, we could say that a tradition offers us not a road, but a shared way of walking.

A tradition in a MacIntyrean sense should not be seen as guided by atavistic traditionalism or by conservatism. On the contrary, a vital tradition 'embodies continuities of conflict [and] is an historically extended, socially embodied argument . . . about the goods which constitute that tradition'[86]. The arguments and conflicts always build on the past, and although the past might try to control the present (through standards, for instance), the tradition is annotated and extended *in the present* by new conflicts and arguments about what excellence means, giving rise to new standards. In other words, a tradition provides guidance, but it is *also* an ongoing discussion about what good practice is and how its internal goods can be best pursued. This call for arguments guarantees a reasonable degree of plurality of views within a given tradition, while maintaining a shared sense of common purpose.

To illustrate, design is rich in agreements and disagreements about conceptions of the good. The pioneers of design were vexed and affronted by the

'dishonesty' and 'fatal newness' of mid-Victorian furniture and the 'ugliness' of the material culture of the late 1800s and the early 1900s. These
were profoundly moral charges: dishonesty, fatal newness, and ugliness
made it impossible for a house to become a proper home, and it was morally
wrong for designers to help bring dishonesty and ugliness into being. All that
was superfluous or purposeless became bad design in a moral sense as well
as in an aesthetic one: 'ornament is a crime', Adolf Loos famously lam-
basted. These condemnations catalysed in the rationalistic and functionalist
ideals of the Modern Movement. Many standards can be found that embody
those ideals; a key one is the minimalist dictum famously popularised by the
architect and designer Mies van der Rohe (1886–1969) 'less is more', which
became, and still is, highly influential to all forms of design[87]*. But critics
emerged and arguments started: life is too complex for simple solutions, the
critics argued, and the goods of (architectural) design did not truly benefit
from the minimalist standard. In 1966, architect Robert Venturi (1925–2018)
proposed: 'Blatant simplification means bland architecture. *Less is a bore*'[88].
The discussion about minimalism continues, of course, and it is a profoundly
moral discussion about the material conditions for living well; what the asyn-
chronous banter between Venturi and Van der Rohe illustrates is how a
tradition transcends its limitations, and how the practice of design is extended through criticisms, arguments, and discussion about the goods of
design and the best ways to pursue them.

What this example emphasises is that while there might be deep disagreement about what constitutes a good dwelling or a good product, the notion
that a building or product is good insofar as it somehow contributes to living
well is upheld. Living well could be seen as an overriding good that enables
rational debates about, for instance, how conducive to the good life minimalism or symbolism are. While Van der Rohe and Venturi agree on the importance of the role dwellings play in living well (the overall good that is
pursued), they disagree on what constitutes living well.

MacIntyre' theory not only permits disagreements but makes them desirable as a way of extending the practice; he argues that 'rebellion against my
identity is always one possible mode of expressing it'[89]. In his view, however, deep disagreements can and need to be resolved, resulting in one conception prevailing once compelling reasons can be given. Contrary to MacIntyre,
I believe that as a long as there is a rough consensus about the overarching
purpose of a practice (in our case, design), settling the internal debates and
reaching agreement on a substantive conception of the good might not a
necessary condition for the continuance of a practice. After all, disagreements over conceptions of the good are 'an inescapable feature of the modern
predicament'[90].

We might take as an example the tension between 'skeuomorphism' and
'flat design' in user interface design throughout the 2010s, where proponents

of both approaches sustained conflicting, albeit internally consistent, rationales about why their approach was the right one. The disagreements can be integrated into the practice as long as the different parties of a debate recognise the other party as pursuing roughly the same overall purpose and upholding a roughly similar commitment to it. In this case that would be achieving more elegant and usable interfaces. In the case of the debate between Van der Rohe and Venturi, the purpose would be enabling people to live well by providing suitable material conditions for dwelling. Modern followers of Van der Rohe and Venturi might think the other party is profoundly mistaken in their means or even in their conceptions of what living well consists of.

To me, these internal disagreements, however radical, seem not to be as important as the similarity in the purposes they pursue. If practitioners pursue similar purposes, they act for the sake of similar things and can be said to have a shared *telos*. In a practice, we said earlier, individual motives become reasons when aligned to the purposes of the practice; if practitioners pursue similar purposes, even if they provide different motives for their actions, their motives can become reasons, and so the disagreement can be lessened[91]*.

In this chapter, we have outlined a virtue ethical perspective for our inquiry. In this approach, we are less focussed on the rightness or wrongness of particular acts than on what people do and want to do with their lives as a whole. Along these lines, virtues are those dispositions and habits that enables us to flourish as human beings. We also adopted Alasdair MacIntyre's account of practices as the conceptual backbone of this inquiry. MacIntyre's account rests on the idea that a good human life is centred on a 'narrative quest' for the good life, during the pursuit of which people acquire and develop the virtues, and engage in different practices to obtain the internal goods of the practices they participate in. In turn, the practices further develop their virtues and bring them closer to their purpose. This quest is not a purely individual pursuit, but is embedded in larger traditions associated with the different practices in which they partake.

In the next chapter, we will apply the MacIntyrean perspective to our analysis of design, with the objective of providing a description of design practice that will serve as a foundation for formulating a plausible overarching purpose of design, and, in turn, ground a regulative ideal that can guide designers in their professional activity.

NOTES

1. The most important milestone in the resurgence of virtue ethics is commonly considered to be the publication of the article 'Modern Moral Philosophy' by G. E. M. Anscombe, in 1958.

Rosalind Hursthouse and Glen Pettigrove, 'Virtue Ethics', Stanford University,https://plato. stanford.edu/archives/win2018/entries/ethics-virtue/.

2. Alasdair MacIntyre, *After Virtue: A Study in Moral Theory* (Notre Dame: University of Notre Dame Press, 2007), 16.

3. Aristotle, *Nicomachean Ethics*, translated by Roger Crisp (Cambridge: Cambridge University Press, 2004), 206.

4. Ibid., 3, 1094a. The first number in the references to Aristotle's *Nicomachean Ethics* indicate the page number (in this case '3') in the edition of the work I have used, which is Roger Crisp's. The second number (in this case '1094a') corresponds to the page number used in the Prussian Academy of Sciences edition of the complete works of Aristotle. Given the quantity of available editions of Aristotle works, both page numbers are referenced. For the benefit of my readers, who are unlikely to be classicists, I am not fully complying with the abbreviated standard format used in classical scholarship.

5. Ibid., 107, 1140a.

6. MacIntyre, *After Virtue*, 149.

7. Geoff Moore, *Virtue at Work: Ethics for Individuals, Managers, and Organizations* (Oxford: Oxford University Press, 2017), 38.

8. MacIntyre, *After Virtue*, 219.

9. Ibid.

10. Julia Annas, *Intelligent Virtue* (Oxford: Oxford University Press, 2011), 74.

11. MacIntyre, *After Virtue*, 219.

12. I build on the example of an architect provided by Moore. Moore, *Virtue at Work*, 39.

13. Annas, *Intelligent Virtue*, 8–9.

14. Shannon Vallor, *Technology and the Virtues: A Philosophical Guide to a Future Worth Wanting* (New York: Oxford University Press, 2016), 19. Italics in the original.

15. Miguel Alzola, 'Virtuous Persons and Virtuous Actions in Business Ethics and Organizational Research', *Business Ethics Quarterly* 25, no. 3 (2015): 293.

16. John L. McMullan and Delthia Miller, 'Advertising the "New Fun-Tier": Selling Casinos to Consumers', *International Journal of Mental Health and Addiction* 8, no. 1 (2010).

17. Annas, *Intelligent Virtue*, 8–9.

18. Aristotle, *Nicomachean Ethics*, 27, 1105a.

19. Fuelfor, 'Standalone Imaging Suite Creating a Patient-Centric Clinic Experience',https://www.fuelfor.net/standalone-imaging-suite-.

20. Aristotle, *Nicomachean Ethics*, 108, 1140b.

21. Ibid., 110, 1141b.

22. Ibid., 114, 1143a.

23. Aristotle, *Politics*, translated by Ernest Barker (Oxford: Oxford University Press, 1995), 10, 1253a.

24. Ibid., 11, 1253a.

25. Almost fifty years of research across multiple scientific disciplines has convincingly shown that *Homo economicus* is not an adequate descriptive model of how real humans behave in the real world. For a critical review of the model from the perspectives of behavioural economics, institutional economics, political economy, economic anthropology, and ecological economics, see Dante A. Urbina and Alberto Ruiz-Villaverde, 'A Critical Review of Homo Economicus from Five Approaches', *American Journal of Economics and Sociology* 78, no. 1 (2019). Paradoxically, although the debunking of *Homo economicus* was spearheaded by experimental psychology and behavioural economics, these fields still retain the model as the benchmark of rationality. Not surprisingly then, in these fields people are seen as irrational *precisely* because they exhibit behaviour that frequently deviates from what the model of *Homo economicus* prescribes. It is telling that Homer Simpson's behaviour is extensively discussed in *Nudge*, the field's most popular volume: Richard Thaler and Carl Sunstein, *Nudge: Improving Decisions About Health, Wealth, and Happiness* (New Haven: Yale University Press, 2008).

26. Moore, *Virtue at Work*, 42. Italics in the original.

27. Hussain Waheed, 'The Common Good', Stanford University,https://plato.stanford.edu/archives/spr2018/entries/common-good/.

28. Amitai Etzioni, *Happiness Is the Wrong Metric: A Liberal Communitarian Response to Populism* (Cham: Springer, 2018), 293.

29. This was point number twenty-four of the *25-Punkte-Programm*, which was the party program of the Nazi Party. See Paul Hoser, 'Nationalsozialistische Deutsche Arbeiterpartei (NSDAP), 1920-1923/1925-1945',https://www.historisches-lexikon-bayerns.de/Lexikon/ Nationalsozialistische_Deutsche_Arbeiterpartei_(NSDAP),_1920-1923/1925-1945.

30. The authors I have in mind are specially those belonging to contemporary communitarianism. For a review, see Amitai Etzioni, 'Communitarianism Revisited', *Journal of Political Ideologies* 19, no. 3 (2014).

31. Hursthouse and Pettigrove, 'Virtue Ethics'. See this source for a treatment of other streams in virtue ethics. For non-Western virtue ethics, see also Vallor, *Technology and the Virtues*, 35–59.

32. Russ Shafer-Landau, *The Fundamentals of Ethics*, fourth edition (Oxford: Oxford University Press, 2017); James Rachels and Stuart Rachels, *The Elements of Moral Philosophy*, ninth edition (New York: McGraw-Hill Education, 2018); Hursthouse and Pettigrove, 'Virtue Ethics'.

33. For example, Gilbert Harman, 'No Character or Personality', *Business Ethics Quarterly* 13, no. 1 (2003); John M. Doris, 'Persons, Situations, and Virtue Ethics', *Noûs* 32, no. 4 (1998).

34. Hursthouse and Pettigrove, 'Virtue Ethics'.

35. Vallor, *Technology and the Virtues*, 21–22.

36. Moore, *Virtue at Work*, 45.

37. Rosalind Hursthouse, *On Virtue Ethics* (Oxford: Oxford University Press, 1999), 36. Elsewhere, she completes this list of virtues: 'Much invaluable action guidance comes from avoiding courses of action that would be irresponsible, feckless, lazy, inconsiderate, uncooperative, harsh, intolerant, selfish, mercenary, indiscreet, tactless, arrogant, unsympathetic, cold, incautious, unenterprising, pusillanimous, feeble, presumptuous, rude, hypocritical, self-indulgent, materialistic, grasping, short-sighted, vindictive, calculating, ungrateful, grudging, brutal, profligate, disloyal, and on and on'. Hursthouse and Pettigrove, 'Virtue Ethics'.

38. Justin Oakley and Dean Cocking, *Virtue Ethics and Professional Roles* (Cambridge: Cambridge University Press, 2001), 79.

39. Ibid., 27.

40. Ibid., 44.

41. David Casacuberta and Ariel Guersenzvaig, 'Using Dreyfus' Legacy to Understand Justice in Algorithm-Based Processes', *AI & Society* 34, no. 2 (2019): 315–16.

42. Martha Nussbaum indicates that principle-based theories do include an account of virtue too, so for her the distinction between virtue ethics versus principle-based theories rests in a confusion. Nussbaum proposes leaving the label virtue ethics behind and speaking of 'Neo Humeans and Neo-Aristotelians, of anti-Utilitarians and anti-Kantians'. Martha C. Nussbaum, 'Virtue Ethics: A Misleading Category?' *The Journal of Ethics* 3, no. 3 (1999). I do not think, however, that doing this will help me guide the readers of *this* volume in understanding the main differences between the approaches. I am a great admirer of Martha Nussbaum and my reasons for not following her on this are not philosophical but pedagogical. The general differentiation between the two distinct approaches (virtue ethics versus principle-based theories) makes the discussion shorter and more accessible, even though it may be a 'the confused story' as she persuasively argues. Also, the categorization into three theories (virtue ethics, Kantianism, and utilitarianism) serve a clear pedagogical purpose and given the character of this volume I prefer to stick to it.

43. Immanuel Kant, *Groundwork of the Metaphysics of Morals* (Cambridge: Cambridge University Press, 1997), 31.

44. John Stuart Mill, *Utilitarianism and on Liberty* (Malden: Blackwell Publishing, 2003), 186.

45. Walter Sinnott-Armstrong, 'Consequentialism', Stanford University,https://plato. stanford.edu/archives/sum2019/entries/consequentialism/.

46. Peter Singer, *Animal Liberation: The Definitive Classic of the Animal Movement*, fortieth anniversary edition (New York: Open Road Media, 2015).

47. See the further reading section at the end of the book for pointers to the literature.

48. Moore, *Virtue at Work*, 35.

49. Neuroscientist Antonio Damasio showed in the early 1990s that patients with impaired emotions because of brain lesions were unable to make good decisions, whereas their overall reasoning was otherwise unaffected by the lesions. Antonio Damasio, *Descartes' Error: Emotion, Reason, and the Human Brain* (London: Vintage, 2006), xxi–xxii.

50. Shannon Vallor discusses this issue in further detail using other examples. Vallor, *Technology and the Virtues*, 7–8.

51. Peter Singer, 'Famine, Affluence, and Morality', *Philosophy & Public Affairs* 1, no. 3 (1972): 231.

52. Philippa Foot, 'The Problem of Abortion and the Doctrine of Double Effect', *Oxford Review* 5 (1967): 7.

53. Vallor, *Technology and the Virtues*, 24.

54. Ibid., 25.

55. Moore, *Virtue at Work*, 47.

56. Beth Crandall, Gary Klein, and Robert R. Hoffman, *Working Minds: A Practitioner's Guide to Cognitive Task Analysis* (Cambridge: The MIT Press, 2006), 131–34.

57. Ignacio Ferrero and Alejo José G. Sison, 'Aristotle and Macintyre on the Virtues in Finance', in *Handbook of Virtue Ethics in Business and Management*, edited by Alejo José G. Sison, Gregory R. Beabout, and Ignacio Ferrero (Dordrecht: Springer Netherlands, 2017), 1154.

58. Cited in Sandra L. Borden, *Journalism as Practice: Macintyre, Virtue Ethics and the Press* (New York: Routledge, 2010), 17.

59. In this book, we are using the third edition: MacIntyre, *After Virtue*.

60. MacIntyre is a widely discussed author. For an exhaustive yet accessible commentary on *After Virtue*, see Christopher Lutz, *Reading Alasdair Macintyre's After Virtue* (London: Continuum, 2012). For a comprehensive bibliography of secondary literature about MacIntyre, see Peter Wicks, 'Secondary Literature',https://www.macintyreanenquiry.org/secondary-chronological.

61. While at the same time adamantly rejecting Aristotle's 'metaphysical biology', according to which 'some men just are slaves "by nature"' (MacIntyre, *After Virtue*, 160.). Aristotle's views on the moral capacities of non-male Greek are evidently wrong: for him women, slaves, and 'barbarians' cannot be virtuous due to their nature, but his biological theory is evidently false, if only due to evidence coming from evolutionary theory.

62. Lutz, *Reading Alasdair Macintyre's after Virtue*, 118.

63. Ibid., 141.

64. MacIntyre, *After Virtue*, 187.

65. MacIntyre develops these ideas in chapters 14 and 15 of *After Virtue*.

66. MacIntyre, *After Virtue*, 190.

67. I thank Angus Robson for pointing out that this distinction was not sufficiently clear in an early draft for this chapter.

68. Dave Lee, 'Microsoft Staff: Do Not Use Hololens for War',https://www.bbc.com/news/technology-47339774.

69. Gerry Lopez, *Surf Is Where You Find It* (Ventura: Patagonia Books, 2008).

70. Alasdair MacIntyre, *Ethics in the Conflicts of Modernity: An Essay on Desire, Practical Reasoning, and Narrative* (Cambridge: Cambridge University Press, 2016), 190.

71. MacIntyre, *After Virtue*, 196.

72. Ibid., 188.

73. MacIntyre, *Ethics in the Conflicts of Modernity*, 190.

74. Ibid., 37. Italics in the original.

75. Moore, *Virtue at Work*, 57–58.

76. Again, I thank Angus Robson for helping me clarify this point.

77. MacIntyre, *After Virtue*, 190.

78. Bryan Lawson and Kees Dorst, *Design Expertise* (Burlington: Architectural Press, 2009), 132.

79. Harry Sapoznik, *Klezmer!: Jewish Music from Old World to Our World* (New York: Schirmer Trade Books, 1999), 107.

80. MacIntyre, *After Virtue*, 219.

81. Ibid.

82. Ibid.

83. Ibid., 216.

84. Ibid., 221.

85. Ibid., 194.

86. ibid., 222.

87. 'Form follows function', a maxim coined by architect Louis Sullivan (1856–1924), would be another standard of comparable importance.

88. Robert Venturi, *Complexity and Contradiction in Architecture*, second edition (New York: The Museum of Modern Art, 1977), 17. Italics added.

89. MacIntyre, *After Virtue*, 221.

90. Andrew Mason, 'Macintyre on Modernity and How It Has Marginalized the Virtues', in *How Should One Live?: Essays on the Virtues*, edited by Roger Crisp (Oxford: Oxford University Press, 1996), 197.

91. I believe I am not falling for the negative 'liberal individualism' that MacIntyre so strongly criticised: a shared purpose inscribes them in a tradition and prevents the individual from being ahistorical and de-situated. I understand that my explanations are far from sufficient to settle the matter, but because the topic much exceeds the scope of this book and possibly the interests of its readership, I will not explore the issue of conflicting positions within a tradition in further detail and refer the interested reader to other sources instead. See, for example, Andrew Mason, 'Macintyre on Liberalism and Its Critics: Tradition, Incommensurability and Disagreement', in *After Macintyre: Critical Perspectives on the Work of Alasdair Macintyre*, edited by Susan Mendus and John P. Horton (Cambridge: Polity, 1994).

Part III

Toward a Practice-Centred Design Professional Ethics

Chapter Seven

Uncovering a Purpose for Design

In chapter 5, we highlighted the need to uncover and reveal the ultimate ends of the practice of design in order to produce a coherent account of an ethics of designing. Now assisted by the virtue ethical notions we acquired in the previous chapter, we undertake that central challenge by aiming to formulate a description of an overarching purpose for design.

Following Aristotle, we will start from things known to us to be true[1]. And what is known to us? One thing that is certain is the excellence that we, practitioners, scholars, and advanced students, recognise in certain outcomes of design. I want to build from this certainty by formulating a plausible description of a purpose that I assume to exist and can be said to have hypothetically and reasonably driven the design of said excellent outcomes.

This sense of excellence is thus our starting point. Through analysis, we will search for this overarching purpose drawing on the different cases of excellent design reviewed in chapter 3; our assumption is that an overarching purpose can be found in and inferred from the different instantiations of excellent design. I am not interested in taking a top-down approach based on formulating a purpose from scratch nor a declaration of what design's purpose *should* be.

My aim is to derive an overarching purpose for design from inductive observation of real-world instances of the practice itself. Because of this, our approach will be bottom-up: we will start by reflecting on specific cases and gradually move toward higher-level generalisations with the goal of formulating a general description of a plausible overarching purpose based on what *already is the case*. It is in this sense that this is a task of revealing and uncovering an ethics of designing rather than creating one.

FROM DESIGN PRACTICE TO OVERARCHING PURPOSE

A professional ethics cannot be solely about reflecting on the ethical dimension of professional activity but needs to be also concerned with providing guidelines for action. In our view of ethics, such a guideline necessitates a conception of right action based on the goods that a particular profession seeks to achieve. A good professional is necessarily concerned with the overarching purpose of their profession. Doctors and nurses, for instance, are concerned with the promotion of health, and although there are huge variations regarding how 'health' is to be understood, few doctors would say that health plays no role in the overarching purpose of their profession. Similarly, teachers are concerned with transferring knowledge or fostering the development of autonomy; that there are serious disagreements about how this is best achieved does not affect the point that these issues are strategic to the profession's overarching role.

Medicine and education have existed as practices, albeit in different forms, for thousands of years, and this enabled their practitioners to engage in shared deliberations and reach tacit and explicit agreements about the overall aims of their practice. Although always temporary and dynamic, these agreements make it possible for practitioners to pursue the internal goods that define the practice by providing them with shared notions of the good. It is difficult to envision how this pursuit can start without a minimally sufficient conception of the good that is being pursued; at some point, at least, *some* conception of the good is required[2].

Practices, as we have seen, are activities that are pursued for the sake of larger purposes than merely performing the activity itself. In design too, it is assumed, a designer designs to contribute to the achievement of some important good; at least, many of the cases of design we have reviewed so far indicate that it does. The pursuit of these important purposes results in goods that are *internal* to design: goods that only design can provide. It should be a truism to point out that many individual designers pursue purposes that transcend themselves as individuals, and that many of these larger purposes are aligned with those of other practitioners. Because of this, design can exist and be sustained as a coherent practice. Needless to say, there is no widely accepted formulation of the purpose of design. Probably due to the youth of design as a profession, this alignment of purposes, however, is not nearly as explicit or clear as in other professions.

In previous chapters, we saw that authors such as Parsons, Margolin, or Buchanan insisted, even if they did so in different words, on the difficulties that design professionals have for articulating and discussing substantive ethical issues. However, a discussion about the grounds, justifications, and aims of the profession is indispensable to sustain a professional activity. As Alasdair Macintyre observes, 'It is in looking for a conception of the good which will enable us to order other goods, for a conception of the good which will enable us to extend our understanding of the purpose and content of the virtues'[3]. Reflecting only about means while neglecting the ends is insufficient for a practice to be maintained and extended.

I will take Richard Buchanan's formulation of the purpose of design and build my arguments upon and around it. Buchanan asserted that design serves human beings in the 'accomplishment of individual and collective purposes. That is, the end purpose of design is to help other people accomplish their own purposes'[4]. This formulation seems quite adequate to retrospectively account for the cases of excellent designing we described in chapter 3; indeed, these cases highlight instances of design that help people accomplish their individual and collective purposes in different ways. However, Buchanan's 'ultimate purpose of design', as he also calls it, appears to be too broad for a design professional ethics; after all, one can help people accomplish their purposes in a myriad of ways. What is more, this formulation, read out of context, could serve to describe the ultimate purpose of most, if not all, professions[5*].

If our goal is to extend our understanding of the purpose of design in order to define guidelines for design action, it is necessary to come up with a formulation of a purpose that is specific to the design profession, and, at the same time, incorporates the general service ideal of professionalism (that is, its public service element). Only after an adequate understanding of the *overarching* purpose behind all that 'conceiving and planning' that design consists of, could we move toward defining guidelines for action. Otherwise, how could these be reasonably put forward without a minimally adequate understanding of the purpose of design?

Having internalised conceptions of the good can contribute to excellence in design. Although it is key for, say, furniture designers to have internalised ideals about what a good chair is, it seems even more crucial (for them and for designers of any discipline, for that matter) to have an internalised conception of what the appropriate ultimate purposes of the profession are. A sense of these ends, of design's overarching purpose, can guide the designer in acquiring appropriate lower-order ideals that fit into the higher-order ideal. The overarching purpose can be taken to be what legitimises and underpins the profession: what it is that designers design for; concurrently, it can be seen to channel and integrate a hierarchy of other lower-order purposes.

Put differently, internalised ideals concerning the overarching purpose of design govern what counts as good furniture; much in the same way a thorough conception of what a good chair is encompasses an understanding of what good back and lumbar support are. Naturally, one could start first with an awareness of lumbar support and then develop a conception of what a good chair is; analogously, one could develop a sense of what a good chair is, and use that awareness as a scaffold to develop a conception of design's overarching purpose.

There is thus a close connection between ideals and purposes. Ideals offer a sense of direction and guide the setting of purposes as well as the evaluation of results. On the other hand, results influence ideals too, as ideals are models of excellence that are constantly under adjustment. We will deal with internalised ideals that embody standards and serve as guidelines for action later on in the chapter.

A Scaffold for Analysing Design Practice

During the next sections, we will work on developing a scaffold that enables us to gain deeper understanding about the ethical dimension of judgment and decision making in design. The scaffold will have three levels:

1. Design as a MacIntyrean Practice
2. A key high-level contribution as the extension of capabilities
3. An overarching purpose centred on contributing to others' flourishing

We will perform a series of analyses to make explicit how the different levels of the scaffold are interrelated. We will do so starting from the practice of design and assisted *by* theory, rather than operating directly *from* theory.

For the first level of the scaffold, our definition of design will be juxtaposed with the notion of practice in a MacIntyrean sense to deepen our comprehension of *design as a practice*. For the remaining levels, our goal is to try to make inferences and provide explanations that could function as an explanation of their ethical dimension. Specifically, for the second level, we will start by uncovering underlaying patterns that can be found in cases of excellent design to describe a *key high-level contribution*. The third level will be based on an account of an *overarching purpose* that could have reasonably been pursued by the designers of those designs.

Starting with the practice of professional design will offer us a truer and more coherent explanation of design goods and purposes than attempting to directly come up with a regulative ideal or a guideline based on principles and theory and try to inculcate them in designers. Asking what goods are achieved when design is performed at its best is another way of asking what type of activities and outcomes are good to be pursued. If cogent, this account may carry normative power, but my intention is not to decree a definitive overarching purpose that designers ought to follow. The goal for the description of the overarching purpose is to catalyse and integrate a series of intuitions and arguments about what good professional design is. In other words, the overarching purpose that will be formulated has a primarily descriptive function, whose aim is to get you to imagine how a designer might have been guided by conceptions of the good and the best in the pursuit of their professional purposes.

DESIGN AS A MACINTYREAN PRACTICE

We explored, defined, and analysed design in the introduction and in chapters 1 and 3. Since then, we gained important new ethical insights that need to be incorporated into the analysis.

In chapter 1, echoing Richard Buchanan, design was described as being about *conceiving and planning the human-made world*. Although this is design's literal goal, it is not equivalent to the practice of design. Clearly, the practice of design *is* about conceiving and planning, but only when it is done in a particular way that is 'consonant with the excellences of the craft', to express it in MacIntyre's words[6]. Put another way, not all ways of conceiving and planning the material world constitute design practice in the sense intended here. For these activities to constitute a practice, they need to be practised in relation to some standards and with some purposes. Practices have standards of excellence that regulate the ways in which the practice is conducted and the purposes that are pursued by their practitioners. Who decides about the standards for the practice? Practitioners, of course, as they are the best suited to recognise the goods arising from their practice and to determine what excellent designing is. MacIntyre asserts: 'Those who lack the relevant experience are incompetent thereby as judges of internal goods'[7]. But practitioners might disagree: as we saw in chapter 5 when we discussed minimalism, there might be conflicting vantage points. One enters a practice that already has an existing tradition, but the existing standards and the very tradition in which these are embedded can be criticised, and so the practice is transformed and enriched.

The Elements of the Practice of Design: Goods and Standards

In line with this, another important point that deepens our understanding of design as a practice is that practices demand the exercise of technical skills, but a set of practical skills, however developed, does not amount to a practice: a practice, in our understanding, 'is never just a set of technical skills'[8]. Thus, a highly skilled designer who *primarily* aims at winning prizes or at getting 'likes' on social networks but does not care for design's internal goods is not truly engaged in the practice of design. This designer, however skilful, is primarily pursuing the external goods that design can offer instead of the internal ones (that is, those goods that cannot be obtained in any other way but by designing). A less proficient designer who truly aims at internal goods can be said to be more deeply engaged with the practice of design than the one who only seeks fame and recognition, regardless the quality of the outcomes. Why do I say that they are not truly engaged in the practice of design? Simply because it is the pursuit of internal goods that constitute the practice in actuality.

To illustrate internal goods with examples: for an editorial designer some internal goods might be choosing a typography, feeling and selecting an adequate type of paper, or, more globally, designing a book that helps illiterate adults learn to read. A product designer might obtain different internal goods, for instance: experimenting with materials, building and testing prototypes, conducting user research, and co-creating with users. A service or interaction designer might obtain these internal goods in framing complex problems or by envisioning the different possible user interactions that a digital service for a hospital has to sustain.

The list of internal goods could be very long, yet a designer that is *primarily* concerned with external goods, such as winning a prize, instead of the pursuit of internal goods, will, alas, fail to attain the internal goods that could otherwise be obtained. This is so even if the designer manages to produce a good outcome or even a technically excellent one that gets them the prize. If a designer primarily cares for an external good, the actual quality of the realised outcome is not conducive to the attainment of internal goods simply because an outcome cannot *become* an internal good *afterwards*, as an external object cannot be internalised[9]. Naturally, in the case of a good or excellent outcome, the designer might feel satisfaction and a sense of legitimate pride, but this is not what internal goods are about.

Oblivious to MacIntyrean nomenclature, graphic designer Paula Scher seems to be very aware of the difference between internal and external goods: 'The accoutrements, and the awards, and my picture in a book don't matter. What matters is the next project'[10]. So, what matters, then? The goods Scher suggests are rather enigmatic but unequivocally internal to design:

> there are all kinds of problems and compromises that I must negotiate. Things that have to be held on to, things that have to be protected to make something move forward. And it's very, very, very hard work. It doesn't have anything to do with fame. It has to do with doing it every day[11].

'Can't a designer post a picture to Instagram?' 'Can't they win awards?' 'Can't they be famous and have their photograph in a book?' Someone might ask. And we must reply that, of course, a designer can obtain many external goods from the practice of design: their picture in a book, fame, money, prestige, and so on. But to say that a designer is truly engaged in a practice, achieving these external goods must be *contingently* attached to the pursuing of internal goods. External goods such as prestige, status, or money are attached to practices 'by the accidents of social circumstance'[12]. Graphic designer Milton Glaser makes the same point: 'I am very happy to have made enough money to live as well as I do, but I never thought of money as a reason to work'[13]. Here I must reiterate that external goods are *indispensable* to sustain a practice as well as the designer's livelihood. External goods deserve further treatment, which is presented in chapter 9.

In some cases, internal goods are obtained *simultaneously* with the practice. For example, when developing a solution, a designer might reach a new understanding of colour or shape, or make successful experimentations with materials that could result in deep enjoyment. These are, undoubtedly, internal goods of design. Conversely, given design's nature, an internal good might also be obtained long after the actual moment of designing. For instance, an internal good is obtained when a designer sees their design solution become a reality in the way they envisioned, or they obtain an internal good when they learn that their design solution actually works. This goes beyond satisfaction. Arguably, to be able to fully experience this as an internal good, the designer (or designers) has to be able to link their present enjoyment (about the design *working*) to their own frame of mind at the time of designing: 'it looks fantastic, just like it was meant to be'. It is hard to see how a designer could achieve (or be *rewarded* with) internal goods if this connection is not there.

Our MacIntyrean framework extends our understanding of design and enables us to sketch some requirements to make it count as a practice in a virtue ethical sense. Design is a practice in the intended sense when it pursues the internal goods of design according to the standards of excellence that design practitioners defined for the practice. These standards embody conceptions of what a good outcome is, but also of what purposes are worth pursuing. Because we are concerned with professional design, these purposes are, in turn, (willingly) constrained by the larger professional goal of contributing to others' wellbeing.

When I say it is practitioners who define the standards for the practice and recognise the goods arising from the practice, I am not claiming that they are the only ones who can judge the *outcomes* of the practice. An outcome of design, at least as I see it, exceeds the practice in which it was created and enters other practices. Professional designers may be the best suited to judge what excellent *designing* is, as practitioners are the only ones who get to obtain this excellence in the form of the internal goods of the practice, but it would be absurd to say that designers are the *sole* judges of excellence in, for instance, cars. Cars are embedded in many other activities and practices, and the people involved in these are, at least in principle, perfectly able to make judgements on cars. Designers, in turn, can revise and adapt their own standards of excellence based on these judgements from outside the practice. To illustrate briefly, car designers use to favour the speed, power, autonomy, thrust, or 'grunt' that is characteristic of internal combustion engines in detriment of electric ones, but are revising and adapting their own standards to make them more in line with laws and general views held within society about internal combustion cars.

The Importance of Good Purposes

There is more to say about the importance of *good* purposes, that is, to purposes worth pursuing. This relates to a point we made in chapter 5 when we mentioned that our individual motives become *good* reasons from a professional perspective only when they serve the purposes of the profession. Having good reasons justifies acting in a certain way or not acting at all. In design, acting is steered by methods, techniques, and skills, which are not ends in themselves, but instruments to achieve meaningful professional purposes. In other words, a designer might have motives for acting in a particular way, but if those motives fail to become good reasons (for example, by ignoring the overall purpose of the profession or by being detrimental to it), acting is not professionally justified.

The role and importance of purposes in determining what counts as a good reason can be illustrated with a simple, though powerful, example: the infamous all-black Schutzstaffel (SS) uniform, which was designed by Karl Diebitsch and produced by The Hugo Boss Company, in Nazi Germany[14]. If we were to analyse Diebitsch's uniform *separated* from the vicious plan to which this design contributed, we could say that it is an excellent uniform: we could praise its elegance and crispness or some other technical aspect. But when we take into consideration the purpose of the SS uniform, what the SS were for, and the absolute evil the Nazis represented, we realise that this understanding is equivocal. The SS uniform can in no way be characterised as excellent, if only because Diebitsch's purpose was the opposite of a good purpose. This extreme example illustrates how, from a virtue ethical perspective, the purpose of an action cannot be separated from the desired outcome of the action. And, importantly for our inquiry, that technical skills, methods, and techniques become valuable in an ethical sense *only* when marshalled for purposes worth pursuing (in our case, professional purposes).

Unlike Nazi medicine, which is not medicine, but torture, Nazi design *is* design; yet a Nazi designer cannot be seen as a good professional in the sense intended here. Alas, this crucial recognition provides little effective guidance as to how to act. What is more, in most situations, evaluating purposes and their conflicting interests is not as transparent as examples involving Nazis. What can be done then?

Virtue ethicists tell us that to know what to do and to act well we can rely on practical reason and the virtues, and the virtues, the qualities, attitudes, and dispositions to act in particular virtuous ways are developed in action. This is why having an overarching purpose is so important, because it can guide our actions by giving us a directionality that is concrete enough as to enable evaluations and assessments of lower-level purposes. In the words of Philippa Foot: 'the wise man knows the means to certain good ends [and knows what these] are worth'[15]. Without a minimally shared conception of the ultimate good that is pursued, no practice can survive as a whole—without it, no goods can be properly ranked or internally assessed. Without it, furthermore, every moral statement becomes an expression of personal attitude or preference.

To end, we said previously that the aim of design is not only 'to conceive and to plan' but to do so according to some standards of excellence, which include consideration and assessment of ends. Why is this so important? Because the result of design as a professional *practice* is not only to generate a good design outcome, but that the designer themselves, as MacIntyre argues, 'is perfected through and in her or his activity'[16]. The point he makes is that practices are the bedrock of morality: it is in practices where the virtues are developed. Put differently: when we seek excellence in the pursuit of legitimate professional purposes, we also grow as human beings.

A KEY CONTRIBUTION: EXTENDING ABILITIES AND POWERS

The goal for this section is to describe a specific key high-level contribution (henceforth, key contribution) that design makes to society *when performed at its best*. To determine that we will rely on a reprise of the analysis we conducted in chapter 3, which will provide us with a conception of the type of internal goods that are obtained in the practice of design, as well as insights into the specific way design serves individuals and society.

Before we continue, a few comments at a meta level need to be made about the type of key contribution that we want to obtain. First, the contribution that we want to extract from these cases should be intrinsically and characteristically connected to all fields of professional design. Second, it will need to be recognised by a reasonable practitioner not only as a strategic element of design, but also as *constitutive* of design. The key contribution need not be seen by all as the single most important contribution that design makes; the only requirement is that reasonable practitioners agree on its overall importance. For example, if we were discussing surfing, 'being one with the wave' would be a good answer, as it is something that most surfers would recognise as constitutive and characteristic of surfing, even by those who might think that surfing has *other* or more important dimensions as well.

The Contribution Design Makes

Piggybacking on the analysis of design we performed in chapter 3, we can now ponder on what the outcomes of the design cases we have reviewed have in common. There we saw, for instance, how design, by enhancing the ease of use and pleasantness of public transportation, enables people to move around their city; something as apparently minor as a map (whether on a physical display or on a screen) can have great impact on the way people (of all types, as well as children and adults with cognitive disabilities) get to their destination easily and with minimal cognitive effort. Or, if one decides to use the car, design is what enables them to know which way to go (either by reading the traffic signs of by following instructions from a GPS navigational system). We can take another example; well-designed voting ballots (or voting machines) facilitate exercising the essential democratic right of choosing our representatives and foster the overall health of the system. Books, magazines, or online platforms help us develop our capacities for practical reasoning, to feel emotions, to voice one's views, to gain awareness of the existence of other ways of being and doing, and so on. Parks, playgrounds, but also videogames, social networks, or cafés and restaurants (they can be designed too!) allow us to play, to exercise our imagination, to have fun, to connect with others, to enjoy ourselves, to celebrate.

Further, we humans carve meaning into objects and build our identities around those meanings, and hence those objects become important *to us*. If you think that it is because of consumerism or advertising that meanings are attached to things, you are profoundly mistaken; it is the other way around: consumerism and advertising simply exploit this human feature to sell us things. We have *always* used things for symbolic purposes: think of the 'talon necklace', eight talons strung together as a necklace by a Neanderthal about 130,000 years ago in what is now Krapina, Croatia[17]. Neanderthals went on doing this until they disappeared, so, not only *Homo sapiens sapiens* but even other archaic human species were bestowing meaning upon things and using them for symbolic purposes.

Good designers know how to work with meanings and cultural and institutional patterns as a 'material' that can be purposively 'moulded', as it were; because of this, designers can be regarded, in the words of Press and Cooper, as 'cultural intermediaries'. Excellent objects, environments, and systems reflect a deep understanding of the symbolic and cultural conditions in which others will interact with and be affected by these artefacts. Along these lines, a project like the redesign of the diagnostic process for breast cancer patients (from a visit to the general practitioner to a diagnosis) is not simply about redesigning the process as if it were a deterministic system like a mechanical clock. Such a project is about dealing with a complex and highly dynamic socio-technical system with the aim of reaching a solution that *works* for the different parties involved, integrating objective aspects (costs, fixed timings, and so on) as well as subjective ones (emotions, expectations, power roles, and so forth). What a project like this achieves is reaching a solution that enables patients and their families to go through a difficult period in their lives the best they can by shortening the wait and thus reducing anxiety, and by providing a mechanism to build trust, to deal with fear, and to feel confident that they are in good hands.

The cases discussed by design scholars Meroni and Sangiorgi in chapter 4, for instance, show how design can be an element of transformation in cities and institutions; that is, those projects permit understanding design as a way of thinking about complex issues, and deciding what to do, and not as the mere formal outcome of a process. The Superblocks Project illustrates how design can be mobilised for integrating citizens in decision and policy making; it also shows that design is a legitimate and potent tool for social innovation. To round up this brief revisitation, we could add that design enables individuals and collectives to understand and delineate 'the shape of things to come', by providing tools and techniques that are suitable for navigating the complexity and the epistemic uncertainty of the future.

To sum up, this analysis highlights the broadness of design's contribution, ranging from stand-alone simple artefacts such as a sign to the planning of services such as health services or complex models of urban mobility. It also makes clear that design can be said to constitute *the backdrop of our existence*, as it plays a key role in shaping the built environment, the networked material and digital world, and the services and products that allow us to do things we would not be able to do without.

Artefacts as Amplifiers

Jeroen Van den Hoven refers to artefacts as 'agentive amplifiers' that allow others 'to access possible worlds that would have been inaccessible without it'[18]. We can illustrate this notion with the example of a pair of headphones and a streaming music service like Spotify, which allows a person to access and enjoy music in a way that was previously impossible. Conversely, because artefacts have a double nature, they could also be seen as 'agentive reducers' that, instead of creating new possibilities, reduce the realm of what is possible, for instance, by impairing the user's ability to be sonically aware of their surroundings. Or by, and this is possibly an unintended consequence, by starving them out of access to more tangible and communal forms of music discovery (such as it occurs in a record store or at the house of a friend), by means of offering an individualistic form of discovery. Artefacts can also 'expand capabilities of some groups, while reducing capabilities of some other group'[19]. Importantly, the capabilities that are expanded need not be human capabilities only, as the example of prosthetic limbs for dogs illustrates.

In conclusion, and restricting the analysis to the human realm, the key feature of design that seems to emerge as an underlying pattern is that design *enables, grants, improves, facilitates, enhances*. The pattern that we find is that all these excellent exemplars of design *extend* in different ways what humans are able to do, know, feel, experience, be . . . that is, typical human abilities, powers, or capacities. For instance, a cosy waiting room extends our capacity for waiting, a good textbook unleashes our capacity for learning, a good DIY tool facilitates our ability to engage in home décor projects, a messaging app enhances our ability to communicate, and so forth. Design empowers us and enables us to do things we would not be able to do without, or greatly reducing the effort that would be necessary, thus augmenting the realm of what it possible and increasing the potentialities of our lives.

The upshot of this analysis is that *a* key high-level contribution that design makes through the conception and planning of the human-made world (seen as a network of agentive amplifiers and agentive reducers) is *extending typical human abilities, powers, or capacities*.

Is This Key Contribution a True Internal Good?

This key contribution embodies an internal good of design (something that designers achieve that can only be achieved through design), but also has instrumental value for other people, as it serves others to attain other goods through those designs. This contribution, while abstract in nature (because of its high-levelness), is quite specific to the design profession, and most professions are unlikely to be able to claim a similar key contribution.

There seems to be an obvious exception: engineering. But this is not surprising; though engineering's methodological approach and overall *ethos* are rather different, design and engineering are 'cousin' professions: there is much design in engineering. Augmenting the realm of what is possible may simply be a high-level contribution pursued by both designers and engineers[20]. While I have nothing to say as to whether this formulation actually fits the engineering profession too, this apparent overlap need not be problematic at all. After all, both doctors and nurses 'care for patients' or 'promote health', but do so from different professional stances. This gives their contribution a particular characteristic that safeguards the uniqueness of the profession; the same can be said of designers and engineers.

DESIGN'S PURPOSE AND THE FLOURISHING OF OTHERS

Building on the key contribution that was just defined, in this section we will try to hypothesise a plausible overarching purpose that could have reasonably been pursued in such cases. We would undermine our inquiry methodologically if we tried to define this purpose 'from above'. We will, however, be guided by the theoretical notions we explored so far. What we seek, then, is to frame these insights within a general description of an overarching purpose. We could also pose the objective of this section as providing an answer to a question: what overarching purposes could those designers of excellent designs have been pursuing?

Here, too, I need to make three comments at a meta level about the nature of the overarching purpose. First, the overarching purpose we formulate must retrospectively account for and serve to reasonably explicate the design cases individually. Second, at the same time, it should match up and explain the cases collectively; for that end we will use the key high-level contribution we defined in the previous section. Third, this exercise is an intuitive and practice-centred way to propose a reasonable higher-level purpose for design, but I am not claiming that the designers were *actually* guided by this purpose.

The Dynamics of Design Purposes

We saw previously that any excellent outcome of design ('from the spoon to the city') can be said to extend people's capacities or powers. But this key contribution cannot be seen as an adequate formulation of an overarching purpose in itself. To see why not, let us briefly consider the dynamics of purposes with the example of the design of a traffic sign.

We need to assume that the designer of such a sign would probably aim to achieve some higher-level purpose than just to finalise the design of a sign; that is, they want the sign to serve some larger purpose. Sticking on this example, we could hypothesise that the designer might set out to achieve a sign that is highly visible and readable under all weather conditions and by people of all ages. This is not yet the type of fully fledged purpose we are looking for, but it gets us much closer. It provides a plausible motive for explaining the designer's decisions: the designer might think that certain colours, typography, size, and shape serve the purpose of the sign. But, still, legibility and visibility are lower-level purposes, and, although important, do not seem to be true ends in themselves; if we want a higher-level purpose, we must still be missing something. We could ask, then, what are readability and visibility good for? Why would our designer want to attain them? Considering the context of traffic, a rather obvious answer is that they want readability and visibility in order to maximise safety. Safety, surely, could be seen as a high-level purpose, as it is, by itself, a strategic good to have. Needless to say, in a practical situation we would not find this strict sequence between higher- and lower-level purposes.

If we want to formulate an overarching purpose, it seems more convenient to infer it not from a collection of concrete cases (the purpose of a sign, of a DIY tool, of a textbook, and so on), but from the key contribution we defined in the previous chapter which already offers and adequate level of abstraction. We can ask, thus, for what purpose would designers want to extend typical human abilities, powers, or capacities?

If we consider designers *as* professionals guided by the regulative ideals of professionalism, an intuitive starting point would be to say, in line with Whitbeck, that they would want to do so to 'promote, ensure, or safeguard some aspect of others' well-being'[21]. In fact, in our view of professions enabling people to live a humanly flourishing life is a key aspect of all professions. It follows from here that for design to maintain its claim to professionalism, it must assist and enable people to secure certain human goods that are crucial to their wellbeing. This is a good start, but we have not yet moved from the overarching purpose of all professions, and we want to go to a more specific level.

An Overarching Purpose for Design

In chapter 1, informed by Ortega y Gasset, we had already linked design (as a general human capacity) to the good life, to wellbeing rather than to mere being. Recalling his insights, the first important point we can make from revisiting our examples of excellent design is that the design profession takes the endeavour of design as a general human activity a step further, extending the scope and accelerating the pace at which societies and individuals attempt to realise their different life plans, helping or enabling them to flourish as human beings[22]*.

Some might wonder in what way the design of a traffic sign, our earlier example, contributes to human flourishing. Indeed, the contribution design makes in this case seems to be rather modest, but traffic signs and other visual cues (such as road marks) guide and control traffic on streets and motorways through a visual language that enable drivers to negotiate their interactions on the road. Signs thus extend our abilities to stay safe on the road and to make the road safe for others by enabling us to know when to exit a motorway, to drive under the maximum speed limit, to be aware of potentially hazardous conditions, and so forth. By directly contributing to the preservation of life, and cognitive and physical health, which are key human goods, traffic signs can be directly connected to flourishing.

Oakley and Cocking argue that a good profession 'is one which involves a commitment to a key human good, a good which plays a crucial role in enabling us to live a humanly flourishing life'[23]. The cases also show that the design profession not only goes beyond the satisfaction of basic needs, but aims at enabling people to make sense of their lives and to shape and reshape the objective and subjective *artefactual* conditions of it. Naturally, this shaping and reshaping is not a one-way street: given its normative dimension, design (and of course, technology in general) also guides and instructs people on how to live, and co-defines with them what living well means. In this way, through the artefacts people constantly interact with, design becomes a constitutive part of their life, of how they act in the world, and it becomes crucial in the definition of the type of person they want to be. After all, one needs a surfboard to be a surfer, a computer to be a twenty-first-century designer, a pair of high-tech shoes to be an Olympic athlete—this is a point Van den Hoven, echoing Ortega, sharp-wittedly makes when observing that one cannot be a samurai without a special sword[24].

This strategic connection to our existence is what makes design so important; yet this importance is not ontological: humans would still exist and be humans without the contribution of the design profession. But focussing on this truism would miss the point. After all, we do not bestow importance on professions because they keep us *alive* in the strict biological sense; we find them important because they are strategically connected with our desire to live well.

MacIntyre's account a practice gravitates around a shared sense of ultimate purpose (a *telos*); practitioners, in his view, seek a similar ultimate purpose, for similar reasons, and they do so assisted by standards that arise in a vital, always developing tradition. MacIntyre's account enables us to extrapolate the common patterns we found in design outcomes to the practice of design as a whole.

If my arguments so far have been cogent, the overarching purpose that we can formulate is that design *seeks to extend typical human abilities, powers, or capacities through the conception and planning of the human-made world so that others can flourish.*

A clarification is in order before we end this section. Seeking to promote others' flourishing immediately opens the philosophical grand question of what counts as one's 'other' and, thus, as a genuine object of care: who or what is this other that we can care about in design. In the context of present design, as far as I see it, 'others' primarily refers to humans and, at times, non-human sentient animals. Admittingly, reasonable arguments could be made for many types of beings and things counting as genuine 'others' and thus as genuine objects of care: naturally other sentient beings (like dogs, chickens, or fish), but also non-sentient beings (like bacteria or plants, for instance) and natural things (such as rivers and mountains), ecosystems and the planet as a whole, as well as human culture (languages, religions, ways of life, and so on).

FROM POWERS TO CAPABILITIES

An important gap can appear between the intended outcome of the design process and the way an instantiated design *effectively* extends typical human abilities, powers, or capacities. I will illustrate this gap with an example involving an urban playground where my youngest daughter plays from late September to early June, but not during the summer months. Why not?

Unfortunately, the playground does not have any shading provisions, which makes it impossible for her to play there during the long, torrid summers we have in Barcelona, Spain. Of course, she *could* go there to play, but not without health and safety risks. Because of this design fault, the playground is mostly deserted during the summer. Because the appropriate conditions for use are not in place, it is as if it the playground did not exist at all.

In other words, having a playground in the vicinity is not enough for children to be *actually* able to play there. This illustrates an important distinction between owning or having access to material means, resources, and potential abilities or powers, on the one hand, and actual 'capabilities' and 'functionings' on the other. This distinction is emphasised by theorists belonging to the so-called Capability Approach, a research programme in economy and philosophy pioneered by Amartya Sen and Martha Nussbaum. The Capability Approach has, since its inception in the late 1980s, accumulated a growing body of highly multidisciplinary literature and has become an influential framework for discussing and investigating issues related to global justice and wellbeing. The approach, which was first linked to design by several design scholars in the mid-2000s[25], can offer us important insights that can enhance our account. The Capability Approach provides a vocabulary that serves to explore the transformation of a potentiality into an actuality as well as the conditions that are necessary for individuals and groups to carry out this transformation. Without aiming to provide a full-blown critical inspection, we will next review a reduced set of its key concepts[26].

Key Terms in the Capability Approach

Conceptually, a 'capability' is more than the potential power, capacity, or ability to do or be, and includes the *effective opportunity* to exercise that power or ability. For instance, a person might have access to a bicycle (a material resource to move around), but a bicycle would *only* truly expand a person's powers to move around if the person has the capability to ride a bike. If the person does not know how to ride a bicycle or lacks access to roads suitable for cycling, then the person's powers are not truly expanded by the bicycle. Conversely, the notion of 'functioning' refers to the *actual* achievements or realised opportunities; for example, having the capability of reading results in the functioning of actually reading. It is often stated that functionings are *beings* (being healthy, being asleep, being tired, being a designer, being a parent, the list is literally infinite) and *doings* (walking, eating, standing, talking to a friend, coding a computer programme, watching a film, solving a puzzle, doing homework, the list is infinite)[27]*.

Capabilities and functionings are interrelated, but are not the same: a capability is a genuine opportunity to achieve a functioning. Philosopher Ingrid Robeyns explains: 'The distinction between functionings and capabilities is between the realized and the effectively possible, in other words, between achievements, on the one hand, and freedoms or opportunities from which one can choose, on the other'[28]. Being literate is a functioning, and the real opportunity to read is the corresponding capability; having one does not entail achieving the other. A person that works two jobs to make ends meet might know how to read, but might be unable to actually read a book because of having too little time or being too tired. A person can be thus said to have the capability to read insofar as they are truly able to get access to material they want to read and have available time to do so.

Capabilities Are Real Opportunities

Both functionings and capabilities are important for people's flourishing, but in different ways. Having capabilities is important because capabilities are indicative of the real choices one has, and having choices is necessary for a good life in the sense that it is in exercising choice that one can exercise the virtues. As Aristotle argued, 'No one deliberates about what cannot be otherwise, or about things he cannot do'[29].

For Aristotle, politicians were not to be concerned with guaranteeing general satisfaction without choice but 'were to aim at producing capabilities or opportunities'[30]. Analogously, it could be argued that the role of design is not to provide functionings that generate satisfaction, pleasure, or convenience, but to create the artefactual conditions that enable and empower individuals and groups to expand their capabilities and achieve desired functionings. This, in other words, can be achieved by creating real opportunities that people can convert into functionings according to *their own life plans*. It is in this space of having open, real opportunities for shaping one's future that a person can flourish. However, this is not to say that functionings are unimportant. On the contrary, the importance of functionings is essential for flourishing too, as actual beings and doings are constitutive of human life. While I have concentrated on a human-centred account of capabilities, it is worth noting that the approach can be aptly applied to animal capabilities as well[31].

When a designer wants to assess and reflect upon the different envisioned trajectories of use for a given design, they could do so by considering the extent to which a design is likely to truly expand capabilities. The notions of capability and functioning provide a larger locus of reflection that goes beyond the artefact and its intended immediate functionality (its potentialities of use). A focus on capabilities and functionings enables a designer to reflect on and account for what people are *effectively* able to be empowered by an artefact, and, in turn, to assess how their designs contribute or could contribute to their flourishing (or be detrimental to it). Along these lines, the primary focus of professional design activity can be linked to the capabilities and functionings that are putatively advanced by design activity[32*].

A Capabilitarian Overarching Purpose for Design

We have seen that artefacts can be conceptualised as agentive amplifiers, but of course, the capabilities they expand might seem to be merely instrumental and unrelated to flourishing. For instance, a car expands one's capability to move, but moving around does not seem directly connected with flourishing as a human person; in most cases one just wants to move around in order to do other things for which moving around is necessary. But that the capability is instrumental to other ends; it does not diminish its importance and its connection with flourishing. For example, a car plays an important role in enabling a person to hold down a job in a city with bad public transportation[33] and in shaping the residential location choices and economic outcomes of low-income households[34]. A car and the capability it expands end up being directly connected with key human goods such as jobs, education, health, and even leisure opportunities. This highlights the need for designers to have a deep grasp of which capabilities are truly relevant and to develop 'systems able to promote and support them'[35].

In the light of all this, the general description of the overarching purpose of design I offered earlier needs thus to be reworded to include these new notions. We can say that *design seeks to expand human capabilities through the conception and planning of the human-made world so that others can flourish.*

This sharpening in focus provides a clearer directionality for the overarching purpose of design; the notion of capabilities can be seen as a robust underpinning to conceptualise design's purpose in terms of contributing to others' wellbeing. Merely aiming at extending abilities or powers may fall short of being conducive to human flourishing; the actual objective is to extend those powers and abilities in a manner that could likely become an effective genuine opportunity to achieve a functioning. The notion of capability captures that dynamic. Consequently, a designer aiming at contributing to human flourishing would primarily focus on extending human capabilities by designing things, spaces, and services that enable people to live 'the kind of lives they have reason to value'[36]. At the same time, the good designer will also be concerned with what the people served by their designs are likely to achieve in terms of functionings (beings and doings) and with the objective and subjective conditions that are necessary to attain those functionings.

For designers and others involved in a design project, all this has important entailments: they get to decide which capabilities and functionings are going to be promoted and enabled, and which ones discouraged or outright prohibited by design. This task is fraught with ethical dilemmas because designers get to *expand* or *reduce* capabilities in a direct or indirect way, expanding the capabilities of some groups, while reducing those of some other group[37]. Electric scooters, for example, expand their users' capability to move around, but at the same time, they reduce others' capability of walking down the footpath undisturbed by risks posed by approaching vehicles. On some occasions, designers might aim at reducing some capabilities by design so that other more important capabilities can be amplified or empowered. For example, safety belts in a car reduce our effective opportunities to move around in the car in order to amplify the capability of riding in it safely (increasing our capability to survive a crash). This purposive reduction of capabilities is not an uncontested thing and is often linked to 'paternalism' (I will come back to this issue in chapter 8).

From Artefacts to Capabilities to Functionings

We arrive to the topic of the trajectory between outcomes of design (artefacts) and realised functionings. The design field adopted terms such as 'usability', 'affordance', 'accessibility', 'learnability', and 'desirability' to account for the variables that can explain the conversion of the usage of a design into a functioning. So, for example, aided with these notions, we can say that because of its extremely poor usability the 'butterfly ballot' that was considered in chapter 3 prevented people from voting for the candidate of their choosing.

Using capabilitarian terminology we can say that these voters had access to a resource (a ballot), but because of its cumbersome design they did not have the capability (the effective opportunity) to use it because it was unintelligible, which obviously resulted in them not being able to achieve the desired functioning (electing their own government). When viewed from the perspective of capabilities, this is not a mere usability problem. Nussbaum argues that 'being able to participate effectively in political choices that govern one's life' is a central human capability[38]; the usability problem becomes thus a serious ill with a clear ethical dimension.

The Capability Approach provides the notion of 'conversion factors' to refer to the factors that play a role in the transformation of resource into functioning; that is, transforming a ballot into being actually able to participate in the election of one's government. These factors can be seen as a combination of internal and external conditions that affect our real capabilities and functionings. Conversion factors can be *personal* (skills, physical condition, disability, etc.), *social* (policies, social customs, class and gender dynamics, etc.), and *environmental* (built infrastructures, geography, climate, available services, etc.)[39]. To come back to the example of the playground, many children are not able to obtain the functioning of playing outside during the summer especially because of environmental factors (the summer heat). But even in the winter, other children might not be able to play there because of *personal* factors (they might have special needs and the playground is not fully accessible); others because of *social* factors (some parents or carers think that playgrounds are dangerous places and do not allow their children to play there while others might simply lack the time to take them there).

Evidently, that it is too hot to play outside in Barcelona from early June to late September is determined by the laws of thermodynamics and the capacity of the human body to cope with heat. But most conversion factors are not primarily a consequence of the laws of nature, but of socio-political systems. Climate and physiology are not the sole relevant factors; a different playground design including shading provisions would enable the children in our neighbourhood to have the *true opportunity* of playing outside by reducing the summer heat to bearable levels.

A different example now. I mentioned earlier that safety belts amplify the capability of riding in car safely. And they surely do, but differently for men and women. Although men are more likely than women to be involved in a car accident, a woman that is involved in one is 47 per cent more likely to be severely injured than a man and 17 per cent more likely to die[40]. This has to do with how cars are designed, and for whom. And with how safety is assessed during regulatory testing[41]. The problem exemplified here shows how the conversion of artefacts (car safety features) into functionings (being able to avoid major injuries or survive a crash) is dependent on the social and gender power dynamics embedded in social institutions, artefacts, and in the designed world as a whole. Conversion factors and design are thus intertwined in the reproduction and consolidation of unjust power dynamics involving race, gender, class, nationality, and other vectors of injustice.

Relatedly, newly designed artefacts are usually embedded in pre-existing socio-technical networks that will also condition and constrain the achievement of actual functionings according to power dynamics. To illustrate we can briefly go back to the One Laptop per Child project. Remember that sponsors and designers failed to consider that an information technology is never a stand-alone system, but is embedded in an ecosystem amid other technologies and services (installation, training, repair, support, etc.). Along these lines, the users of the laptop (especially children and their teachers) were unable to convert the resource (the laptop) into actual functionings (learning, playing, socialising, etc.) because important conversion factors were not adequately considered (the need for teacher training, creating software and content, delivering maintenance and support, etc.)[42]. The upshot of all this is that conversion factors (whether social, cultural, political, institutional, or economical) are important because they constrain and condition the transformation of artefacts into capabilities and functionings, in other words, into actual true opportunities.

REFLECTIONS AROUND THE *TELOS* OF DESIGN

In this section, I want to introduce several observations on the overarching purpose that was formulated above.

Some Observations and Challenges

The overarching purpose and the scaffolded account as a whole is presented as a reasonable object of deliberation around the overall telos of design. Given the plurality of views within the fields design, and the breadth of the range of disciplines, it is almost certain that reasonable designers and scholars of design will reject this formulation for the overarching purpose because it might conflict with other formulations that they find more persuasive or with their intuitions on the matter[43*]. Some might accept it as a valid purpose while disagreeing with its putative overarching character: they might say that contributing to flourishing is an important purpose worth pursuing, but not the overall *telos* of design. They could claim, for instance, that design's overarching purpose is to contribute to human rights or to social change or transitions to more sustainable futures. To me, all this could also be reasonably seen the other way around. Human rights can be seen as the minimum necessary conditions for being able to flourish as human beings. Similarly, one could advocate social change or 'transitions' toward more just and sustainable societies because a more just and sustainable future is likelier to provide better conditions for flourishing. Does promoting human rights or designing for transitions provide truly higher-level purposes or are they just means toward flourishing? I am inclined to choose the latter alternative over the former.

Indeed, claiming that a purpose is overarching implies claiming that there is no other higher-level purpose; this is, no doubt, a strong claim. Yet it would be absurd to expect that *everyone* in the field of design would agree with the claim of overarchingness; this would be an implausible condition to set. However, I propose instead a more plausible way of dealing with the possible disagreement: a purpose can be considered overarching as long as it can be said to encompass lower-level purposes and is not less compelling than other possible candidates to be the ultimate purpose (in other words, as long as there are no inescapable reasons for rejecting it). This opens up a possible scenario in which several formulations for the ultimate *telos* of design might exist that are equally compelling; in this case, the different purposes could be temporarily accepted jointly until they can be somehow ranked in order, which may not be possible to decide[44*].

Other critics might consider that 'seeking to expand human capabilities through design so that others can flourish' not only is not *the* overarching purpose of design, but it is not even a legitimate purpose for design *at all*. In this case (assuming the critics are not merely questioning the wording of the purpose, but its substance), I do not see how this objection could be resolved. The lack of acceptance of the key role human flourishing has for design would highlight a clash so deep in the conception of design that it would suggest that no formulation of a shared telos could be found between these antagonistic positions. For example, somebody who views design *exclusively* as a means for business to achieving and sustaining economic growth, and not more than that, is unlikely to be persuaded by my arguments. In this case, good positive reasons could possibly be provided for each of the positions, but these reasons do not *override* the reasons that can be given for the other position. We would be lacking the common ground needed for arguments to work, maybe finding ourselves in an irresolvable 'deep disagreement'[45].

Admittedly, the issue of the ultimate purpose of design, just like most things in ethics, will not be definitively settled, but a description of an over-arching purpose can help designers reflect upon and improve on their own pre-existing internalised ethical ideals or on the lack thereof, if only by rationally engaging in a reflexion around the matter. Nevertheless, the purpose is formulated here as a thesis for discussion and is necessarily tentative and temporary: purposes cannot be fixed forever, as practices and the communities in which those practices are embedded are not static and change with time. In line with this, even if the purpose is accepted as a plausible purpose for design to have, a new higher-level purpose could arise in the future.

Is This Purpose Ever Plausible?

Although pursuing a purpose like this would not be an easy feat, there is nothing in it that makes it impossible *in principle*. But even if it were, the impossible could serve as an ideal. Perfection might not be possible in practice at a particular time, but striving to get as close to it as we can is something that can guide what we do, and even result in pushing the boundaries of what seems possible.

My formulation of the *telos* is pluralistic about design; that is, it can accommodate a diversity of views and values that are present in the field of design. To illustrate briefly, it is neither in favour or against minimalism, expressivism, functionalism, or any other *-ism* in design. It also allows for a designer to have many other types of lower-level purposes and ideals of excellence, such as upholding functional or technical standards, meeting business goals, honouring confidentiality, or striving for elegance, ingenuity, or beauty.

At the same time, the purpose highlights the orientation toward others' wellbeing, which legitimises design as a profession. Yet the formulation of the overarching purpose is also pluralistic about flourishing: that is, it is agnostic to the designer's own views of the good life and should be able to accommodate a radically broad spectrum of conceptions of what it is to truly flourish as a human being[46*]. At the same time, while making an explicit commitment to a crucial human good, this formulation avoids placing the professional designer as being ultimately responsible for (designing) people's flourishing itself. Such standards of excellence would be haughty. And quite possibly ridiculous.

Lastly, on all three levels of the scaffold we developed, there is a connection between design, purposes, and key human goods that play a strategic role in human flourishing. This means that the account, although pluralistic about design and about wellbeing, is grounded in a view of professions that is primarily concerned with human flourishing, which entails that any other ideals that a designer might uphold (aesthetic, commercial, or technical ideals, for instance) cannot conflict with this primary concern for human flourishing.

The Challenge of Anthropocentrism

This focus on human flourishing might be characterised (and disregarded) by diverse scholarly perspectives as being *anthropocentric*. In this context, the charge of anthropocentrism carries negative connotations associated with the magnification of the importance of humans in the world. In the strongest sense, anthropocentrism holds that humans are the most significant entity in the world. Though in tempered form, the idea can be traced back to the Bible and is best embodied in Francis Bacon's idea of the dominion of humans over nature.

For its contemporary critics[47*], the failure to grasp the animal character of human life, and the interconnectedness of human and non-human life, results in global warming, environmental destruction, and the decimation of natural species, which is a poignant consequence of seeing the whole world (and even the universe) as a resource to be exploited to advance human ends; the outcome of which is today's climate and environmental emergency.

The critique against anthropocentrism is a legitimate one. And granted, this inquiry could indeed be argued as a form of anthropocentrism because I have defined professions and practices in a way that seems inescapably anthropocentric. But this focus could be also reasonably seen as *anthropological*, rather than anthropocentric: the central actors in a profession are humans. Also, practices in general, and design in particular, are intrinsically human endeavours. It is not that far-fetched on my side to suggest that it makes much sense to unambiguously frame an inquiry into the purposes of a profession in terms of human interests, values, and experiences.

At the same time, it does not necessarily follow from this framing that there is no inherent value in the natural world, nor does it follow that humans should be considered the most important species; it much less follows that humans are self-sufficient beings that are not a part of the natural world. As far as I see the matter, the elements of anthropocentrism that are standardly charged against by the critics are not present in my account.

What is more, in the philosophical anthropology we have been discussing, it seems difficult to imagine that a person would be able to live a full flourishing life without being emotionally and physiologically linked to the natural world as a whole. Along these lines, human flourishing cannot ever be truly seen as wholly separate from the welfare (and in some cases flourishing) of other beings and natural things; nor can the natural world merely be seen as a requisite or resource for human existence. At the same time, from the vantage point of the present, it is hard to imagine a type of concern for the non-human world that is not necessarily mediated by our human values, vocabulary, desires, beliefs, and experience.

This has serious implications for a designer that realises that human flourishing cannot be achieved with disregard for the non-human world. The intrinsic value of other beings and natural things and their needs must be taken seriously in design activity. The most obvious way to do so at a professional level is to internalise the notion that natural life has intrinsic value and is more than instrumental. The externalisation of this can manifest itself in the care for other beings and natural things by incorporating this notion as constitutive of the criterion of appropriateness of a design solution so that it is in balance with the planet's overall capacity to support human and non-human life, and is at least not detrimental to it. For a designer this could mean, but is not restricted to, seeking to implement and follow one of the many ecologically sound design methods and processes that are available[48*].

Alas, a full response to the critique would be to reconsider the overall goals of professions, but proposing a *post-human* professional ethics is not what I intend to do here. The charge of anthropocentrism is far from irrelevant, and this response might seem insufficient to some, but it is as far as I can go within the umbrella of professional ethics and current understandings of professionalism. The problem of anthropocentrism opens up important ethical, metaphysical, and anthropological topics like the relationship between humans and other beings and nature, and the rights and the status of personhood pertaining to living and inanimate things; these topics much exceed the scope of our inquiry and are the subject of ethics at large. As Jeffrey Chan argues, 'this dilemma is not entirely up to design to resolve. After all, there is still little consensus today on how to value non-human species independent of anthropocentric values'[49]. I hope, however, that the treatment of this objection, although not in the least resolving the issue, might allow us to move forward.

To sum up, the overarching purpose of design that is proposed here—this *telos* for design—is not a representation of what the design profession currently is as a whole, but of *what it can be at its best*. When a designer embarks on the pursuit of a good professional purpose, in a manner that is 'consonant' with the excellences of the profession, they get closer to achieving the goods of design and the telos that is being pursued. In addition, it is on this journey toward the good where they gain virtues such as responsibility, courage, integrity, and constancy, developing the skills for practical reasoning that are necessary to cope with the difficulties and constraining conditions they will encounter along the journey.

REGULATIVE IDEALS AS INTERNALISED GUIDELINES FOR ACTION

In our working definition of a profession, we saw that they are undertaken to promote, ensure, or safeguard some aspect of others' wellbeing and also involve a commitment to key human goods that enable a person to *flourish* as a human being.

Every profession adds, in turn, its special contribution to this general purpose, which is, just like any purpose, *ideal* in nature. Although teachers educate by, for instance, fostering certain skills, this is not the final end of education. Teachers *ideally* pursue larger purposes for which those skills are instrumental. For instance, they prepare students for membership of the community or to develop autonomy in reasoning (for example, to be able to critically reflect about one's life). Similarly, designers do not just select colours or materials or make prototypes, which could be an immediate goal of design activity. As we saw in the previous sections, what they *ideally* do is to extend typical human abilities, powers, or capacities through the conception of material and immaterial artefacts. To this end, they attempt to realise outcomes that are, for instance, pleasurable, safe, just, useful, and so on and so forth.

Our analysis indicated that if designers want to achieve excellence in their profession, it is indispensable for them to commit to realising this substantive ideal, as well as to be concerned with technical goals, which are not true ends in themselves, but instrumental in the attainment of the larger purpose. Further, besides professional achievements, the pursuit of one's professional *telos* enables designers to achieve actual personal accomplishments, however modest, along the way.

Regulative Ideals in Design Practice

But how can a designer operationalise all this? How can they translate the highly general overarching purpose into discrete actionable goals? In other words, how can they decide what to do in their everyday jobs? And are there any guidelines they can follow? Well, virtue ethics does not provide universal guidelines for action in the way Kantianism does, but it does offer accounts of how the virtues can guide behaviour by providing a specification of 'right action'.

One of these accounts gravitates around the notion of 'regulative ideal', which we introduced in chapter 6. We mentioned that this notion refers to the internalised conceptions of correctness and excellence that allow a person to attune their motivation and conduct to that standard[50]. The ideal is not simply declarative or aspirational, but 'regulative' in the sense that it regulates our motives and behaviour. So, for example, a designer with an internalised conception of what good typography is can be guided by the regulative ideal so that their behaviour and motivations as a designer are aligned with their idea of what a good designer would do.

Because a regulative ideal operates based on internalised conceptions, we might not be fully aware of its guiding effect. The example involving Frank Gehry, in chapter 1, can serve to illustrate this. Gehry has an obviously well-developed conception of what excellence is in architecture that enables him to know that something in the model of the building he is developing is 'weird'. Even though he is unable to put into words why he does not like it, he *knows* that the model is weird because it conflicts with his regulative ideals; that is, with his conceptions of what good architecture is supposed to be. It is only after a deep rumination that he is able to very tentatively explain that he does not like it because it seems a little 'pompous' and 'pretentious'. The same case provides another illustration of the workings of a regulative ideal: when Gehry is working on another side of the model, the internalised regulative ideal informs and guides him in finding the notion of 'crankiness' to anchor and guide his creativity in the search for the specific type of 'crankiness' he wants for his building. That he is not able to get it right at the first attempt does not mean that regulative ideal is not functioning; on the contrary, it is precisely the regulative ideal that tells him that he is not succeeding. Importantly, the regulative ideal cannot at all be reduced to the mere notion of 'crankiness'; the regulative ideal is what indicates to Gehry that 'crankiness' in needed to achieve excellence.

In chapter 6, we mentioned that regulative ideals are based on standards of *correctness* (rules and principles) and of *excellence* (virtues, values, and other considerations beyond the correct or incorrect). One could, for instance, codify standards of correctness for basic aspects of usability or the durability of the materials used in architecture, but surely no standards of correctness for 'cranky' architecture could be laid down. Standards of excellence are much more elusive and are directly related to the contextual decisions for which practical reasoning is indispensable. The standards of excellence regulate our behaviour, but they do so in a way that cannot be codified because what constitutes right action needs to be assessed on a case-by-case basis and cannot be determined beforehand.

Regulative ideals about the outcomes of design could be said to be frequently structured around the notions of usefulness, usability, aesthetic quality, desirability, and sustainability. For instance, when designing a chair, a designer might be guided by a regulative ideal consisting of a chair having to be comfortable and adequate for people of all sizes and shapes. Needless to say, a good designer would also be informed by other regulative ideals that would direct them toward consulting the design literature (to look for anthropometric measurements or other chairs that could be used as referents, for example), to researching the way people interact with chairs, to having their designs tested for usability with real users, and so on. When the regulative ideals possess codifiable elements that inform a designer about what would count as *correct* forms for a chair, a designer has a clearer path to reach a threshold of correctness.

As we saw previously, regulative ideals also have uncodifiable elements that guide designers in achieving excellence. In this case excellence might be achieved by, for instance, a particular way of doing user research or by developing a new sophisticated understanding of what the purpose of a *good* chair is; the list *is* endless. Standards of excellence are necessarily broad because they focus on the activity and its outcomes *as a whole*, taking into consideration the overall goals and purposes and their systemic features. For instance, 'crankiness' is not the purpose Gehry ultimately seeks to achieve; he is possibly interested in that 'crankiness' as a way to engage people feelings, to shock, to confuse, and so on. In this way, a regulative ideal guides actions without *necessarily* becoming the purpose of the actions it regulates[51]. For instance, a regulative ideal that embodies ideals of minimalism or symmetry can guide a designer in the conception of a basic layout for an interface, but their purpose may be to design a good product page for an online supermarket, rather than to achieve minimalism or symmetry per se.

By contrast, in other cases the regulative ideal and the purpose may be overlapping. So a designer working on the design of a schoolbook may be guided by ideals of readability and usability, which could also become the designer's immediate purpose when choosing typefaces, and making decisions about the size of headings or about the page layout, if what they explicitly seek is to design books that are readable and usable.

Regulative Ideals and Character

The distinction between standards of correctness on the one hand, and of excellence on the other, mirrors the difference we made earlier between principle-based ethics and virtue ethics. It is not a coincidence, then, that the standards of excellence embodied in regulative ideals are closely connected to one's character. In our discussion of virtue ethics, we saw that virtue was about acting, feeling, and reasoning in a certain way, out of certain motives that incorporate an appropriate sense of the virtue in question. So, for example, one acts honestly when one acts on motives that incorporate an appropriate sense of what honesty is. To have internalised a regulative ideal around honesty is to have a sense of honesty that guides one to act honestly. The connection between character and right action is thus in the possession of 'certain standing commitments or normative dispositions, which need not always be consciously formulated or applied, but which will govern and shape their motivations and actions'[52]. Character is thus about having the dispositions to act consistently with a certain regulative ideal one has internalised.

Let us bring this closer to design again by way of an example; imagine a designer working for the American furniture company Herman Miller, which is renowned for the attention it pays to the customers' wellbeing and for striving to curtail its environmental footprint. A Herman Miller designer, arguably, would have internalised, on the one hand, a series of regulative ideals that would guide them in designing, say, office chairs that are ergonomic, beautiful, sustainable, and reliable, and, on the other hand, they might seek to design chairs that substantially contribute to people's wellbeing (to the extent an office chair can do that). An excellent furniture designer would know, and have internalised as a regulative ideal, that human bodies have evolved during eons of natural selection to be moving around, and that sitting for too long over the course of a workday is seriously unhealthy. One could plausibly imagine that the designer Brian Alexander was guided by regulative ideals like these (among many more, of course) when working on the 'Renew Sit-to-Stand' desk, which is an example of an excellent design that facilitates switching between sitting and standing. If this designer has also developed the capacity for practical reasoning, they might reach design solutions that afford not only sitting, but also standing and moving. Plausibly, the regulative ideal, assisted by the designer's own capacities for practical reason, guided them in thinking through the different design alternatives and in acting out their professional purpose. It could be argued that the ideals, albeit partially, become embodied in a chair that is comfortable, prevents injury, facilitates movements, and so on.

The notion of character also serves to explain why Alex, the designer in the story from chapter 5, was conflicted about designing ads that targeted ex-gamblers and portrayed gambling as a way out of poverty. They had internalised conceptions about what constitutes *honest* advertising, and about professional *integrity*; these notions allowed them, under some strict conditions, to create ads for state-run lotteries in the past without compromising their professional integrity. Conversely, the required new ads conflicted with different regulative ideals Alex had internalised, and that is why they were acting virtuously (by showing integrity and consistency) when questioning the new ads and acting upon this conflict.

Why is the notion of regulative ideal important for our inquiry? It is important because if we are concerned with guidelines for professional action in the field of design, it might be productive to think about them in terms of plausible regulative ideals. Regulative ideals offer a way to include codifiable principles and norms ('do not combine more than two typefaces', for instance), as well as the more ambitious notions around excellence in design (for instance, what an excellent layout is), virtue, practical reason, and flourishing. In a way, regulative ideals work as our own internalised code of conduct; they 'operate as guiding background conditions on our motivation'[53], and can thus function as a guideline toward right action by embedding conceptions of what constitutes good design as well as good professional behaviour.

In our understanding of professional ethics, providing ethical guidance is not about providing guidelines that instruct the designer to choose such and such options or avoid doing such and such things, nor is it about offering a clear-cut decision making procedure that designers can use to resolve ethical dilemmas. Ethics, unlike mathematics, as Aristotle famously posed, is concerned 'with particular facts, and particulars come to be known from experience'[54]. The ethical guidance that can be provided here aims to contribute to the development of one's character as a designer; a book can only go so far, and this is a task that the designer themself will have to carry out. Along these lines, developing ethical expertise is in a sense partly about developing, acquiring, and internalising regulative ideals that govern the different spheres related to one's professional activity.

Certainly, *some* general guidance can be offered here: a horizon as a starting point of sorts, rational arguments around a particular telos for design, reasons to see things in a particular way, a vocabulary and a conceptual apparatus to think about one's personal and professional life. All this, in turn, provides a motivation for not putting these aspects of one's life in separate compartments so that life can be experienced as an integrated and coherent whole, and not as compartmentalised fragments.

We have covered a lot of ground in this chapter and obtained a formulation for an overarching purpose structured around promoting others' capabilities through design. We also reviewed how regulative ideals can guide designers in their everyday activities without the need for following codes of ethics of declarative guidelines. In the next chapter, we will deepen our account of design professional ethics by exploring the notion of responsibility and its relation with the overarching purpose and the ideal of professionalism.

NOTES

1. Aristotle, *Nicomachean Ethics*, translated by Roger Crisp (Cambridge: Cambridge University Press, 2004), 6, 1095b.

2. Alasdair MacIntyre, *After Virtue: A Study in Moral Theory* (Notre Dame: University of Notre Dame Press, 2007), 219.

3. Ibid.

4. Richard Buchanan, 'Design Ethics', in *Encyclopedia of Science, Technology, and Ethics*, edited by Carl Mitcham (Detroit: Macmillan Reference, 2005), 507.

5. However, when read in context, Buchanan's formulation becomes absolutely specific to design, and any minimally charitable reading yields many valuable insights about specific purposes. But, alas, no more specific formulation of design's purpose is presented in his entry that can only be valid for design.

6. Alasdair MacIntyre, 'A Partial Response to My Critics', in *After Macintyre: Critical Perspectives on the Work of Alasdair Macintyre*, edited by John P. Horton and Susan Mendus (Notre Dame: University of Notre Dame Press, 1994), 284.

7. MacIntyre, *After Virtue*, 189.

8. Ibid., 193.

9. Alejo José G Sison, 'Revisiting the Common Good of the Firm', in *The Challenges of Capitalism for Virtue Ethics and the Common Good*, edited by Kleio Akrivou and Alejo José G. Sison (Cheltenham: Edward Elgar Publishing, 2016), 108.

10. Debbie Millman, *How to Think Like a Great Graphic Designer* (New York: Allworth Press, 2007), 51.

11. Ibid., 50–51.

12. MacIntyre, *After Virtue*, 188.

13. Millman, *How to Think Like a Great Graphic Designer*, 32.

14. Adrian Gilbert, *Waffen-SS: Hitler's Army at War* (New York: Da Capo Press, 2019), 13.

15. Philippa Foot, *Virtues and Vices and Other Essays in Moral Philosophy* (Oxford: Oxford University Press, 2002), 5.

16. MacIntyre, 'A Partial Response to My Critics', 284.

17. A. Rodríguez-Hidalgo, I. Morales, A. Cebrià, L. A. Courtenay, J. L. Fernández-Marchena, G. García-Argudo, J. Marín, et al, 'The Châtelperronian Neanderthals of Cova Foradada (Calafell, Spain) Used Imperial Eagle Phalanges for Symbolic Purposes', *Science Advances* 5, no. 11 (2019).

18. Jeroen Van den Hoven, 'Human Capabilities and Technology', in *The Capability Approach, Technology and Design*, edited by Ilse Oosterlaken and Jeroen Van den Hoven (Dordrecht: Springer, 2012), 35.

19. Ibid.

20. See Bucciarelli's extraordinary account of the way engineers go about designing: Louis L. Bucciarelli, *Designing Engineers* (Cambridge: The MIT Press, 1994).

21. Caroline Whitbeck, *Ethics in Engineering Practice and Research*, second edition (Cambridge: Cambridge University Press, 2011), 77.

22. It is worth noting that a previous wave of acceleration started during the Neolithic Revolution, when artisans appeared due to the development of technology, especially in the fields of metallurgy, ceramics, and woodworking. This acceleration continued across the globe and gained momentum during the Middle Ages until it was replaced as an accelerator of change by the Industrial Revolution. Evidently, that artisanship was replaced as an accelerator does not at all entail that artisans disappeared; they still form a very significant social group in contemporary societies of all economic levels. Josef Ehmer, 'Artisans and Guilds, History Of', in *International Encyclopedia of the Social & Behavioral Sciences*, second edition, edited by James D. Wright (Oxford: Elsevier, 2015).

23. Justin Oakley and Dean Cocking, *Virtue Ethics and Professional Roles* (Cambridge: Cambridge University Press, 2001), 74.

24. Van den Hoven, 'Human Capabilities and Technology', 33.

25. Ezio Manzini, 'Design, Ethics and Sustainability. Guidelines for a Transition Phase', *Cumulus Working Papers Nantes 16/06* (Helsinki: University of Art and Design Helsinki,

2006); Carla Cipolla, 'Sustainable Freedoms, Dialogical Capabilities and Design', *Cumulus Working Papers Nantes 16/06* (Helsinki: University of Art and Design Helsinki, 2006); Andy Dong, 'The Policy of Design: A Capabilities Approach', *Design Issues* 24, no. 4 (2008); Ilse Oosterlaken, 'Design for Development: A Capability Approach', *Design Issues* 25, no. 4 (2009). See also the further reading section at the end of the book.

26. Neither will I attempt to find common theoretical grounds between the Capability Approach and the MacIntyrean account beyond their common focus on *eudamonist* wellbeing. This connection would necessitate a very extensive scholarly discussion that is not central to our purposes.

27. Although these examples illustrate 'desirable' functionings, it is important to note, however, that functionings and capabilities are in themselves morally neutral. They can be 'unequivocally good (e.g., being in good health) or unequivocally bad (e.g., being raped)' or of a not so straightforward moral evaluation (travelling or being a parent). Ingrid Robeyns, *Wellbeing, Freedom and Social Justice: The Capability Approach Re-Examined* (Cambridge: Open Book Publishers, 2017), 42.

28. Ibid., 39.

29. Aristotle, *Nicomachean Ethics*, 107, 1140a.

30. Martha Nussbaum, *Creating Capabilities: The Human Development Approach* (Cambridge: The Belknap Press of Harvard University Press, 2011), 125.

31. Jozef Keulartz and Jac. A. A. Swart, 'Animal Flourishing and Capabilities in an Era of Global Change', in *Ethical Adaptation to Climate Change: Human Virtues of the Future*, edited by Allen Thompson and Jeremy Bendik-Keymer (Cambridge: The MIT Press, 2012).

32. Design itself can be conceptualised as a capability: the capability to shape one's environment, for instance. But this is not the main intended sense. For a discussion of design as a capability see Dong, 'The Policy of Design', and Ezio Manzini, *Politics of the Everyday* (London: Bloomsbury, 2019), 47–48.

33. Alana Semuels, 'No Driver's License, No Job', *The Atlantic*,https://www.theatlantic.com/business/archive/2016/06/no-drivers-license-no-job/486653/.

34. Rolf Pendall, Christopher Hayes, Arthur George, Zach McDade, Casey Dawkins, Jae Sik Jeon, Eli Knaap, et al, *Driving to Opportunity: Understanding the Links among Transportation Access, Residential Outcomes, and Economic Opportunity for Housing Voucher Recipients* (Washington, DC: The Urban Institute, 2014).

35. Manzini, *Politics of the Everyday*, 48n.

36. Amartya Sen, *Development as Freedom* (New York: Alfred A. Knopf, 2000), 10.

37. Van den Hoven, 'Human Capabilities and Technology', 35.

38. Nussbaum, *Creating Capabilities*, 34.

39. Robeyns, *Wellbeing, Freedom and Social Justice*, 45–47.

40. Caroline Criado Perez, *Invisible Women: Exposing Data Bias in a World Designed for Men* (London: Vintage, 2019), 186.

41. Ibid., 187.

42. Kenneth L. Kraemer, Jason Dedrick, and Prakul Sharma, 'One Laptop Per Child: Vision Vs. Reality', *Communications of the ACM* 52, no. 6 (2009).

43. I refer explicitly to designers and scholars of design because it is primarily up to practitioners to decide what their practice is for. Naturally, the formulation could also be rejected by anybody on grounds relating to the general cogency of my arguments.

44. This scenario opens up a hugely complex discussion related to, among other things, unresolvable conflicts between values. I will leave this issue untouched, as it is not central to my arguments and way beyond the scope of the book and the interests of its readership.

45. Robert Fogelin, 'The Logic of Deep Disagreements', *Informal Logic* 25, no. 1 (2005).

46. To illustrate, although we are clearly committed an *eudaimonic* view of wellbeing, the overarching purpose could be seen as a legitimate one by somebody holding a different understanding of wellbeing such as desire-fulfilment perspectives. See Roger Crisp, 'Well-Being', Stanford University,https://plato.stanford.edu/archives/fall2017/entries/well-being/.

47. There is a broad non-homogeneous movement of cultural and philosophical critique (often referred to as 'post-anthropocentrism' or 'post-humanism'), which challenge human specialness. Authors like Rosi Braidotti, Timothy Morton, and Graham Harman blur the dis-

tinction between humans and non-humans and broaden the range of what could count as a genuine 'other'. Beware: this type of post-humanism should not be conflated with another, fully unrelated conception of 'post-humanism' that is broadly concerned with the use of technology to eradicate undesirable biological human features and enhance desirable ones (aging, dying, limited cognitive capacities, etc.).

48. For different perspectives, see Stuart Walker and Jacques Giard, eds., *The Handbook of Design for Sustainability* (London: Bloomsbury, 2013); Conny Bakker, Marcel Den Hollander, Ed Van Hinte, and Yvo Zijlstra, *Products That Last Product Design for Circular Business Models* (Amsterdam: Bis Publishers, 2014); John Ehrenfeld, *Sustainability by Design* (New Haven: Yale University Press, 2009).

49. Jeffrey K. H. Chan, 'Design Ethics: Reflecting on the Ethical Dimensions of Technology, Sustainability, and Responsibility in the Anthropocene', *Design Studies* 54 (2018): 193.

50. Oakley and Cocking, *Virtue Ethics and Professional Roles*, 25.

51. Ibid., 27.

52. Ibid., 28.

53. Ibid., 44.

54. Aristotle, *Nicomachean Ethics*, 111, 42a.

Chapter Eight

The Full Circle

From Responsibility to Action

We ended the previous chapter connecting the purpose of design to the promotion of others' capabilities, and providing an explanation of how regulative ideals can guide designers in their professional activity. In this chapter, we will explore how all this can be converted into design artefacts that are conducive to others' wellbeing. To do so, we will once again consider the virtues by exploring the notion of responsibility, continue with care as a core element in responsibility, and end with an analysis of practical reason as the virtue that converts a willingness to care into actual instantiations of design.

DESIGN AND THE VIRTUES

From the virtue ethical perspective adopted here, our life is made up of different practices in which we participate and pursue the internal goods they have to offer. This engagement becomes one's narrative quest: the process through which we discover, frame, and reframe our purposes in life. Irremediably so, we are born with a past, and our own story and narrative quest are threads woven into the blanket of the larger narratives of the communities and practices to which we belong. Naturally, one could burn one's bridges down, deny prior relationships, and start afresh with a blank slate, but even then, one's quest would be bound to be, once again, intertwined in other interlocked narratives.

Design, just like the other practices with which they engage, rewards designers with internal goods, which constitute the characteristically desired results of a practice. These internal goods are not just goods worth pursuing but constitute and legitimise the very practice that produces them. As discussed in the previous chapter, the purpose for the design profession could be associated with fostering others' capabilities.

Besides design, designers engage in other practices as well, such as family life, friendships, sports, hobbies, volunteer programmes, activism, and so on. But it is the practice of professional design that engages them during a large part of their time. When practitioners pursue the internal goods of a practice, provided they do so according to its standards, they attain and develop virtues (excellences) in their journey toward the good. In this journey, not only is the practitioner's technical ability and overall expertise improved and extended, but the practitioner themself grows as a person as they encounter and overcome obstacles, harms, and distractions as they pursue the internal goods.

In this way, practices provide practitioners with opportunities to achieve their professional purposes, as well as to extend not only their technical abilities but also to flourish as persons when fully engaged in their work in the pursuit of the goods of the practice. As Milton Glaser illustrates,

> much of my satisfaction and happiness in life comes out of my relationship to my work. And I still have the feeling that I have enormous opportunities and possibilities. There is always so much more to understand about the nature of communicating and design and color. You reach a point in your life when you realize that you know nothing about color or shape! [1]

Design is such a generous practice that even one of the most celebrated graphic designers in contemporary history is able to find new ways of understanding colour or shape after a career spanning seven decades. But that is not the central point here. The important issue has to do with the way in which Glaser approaches the practice; a way that allows him to remain curious and eager to find new understandings about colour and shape (things he understands better than most people on this planet). This way is enabled by the virtues Glaser possesses next to his technical skills and knowledge; it is the virtues that make it possible for him to learn new things. The virtues of curiosity and intellectual humility, for instance, enable him to accept that there are still important things about shape or colour that need to be understood; or patience and perseverance, which aid him in accepting and carrying on with his learning process.

The professional activity and the virtues are thus interrelated; seeking and obtaining internal goods have profound consequences for us that go beyond the professional realm and affect us as whole persons. Moore writes that the internal goods we achieve 'lead to *the* good for ourselves and hence to us fulfilling our own purposes in life'[2]. For example, when a designer works through a difficult assignment, besides the necessary intellectual and technical skills, they will require and develop particular attitudes and disposition that will enable them to negotiate through the difficulties. For Macintyre, the virtues tend 'to enable us to achieve those goods which are internal to practices and the lack of which effectively prevents us from achieving any such goods'[3]. For instance, imagine two designers who separately find themselves in a similar position: after generating a minimally acceptable but rather mediocre design proposal, they come to a dead end in their projects: every new idea they have is a bad idea. Although both designers possess the intuition that the minimally acceptable design proposal should be taken a step further, one of them gives into frustration and abandons the search for improvement, settling for a mediocre solution in the hope that the client will accept it. Conversely, the second designer carries on with the search; they cannot bear to deliver the mediocre solution to the client; they seek advice from others, they apply new techniques to come to new insights, and they keep engaging in deliberate practice with the intuition that the effort will pay off and a better solution would be found. The upshot of this is that the second designer exhibited the virtues of honesty, patience, perseverance, creativity, and humility; these virtues would be strengthened by having carried on, by having persevered, having been patient, and so on. And this could be so regardless of the quality of the solution that they actually reached (a mediocre result could be due to a bad brief or other external constraints).

I have neither the space nor the intention to make an inventory of virtues that are relevant to design. Rather, what I want to do here is to concentrate on giving an account of one particular virtue that stands out above all others: the virtue of responsibility. Responsibility understood as a virtue can serve to structure an account of what it means to be an ethical professional designer.

RESPONSIBILITY AS A VIRTUE

It would be a truism to say that all professions, by their being connected to key human goods, require responsibility from professionals. Responsibility, as we saw in chapter 2, has different meanings. The sense that will concern us here is 'responsibility as a virtue'. Here, responsibility is not understood as having to do with blame or praise (in the sense of attribution as when a person is *held* responsible), nor as a relation between a person and a state of affairs or act (in the sense of obligation as when a person is responsible *for* something), but as a characteristic or character trait of a person. In the virtue ethical sense that will concern us here, a person is not necessarily responsible for something, but simply *is* responsible; a person is responsible in the same holistic way they can be funny, courageous, or well-read.

According to ethicist Jessica Nihlén Fahlquist, there are three key components of responsibility as a virtue: care, moral imagination, and practical wisdom. This notional triad will serve as points of departure to deepen our discussion on responsibility as a virtue and its entailments for design professional ethics.

Responsibility and Care

The central aspect to grasp about responsibility as a virtue is that being responsible is a feature of the person and not about having or meeting a set of isolated obligations. Naturally, a responsible person may enter an obligation and be responsible *for* something, but although the notion of obligation is surely relevant, being responsible in the virtuous sense exceeds the sense of duty and the obligations one has entered. In fact, one can be responsible in the virtue ethical sense and not have an obligation in the strict sense at all. Nihlén Fahlquist explains: 'to be a responsible person is to feel responsible and to feel responsible is to care about others' wellbeing and how one's actions affect fellow human beings'[4]. A responsible person feels responsible because they *care*, not necessarily because they have an obligation toward them (they might have none). In the virtue ethical sense, the obligation is less important than the feeling of responsibility and the readiness to care that ensues from that feeling. All of this will become clear as we go along in this section.

A person that is 'responsible' in the intended sense is personally invested in what they care about. I will explain. Philosopher Harry G. Frankfurt posits that *we identify* with what we care about[5]. When our object of care is harmed, we experience that as a personal loss; analogously, when our object of care flourishes, we see that as a personal gain. And what would responsible professional designers care about? In the previous chapter, we defined an overarching purpose for design that was connected to purposes such as promoting others' wellbeing, expanding human capabilities, and, ultimately, promoting human flourishing. So one likely object of care could be all these 'others' who interact and are affected by the outcomes of design. But I want to claim that design as a profession is itself *also* a reasonable object of care for a responsible designer to have.

The notion that design as a profession (and naturally as a practice) plays an important part in the designer's own quest to lead a good life can be seen as providing motivation for acting responsibly. On the one hand, it is because design personally matters to us that we are driven, at least to some degree, to acting responsibly. On the other hand, regulative ideals explain how a designer can be guided in a particular direction by a sense of responsibility (indicating *that* we should care and *what* we should care about). Acting responsibly, to reiterate, includes meeting the obligations we may have (for instance, a deadline or respecting the client's confidentiality), but it is not restricted to that. It follows from all this that professional designers would be personally affected when their object of care (design) is advanced or hindered, hence seeking to achieve the overarching purpose of design and the results of design activity mattering to them.

Although the status of emotions as a valid source of moral knowledge is a hotly and widely contested issue in philosophy, the character-based perspective of virtue ethics leaves plenty of room for considering them as a relevant input for moral reasoning. This allows for a different dynamic of ethical reasoning than what we find in the extreme rationality of narrow ethics and principles such as the principle of utility and the categorical imperative, which eschew emotions. Responsibility as a virtue is expressed by asking oneself 'what choices are compatible with or reinforce desirable aspects of one's personal identity. Questions like "Could I bear to be the sort of person who can do that?" are foremost'[6].

Remember the interface designer that was troubled about including a field for gender in the design of a form; they had to negotiate with their manager or client and argue why it was a bad idea to include said field. To them it was a bad idea but not for technical or aesthetical reasons, their concerns were ethical ones (perhaps they simply felt uneasy about it); the upshot is that they *cared* and did so because they imagined that some users encountering the field might feel aggrieved by it (even if most would perhaps not be). The troubled designer cared because a good design would not put people in this uncomfortable situation, and even if people did not worry too much about the field at all, it was not a good thing to be asking gender-related questions, as information about gender was in no way necessary for the correct functioning of the service. Through this example, we see how a designer can feel responsible and care, without having entered into an obligation in the strict sense.

Caring as Personal Involvement

In the previous chapter, designer Paula Scher was quoted alluding to the elusive things that have to be 'held on to and protected' during a negotiation. It is common knowledge that every designer constantly negotiates and compromises in their daily work about technical, formal, or aesthetic issues. But what is important to note from Scher's words is that she is not merely *wishing* to hold on to those elusive things: these things are really important to her. She makes these things especially her own; they truly matter to her because she identifies with them. Moreover, because she deeply cares about them, it would be a loss to her if what she cares about was diminished. And it is pertinent to note that this is not a 'zero-sum game', where she wins and the client loses; it would not be reasonable to believe that Scher's 'ellusive things' are matters that could be detrimental to the interests of her clients.

Becoming a virtuous designer means learning to recognise, appreciate, and care about these elusive things. At its most general, the virtue of responsibility is expressed in truly caring for certain things because they are seen to be directly connected to the attainment of the internal goods of design. It is because of this that design matters to the designer: what they do as practitioners is an important part of their narrative quest, of how they want to be as persons.

To illustrate further with an example, many cities across the world have seen the introduction of electric pay-as-you-go scooter sharing programmes whereby a rider can find one and unlock it with a smartphone app. Unlike some shared bicycle services, which require parking them in docking stations, these scooters do not usually have specially allocated parking spaces and are often left in places where they obstruct the footpath causing localised problems, for example, for parents with baby strollers or people with disabilities.

In the strict sense, the designer is not responsible for the rider's behaviour, as it is the rider who is responsible (in the sense that they have the obligation) for parking the scooters out of the way, a clause that is commonly included in the service's terms of use. Nonetheless, a designer can *feel* responsible for the situation, even if they are neither directly causing it nor formally responsible for making sure the rider meets their obligation. Along these lines, a responsible designer would care about more than just what they can be directly held accountable for, and proactively assume a forward-looking responsibility to design the service in such a way that other footpath users were not disturbed or harmed by misbehaving riders (an unintended consequence of the service). It is important to note at this point (and we will develop this argument when we discuss empathy in the next section) that this attitudinal or dispositional sense of responsibility requires neither warm feelings, sympathy, nor empathy toward others affected by the service.

The designer could deal with this forward-looking responsibility in several different ways: for instance, by designing adequate returning protocols that take into consideration the interests of other parties involved in the usage of footpaths. They could use nudges or other behavioural clues to steer user behaviour, for example by gently reminding the rider to care for other users and directing them to park the scooter in an adequate location. They could also choose a so-called ethics by design approach that makes it difficult or impossible for a user to leave the scooter in the middle of the footpath; for example, by using sensors or other technological solutions that check whether the scooter is placed in an acceptable place or refusing to accept the return until the scooter is located in a suitable spot. A different approach would be to avoid prescribing a single 'right' way of acting and to consider things more systemically and in less adversarial terms than framing the issue as the convenience and safety of the riders versus those of other stakeholders. This systemic approach would be more desirable than nudges from a virtue ethical perspective, if only because it increases the space for moral agency and is not focused on mere compliance with norms.

But designers often find themselves in a position in which taking a systemic approach is not a real alternative, and their task is more limited in scope as they are hired to work on a more limited commission. But this constraint does not entail that the designer needs to be less responsible. For ethicist Garrath Williams, responsibility as a virtue 'represents the readiness to respond to a plurality of normative demands'[7]. This readiness is bounded by the true *ability* the person has to responsibly deal with those demands; for Williams, ability consists of internal capacities such as imagination, perseverance, or judgement, as well as the external institutional dimension of one's circumstances[8]. These internal and external elements determine the effective freedom the designer has to realise their willingness. To put it differently, they need to find themselves in a *design situation* that truly opens up opportunities for action.

Because of all this, a responsible designer will seek ways to be less constrained by the circumstances when they conflict with their regulative ideals. The designer of the scooter service might prefer to approach the service from a systemic perspective, but their effective freedom to act upon this intention might be constrained by the actual possibilities that are realistically available to them. Although their preferred goal would be to expand capabilities, a responsible designer working in a project whose scope is too narrow will seek to *at least* not diminish the capabilities of other users of the footpath (for instance, by caring for the safety of pedestrians and people with disabilities), even if they are not able to design with a systemic approach as they would want.

Ezio Manzini proposes the related notion of 'field of possibility' to refer to 'everything that the nature, culture, economy, and technology of a given society allow to be done in a given place and time. It includes the spaces of freedom within which each subject can theoretically move'[9]. The subject, a designer in our case, can increase the range of available options at their disposal in two complementary ways. First, by extending their field of action by improving their personal skills, interests, and creativity. Second, by widening their 'field of possibility by reducing the technical, regulatory, financial, and cultural limits of the system with which subjects interact'[10]. The first way is clearly internal to the subject, but the second should not be seen as purely external (though it might appear to be that way); the subject aiming to increase its field of possibility works from within to modify it by collaborating with others.

Validity and the Designer's Stance

Another way of thinking about what is truly possible is to think about what counts as a 'valid' design outcome. I propose that validity could be understood as reaching a threshold that minimally meets the (ongoing) requirements and expectations of the relevant stakeholders that are directly involved in a design project with regard to the expected design outcome that is sought. The validity of an expected design outcome is important because it is a key factor that bounds the solution space that is available to designers within a given situation (budget or time frame are other important factors). Determining validity is the result of a series of complex negotiations between different parties that are involved in the project (project commissioners, project or product manager, development engineers, marketing department, and so on).

Validity is strongly linked to the notion of 'appropriateness' that was discussed in chapter 1. Validity and appropriateness are often congruent but not necessarily so; a solution can be valid but not appropriate and vice versa. What counts as a 'valid' solution depends on an agreement between stakeholders with decision making capacity, whereas 'appropriateness' refers to being adequate in the context of the problem or audience; 'validity' has thus a narrower sense than 'appropriateness'. So a valid solution could be still deemed inappropriate (or just partially appropriate) when considered overall (for instance, because it is unsustainable or manipulative).

To illustrate, imagine that a designer proposes the design of a systemic mobility solution that avoids a 'techno-fix' such as using sensors and artificial intelligence as to force the rider to act in a pre-established way. The designer believes a techno-fix is not a fully appropriate solution as it unnecessarily restricts the rider's agency by making 'good' behaviour mandatory. If the client contemplates said 'techno-fix' as the only type of valid expected outcome, then the systemic solution, which the designer believes to be more appropriate, would not be deemed valid. To go ahead with the systemic solution, the designer would need to gain buy-in from the client first. Naturally, the designer could negotiate a compromise or settle for a 'techno-fix'; in any case, a consensus about validity needs to be reached.

The upshot of all this is that a responsible designer would not only be concerned with the design of the outcome itself (the artefact), but would also be aware of what bounds and constrains them, which would also have them invest a lot of effort in extending the realm of what counts as a valid expected design outcome. In practice, this effort consists primarily in engaging in negotiations and persuading others (the client, project managers, etc.) into a consensus about the validity of the approach or perspective the designer is advocating, and in fostering the mindset that is necessary to be open to accepting it.

The narrower the conception of what counts as a valid expected outcome, the more constrained the space of true possibilities, and, in turn, the designer's true options for responsibly dealing with the legitimate plural demands they face. From an ethical point of view, the primary goal of this negotiation is to ensure the creation of a sufficiently broad space for responsibly reaching what the designer considers to be an appropriate outcome that is at the same time a valid one. In a sense, the negotiation about validity is also a negotiation about the conditions for exercising design responsibility.

Imagine, as an example, a designer that receives a brief with a narrow understanding of mobility asking for the design of a scooter service that is convenient and usable for the end-user. The responsible designer, wanting to extend the realm of possibilities, might attempt to highlight to the client the capacity design has to go beyond the low-value end (making things merely convenient or usable); they might make a case for design's core value as a process for meaningfully dealing with complex problems with the result of conceiving and formulating new ways of being. In our case, these new ways of being could be systemic solutions for urban mobility that extends human capabilities for moving around.

This could be critiqued as being arrogant. Why can't the designer just do as requested in the design brief? 'The client simply asked for a scooter service, geez!' a critic could lambast. Naturally, a designer *could* do that, but a responsible designer would not because they have developed a willingness to engage in this negotiation so that they can appropriately deal with the multiple demands that need to be weighed. This does not necessarily mean that the designer should try to convince the client to forget about scooters and go for a full-blown mobility service that rethinks urban mobility from scratch. This would be absurd.

A more reasonable designer could, realising the danger that unattended scooters represent for other footpath users, start by challenging the exclusive focus on the end-user's convenience to include the safety of other footpath users as well. A responsible designer would thus attempt to go beyond the original request, and question and extend its premises whenever necessary in order to pursue what design can do and be at its best[11*]. In some design disciplines such as user experience design and service design, the brief formulated by the project commissioners is standardly reinterpreted and reformulated by the designer (or the design firm to which they belong) to convey a broad set of initial hypotheses, objectives, and methodological approaches; in the Spanish and Italian design culture, this reformulation is traditionally called 'contra-brief' ('counter-brief'). So far, I have referred to reformulations that occur at the initial stage, but the same can be said about a situation deep in the design process when new aspects from the brief emerge that call for being challenged; this could happen as a consequence of conducting user testing, for instance.

Regardless of when this reformulation occurs, it is a product of a value-laden analysis, not a 'neutral' one. Philosopher and computer scientist Paul Dourish poses the notion of the 'designer's stance' to refer to the designer's own conception of what they are doing, to the particular perspective they adopt regarding the role they play in a given design situation.

A responsible designer adopts a *caring* stance, which, as I argued earlier, includes a sense of being truly invested in the outcome that much exceeds what one directly causes or the obligations one has entered. This stance also exceeds the set of obligations that arise from professionalism (that is, its public service element). When a responsible designer cares about something they seek to act as to ensure that 'the caring actually occurs'[12]. This is why a responsible designer seeks to broaden the space for responsibility: it is because it personally affects them if the object of care is benefitted or harmed; it has to do with a sense of integrity and identification with what they care about. It is a way of self-enactment.

The willingness and readiness to care about others is only the first step, it is by extending the space of possibility that they are able to transform this caring *about* into caring *for*. To reiterate, the responsible designer (in the virtue sense) is not moved by duty, but because they care, and they care first and foremost because caring is a constituent part of who they are and who they aspire to become. Caring is characteristic of how they engage in the pursuit of the internal goods of design, and, more generally, in the narrative quest of becoming who they want to be. As I argued previously, caring in this sense is not *necessarily* (though it can be) selfless, altruistic, or out of empathic concern for others, as we will see shortly; it is, however, unquestionably personal.

EMPATHY AND MORAL IMAGINATION

Philosopher Nel Noddings argues that 'we can . . . "care about" everyone; that is, we can maintain an internal state of readiness to try to care for whoever crosses our path'[13], as the example of the scooter illustrates. However, caring in the intended sense is not only caring about but caring for: *caring is an actuality not a possibility*[14]. A responsible designer cares by realising that other users of the footpath may be hindered or even put in a risky situation by riders leaving the scooters in the middle of the footpath; that these footpath users are likely to be complete strangers is fully irrelevant to a responsible designer.

Nihlén Fahlquist argues that we are able to care thanks to an ability to empathise with others and to exercise moral imagination[15]. Along these lines, Noddings argues that 'caring involves stepping out of one's own personal frame of reference into the other's'[16]. Empathy is what enables the designer to emotionally grasp how the different parties involved in a situation would be affected by the designer's actions. This grasp, which Noddings calls a 'feeling with'[17], is not about projecting oneself into the other by asking 'How would *I* feel?' in order to understand them. On the contrary, it is about receiving, not projecting: 'I receive the other into myself, and I see and feel with the other. I become a duality'[18]. This commitment to receiving the other's feeling is what enables us to truly put ourselves in the other's shoes.

Design researcher Indi Young offers an alternative understanding of empathy that is inscribed in the framework of user-centred design. Here, 'empathy' is understood as a cognitive ability rather than as an emotional one. It allows one to gain insights into the point of view of others in order to understand what makes different people tick. She refers to this capacity as 'practical empathy', which is 'about understanding how another person thinks—what's going on inside her head and heart'[19]. Although she offers compelling reasons to adopt the practical methods she proposes, she is not equally successful in showing in what way one could 'comprehend another person's . . . emotional states'[20] without resorting to one's emotions. Naturally, one could understand *that* somebody feels sad and *why*, but one would have a difficult time truly comprehending the very idea of sadness without involving one's own emotional experience of it.

Be that as it may, empathy (whether conceived as a cognitive skill, a capacity, an attitude, or an emotional state) is important because it is intertwined in the process of moral imagination, which business ethicist Patricia Werhane defines as 'the ability to discover and evaluate possibilities within a particular set of circumstances by questioning and expanding one's operative mental framework'[21]. On the one hand, moral imagination involves envisioning how a decision one makes might harm others. On the other hand, it is the development of a moral outlook that will sustain envisioning alternative scenarios that will seek to avoid or minimise those potential harmful or less desirable outcomes. In short, moral imagination consists of two activities: first, gaining an empathic awareness of what is going on from the ethical perspective; and second, being able to envision and evaluate possible courses of action from different perspectives. It is not about generating wild speculations but feasible possibilities and ideals that *could* become a reality; Werhane asserts that 'such possibilities have a normative or prescriptive character; they are concerned with what one *ought* to do'[22].

All this is an important departure from conventional ethical theories such as Kantianism or utilitarianism that are ruled by the application of rational principles and exclude feeling and imagination in the formation of moral judgements.

Is Empathy a Necessary Motive?

Care theorists have pointed out that empathy is a 'valuable—or even essen-tial—tool for developing our understanding of others and enabling us to determine what the best thing to do is in real world situations'[23]. Fully acknowledging this valuable role, I want to build on the suggestion I made in the section titled 'Responsibility as a Virtue' and argue that empathy is not *necessarily* the designer's primary motive for caring. To express it the terms of our previous example: it is not necessarily from a motive of empathy that a responsible designer cares about the users of the footpath that may be put in a risky situation by a scooter rider.

Let us explore a new example that will serve to substantiate this claim. Design scholar Laura Forlano recounts how she was kept awake at night by her artificial intelligence (AI)-driven insulin pump (used for Type 1 diabetes) after she started using it. Like every AI system, the pump needed initial training and calibration, but the pump required calibration up to twenty-five times on certain days, and even eighteen months after having started using the device, Forlano was still being woken up at night for 'a 3am calibration' (and for a myriad of other technical issues such as low batteries). People with Type 1 diabetes 'are already burdened with additional work in order to go about their day-to-day lives. . . . A system that unnecessarily adds to that burden while also diminishing one's quality of life due to sleep deprivation is poorly designed, as well as unjust and, ultimately, dehumanizing,' she con-cludes[24].

Imagine being the designer in charge of redesigning this insulin pump. There is one scenario in which empathy is indeed the motive for caring. In this scenario, you empathise with the person using the pump and 'feel with' them what it is to be constantly woken up during the night and disturbed by constant alerts during the day. This empathy is your primary motive for caring (or willing to care). A different scenario does not require empathy. Here, you still care just like in the previous scenario, but you care because of a different motive than empathy: you care because your profession and achieving its internal goods matter to you and you feel personally responsible for how the outcomes of your work affect the user. I will develop this idea in the following.

Empathy has been defined as 'an affective response more appropriate to another's situation than one's own'[25]; we see, however, that in the second scenario your affective response is not primarily about *another's* situation, but about *your own*. You care about the user because you find it unacceptable that a person gets woken up in the middle of the night to replace batteries. As the designer in charge of the redesign, you feel a sense of (forward-looking) responsibility over the pump and outright reject the possibility of a user being woken up every hour to recalibrate the device. When you speak to your client or project manager, you try to make the point that this is a major design failure that needs to corrected before the product is made available to the public.

In the second scenario, as opposed to the first one, you feel responsible and willing to care because you care about good design: an insulin pump that behaves that way is poorly designed. The point here is that empathy is not necessary to arrive to this conclusion. The pump is also unjust and dehumanising, as Forlano rightly claims. And all this matters to you as a professional designer because the pursuit of good design matters to you *personally*. You consider that unusable, unjust, and dehumanising design is bad design; thus, it is something that must be avoided at all costs.

The overarching purpose that we proposed in the previous chapter tells us what is wrong with the pump from an ethical perspective: although the pump increases the person's capability for good bodily health, at the same time it harms and diminishes other central capabilities such as psychological health and being able to use the senses, to think, to reason, and to rest (Forlano explains that she felt 'groggy and delirious' after being awakened no less than five times by various alerts)[26]. This roughly tells the designer that the design for the pump in its present form is far from what it could be at its best. Second-order regulative ideals can also guide a responsible designer toward caring for users by providing standards of excellence (for instance, a design should treat the user 'humanely') and standards of correctness (for instance, a pump should not wake up its user to replace batteries; especially if it is driven by AI, it should alert the user about this at a more reasonable time). Interestingly, in both scenarios, whether based on empathy or on regulative ideals, the overall direction for the redesign could be the same: redesigning the pump device as to eradicate or at least minimise the burden the device adds to the person with diabetes, while of course enabling them to effectively control it.

The second scenario also illustrates how a designer can exhibit a readiness to care without empathy being the motive behind it. A responsible designer would care for the user because good design matters to them personally, as does seeking to achieve what design can be at its best (that is, extending human capabilities).

Some Problems with Empathy

There is a cautionary warning that can be made about having empathy as the motive that prompts us to care. Empathy is unreliable as a starting point because people can develop tolerance for faulty designs and put up with ugly, dangerous, unjust, unsustainable, and non-inclusive material and immaterial artefacts. This should not be surprising: in all spheres of life, we all occasionally bend and adjust our aspirations and expectation in light of experience.

Social scientists call this phenomenon 'adaptive preferences formation': people non-consciously adapt to the life they have and adjust their wants to what seems possible[27]. In other words, empathy might not always be a reliable prompt for care and responsibility because when people are happy and satisfied with dangerous or unjust designs it might be difficult for a designer to start to empathise because people would seem to be doing just fine.

As capabilitarian scholars have taught us, a 'person may be in a desperate situation and still be contented with life if she has never known differently'[28]. So looking at happiness or satisfaction of preferences might also hide important dimensions of analysis[29]. Imagine a designer interviewing a person with diabetes who is not that much bothered by the constant alerts and the 3 am calibrations of the insulin pump we discussed before. Because sleeping is something that everyone does, the designer might still develop empathy by projecting how they would feel in the same situation, which is the total opposite of 'feeling with' others that we discussed earlier. This type of empathy that is not based on true encounters with others but in self-reference can be a slippery slope that paradoxically leads to even more self-referential decisions, as research shows[30].

Conversely, a responsible designer might start to empathise (and care about) *because* they realise that the device harms capabilities (psychological health, clear thinking, resting, and so on); it is because they are guided by regulative ideals centred around relevant capabilities that they are able to perceive that a design may be unjust *even* if the user does not complain about it. Naturally, this designer might develop empathy as their understanding of what is truly wrong with the pump increases. We see, thus, that thinking in terms of extending capabilities (or at least preserving them) can be a more reliable guide for professional ethics than empathy alone.

The upshot of all this is that whereas responsibility as a virtue is directly connected to a disposition to care about and a readiness to care for others, it seems reasonable to claim that empathy needs not be the primary motive for caring. It might also be reasonably claimed that truly caring for the telos of the practice can also provide a plausible motive for caring about and for others. This can occur insofar as the practice and its internal goods truly matter to the designer, and seeking to attain these goods is constitutive of the person they are or aspire to be. It is in the quest of pursuing the internal goods of design that the designers are furnished with 'increasing self-knowledge and increasing knowledge of the good'[31].

To illustrate, for a designer acting responsibly starts with becoming aware that a device that wakes up its users at 3 am to replace the battery is bad design, and it is bad design because it prevents people from sleeping, which is a central capability that affects other central capabilities (being able to think clearly, for instance). This realisation provides the designer with sufficient motivation to care for the situation and to feel the responsibility to act.

It is, thus, in the pursuit of the telos of design that the designer becomes empathetic and develops the emotional ability that is necessary to instantiate the readiness to care into actual designs that extend others' capabilities and contribute to their flourishing.

To reiterate, *empathy is clearly important*. My intention for this section is not to deny its key role in responsibility as a virtue, and more generally in ethical reasoning. But as I argued previously, empathy may be unreliable at times. My goal is to present an account of how designers can care for human capabilities and flourishing without necessarily acting from a motive of empathy. This difference is important because it highlights the role professional regulative ideals play. My account is structured around the idea that the importance a profession has for the designer *as a person* can operate as a sufficient motive for acting responsibly (which includes both a readiness to care about, and a disposition to convert this readiness into actual care). But, having explained motivation and readiness, we are only halfway along on the road to explaining responsibility as a virtue; in the following section, we will turn to what the notion of practical reason means for the process of transforming this readiness into an actuality.

DESIGN COMPLEXITY AND PRACTICAL WISDOM

Having a disposition to care about how our designs affect others requires being open to notice that something is the matter before knowing *what* is the matter. But there is more. Just like with any other virtue, responsibility is not a mere disposition; a virtue consists of a behavioural element that aids the designer in converting a readiness into concrete actions. Responsibility is ultimately about seeking to actualise that willingness to care and convert it into an actual design[32*].

Because responsibility as a virtue calls for going 'all the way' from 'caring about' to realisation, it places limits on the number of things a designer can undertake; if only due to operational reasons (most unarguably, time), they cannot realise all the things they care about. At the same time, the designer faces general project constrains (budget, technologies, and so on) and needs to deal with a large number of people who are involved and have a say in the design process. It is because of all this that 'to be responsible, one needs to care in a relatively focused and effective way'[33].

The virtue of responsibility is typically called for in complex situations where plural demands are made on the professional, which highlight the need for a negotiated response. Design scholars Adam Thorpe and Lorraine Gamman offer an eloquent description of the dynamics in which a 'responsive' designer operates (we will return to the notion of 'responsivity' in the next section):

> Appreciation of context requires designers and other actors to be mindful of competing resource requirements, goals and needs, and to be able to consider and decide which factors are to be prioritised in the design response (given that drivers are sometimes contradictory) [34].

What is needed thus is more than simply weighing pros and cons in a linear fashion according to one dimension, but to find a way to form an appropriate response by balancing specific circumstances, contexts, and conflicting interests that may be incommensurable to each other. Rather than washing away these tangles by attempting to apply a moral algorithm, responsibility as a virtue gives the designer what philosopher Larry May describes as 'a wide discretion concerning what is required to be a responsible person' [35].

Being responsible asks, in turn, for being able to accept the inherent unresolvability of certain dilemmas and the intrinsic uncertainty of some situations. Business thinker Roger Martin speaks of 'integrative thinking' to refer to the 'metaskill of being able to face two (or more) opposing ideas or models and instead of choosing one versus the other, to generate a creative resolution of the tension in the form of a better model, which contains elements of each model but is superior to each (or all)' [36].

To respond appropriately to a situation, the designer needs to produce this type of 'integrative thinking', which requires the right intellectual and emotional mindset. In most cases *there is no one single best response that can be produced*, as what is best will depend on how the different criteria are balanced (and different reasonable designers will balance criteria in different ways). Conceivably, in some situations the best course of action might be *not acting*. Designers might do well to realise that they, at least the proficient ones, are already able to think like this in their daily work when they deal with design problems, which are commonly not only complex and ill-defined, but often also indeterminate and even 'wicked'. 'Integrative thinking' is a key feature of design thinking; designers just need to learn to extend the scope to include the ethical dimension of design.

Making Wise Decisions

The ability that is required to effectively and virtuously deal with a complex situation is called 'practical wisdom' (*phronēsis*, in Greek), a notion we encountered briefly in chapter 6. From a philosophical perspective, wisdom can be understood as a virtue that 'reinforces theoretical reason in its intellectual recognition of reality' [37] allowing us to better comprehend the world. Wisdom is about the knowledge and judgement about the meaning of life, and how to live well and flourish.

Practical wisdom is concerned with determining what is good and right in each situation. According to education scholar Stephen Kemmis, it 'consists, first, in a preparedness to think critically and understand a given situation in *different ways*, and not to accept immediately that the situation is what it appears to be'. It also entails a preparedness for being open to wholly *'new ways of understanding a situation'*, and to being open to *'experience itself . . .* by trying out a new way of being in the world'[38].

Some decision theorists describe a similar process as 'sensemaking': the mental effort to gain awareness and understanding of a given situation or event. Sensemaking is the constructive, iterative task of fitting inputs from the world into frames of understanding in order to interpret the world and give meaning to one's experience[39].

Besides this capacity to think critically about situations and what they are, for Kemmis, the person that has the virtue of practical wisdom has the capacity to:

> think practically about what should be done under the circumstances that pertain here and now, in the light of what has gone before, and in the knowledge that one must act (and that even not acting, or not appearing to act, may be the right action)[40].

These characterisations and definitions resonate with many of the descriptions of how expert designers think that have been generated in the last three decades, which we briefly reviewed at the beginning of the book when we touched upon the reflective, situated, and interpretive nature of design. What was said of design problems is also true of situations that call for practical wisdom; these situations are normally *fuzzy* and are understood and framed differently by different people, whose own different internal configuration of values determine a different balance of conflicting elements, and a possibly different response to the situation.

Practical wisdom is what enables the designer to transform care and moral imagination into actions by prioritising their caring efforts and judiciously balancing the different concerns[41], directing the designer toward their purpose. Drawing in their professional expertise and practical wisdom, the virtuous designer aims to make a judgement about what to do; a judgement that, while being contingent on the situation's many variables, is at the same time consistent with the different regulative ideals that guide their behaviour. In this way, a key component of practical wisdom is self-reflectively examining whether one has the best reasons for choosing a given course of action. We see how for a responsible designer, a design decision *always* involves ethical considerations either implicitly or explicitly; ethics and professional design are not on different layers—they are glued together by practical wisdom. When a responsible designer makes a design decision, they are also considering the relevant ethical aspects of a situation, as well as the technical, aesthetical, operational, and technological ones.

Along these lines, many design methods and techniques could be seen to foster practical wisdom for design. They inform and enable the designer to think critically about a design situation and aid them in understanding it and in producing an appropriate response. To illustrate in the terms Kemmis uses to describe practical reason: traditional user research methods such as interviews or surveys aim to inform *different ways* of understanding a situation; similarly, methods like 'Personas' or SWOT analysis serve a similar goal[42]. Approaches like 'frame creation',[43] 'co-creation'[44] practiced at the early front end of a project, and participatory research methods like ethnography seek to provide wholly *new ways* of considering it. Scenario-based design, speculative design, and co-design methods such as the 'diffuse design strategies'[45] discussed by Manzini offer designers (and non-designers) ways to *experience the world* in new ways. In this manner, practical wisdom becomes intertwined with design methods, rather than something we exercise apart from them.

To go back to a previous example, the designer that is hesitant about whether to include a field for gender in a form tries to decide and balance not only the possibly conflicting interests of the different parties involved (most notably, the client's and the end-users'), but also legal regulations, technical, functional, or aesthetic standards, guidelines, and best practices. What is more, the designer also needs to integrate and balance other considerations that are truly important to them as a person, which arise from other roles the designer has in all the spheres of their life (as a parent, a citizen, a friend, a colleague, an employee, and so on). Remember Alex, the designer that was troubled by designing for an online bookmaker targeting people in lower-earning brackets, ex-gamblers, and young people. In both cases, what needs to be negotiated is not only others' demands, but one's own demands as well.

Alex was troubled by collaborating with an online bookmaker precisely because they sensed that associating gambling with sex, and presenting it as a way out of poverty, is a matter that calls for moral deliberation. The ethically salient features of a situation are often highlighted by an emotional reaction; so, for instance, uneasiness, outrage, and indignation can indicate the presence of injustice or the risk of harm to others. Aristotle famously argued that our emotions are central to virtue: 'to have them at the right time, about the right things, towards the right people, for the right end, and in the right way, is the mean and best; and this is the business of virtue'[46]. Naturally, an emotion is not in itself proof that something is truly *wrong* from a moral perspective; we may be disgusted at something or somebody, but it may very well be that it is our emotions which are misplaced. As they often are! Nevertheless, affective responses are useful because they can ignite a moral reflection.

In some cases, our affective response might result in the immediate stipulation of a course of action *without* the need for conscious deliberation. This can happen when people lacking practical wisdom simply act without thinking about what they are doing; in this case we would call this person reckless and careless. Conversely, it is also possible that a person reacts without conscious deliberation because their practical wisdom is strong, which enables them to assess the situation and act intuitively[47]*. In this case, behaviour flows naturally and the actions fit the demands of the situation without the need for deliberating. When wise people deal with typical situations, they are not 'making' decisions but carrying out actions that are likely to be successful. In contrast, when situations are novel or atypical, expert deliberative judgement becomes truly important[48].

As we see from the previous examples of the hesitant and the troubled designer, the different concerns that they need to juggle with are not only design related. The relevant concerns might come from other practices to which they belong and from notions about what is good for them, for others, and for the planet as whole. A responsible designer aims to balance the conflicting concerns *precisely* because they seek to resist living a compartmentalised life, where each sphere of activity is separated out. Practical wisdom is thus tied up to our personal and professional purposes. Along these lines, bioethicist Daniel Hall argues that practical wisdom is 'explicitly value laden because it functions only in relation to thick notions of what human beings are meant to be and become'[49].

To sum up, practical wisdom consists of several intertwined elements: noticing what is important in a given situation with respect to the good, deliberating on what needs to be done, reflecting on ways to carry that out and what possible consequences could ensue, and adopting a reasoned course of action. All this prepares us to take moral responsibility by binding together care and moral imagination. In the words of Kemmis, 'the virtue of *phronēsis* is thus a willingness to stand behind our actions'[50].

TWO OBJECTIONS: 'RESPONSIVITY' AND PATERNALISM

I have just presented how responsibility as a virtue can serve to structure a professional ethics for design centred on practices and internal goods. In this section, I will discuss two possible challenges to this account: first, an objection to the very possibility of responsibility in design, and second, an invalidation of the account for being paternalistic.

Can a Designer Ever Be Responsible?

Considering the way designers are subject to compromise, Thorpe and Gamman question whether a designer can be responsible at all. Their challenge affects both market-led design and what they call 'socially useful' design. Their attack on the very possibility of responsibility in relation to market-led design emphasises the designer's lack of control over resources 'to set and fulfil their own agendas', which leads to results that serve the interests of the market over those of society[51]. Responsibility is questioned in 'socially useful' design on grounds of the compromises that designers need to make over both processes and products with other stakeholders[52].

Instead of responsibility, they argue for the notion of 'responsivity' to account for what is necessary for appropriately dealing with complex design situations. This type of situation is 'characterised by competing and often contradictory drivers, and scenarios in which there are multiple "correct" answers to design problems, depending on context and stakeholder perspective'[53]. A 'responsive' designer is 'one who is acting to effectuate societal change with available collaborators and resources, and settling for the best that can be achieved in a particular context'[54]. Along these lines, Ezio Manzini describes a 'responsive' person as someone who acts in informational and systemic complexity and is ethically sensitive to the context[55].

Thorpe and Gamman's arguments and their description of overall design dynamics (included in the section titled 'Design Complexity and Practical Wisdom') nicely fit within the account of responsibility as a virtue that is defended in this chapter; it also matches the claim that a designer needs to balance a 'plurality of normative demands'. However, some might reject this similarity and posit that Thorpe and Gamman explicitly present responsivity as something opposed to responsibility when they find the 'pluralism and adaptability of the designer's role to be crucial and one of *responsivity* rather than *responsibility*'[56].

Yet Thorpe and Gamman's rejection of responsibility might rest on a different (though not erroneous) understanding of responsibility than the one I have been arguing for in this chapter. They somewhat equate responsibility with 'control over resources that would allow them to set and fulfil their own agenda'[57]or 'agency over the outcomes and impacts'[58]. In my view, the dichotomy these authors highlight need apply only to responsibility as obligation and as attribution. Accordingly, the opposition they stipulate between responsivity and responsibility does not include responsibility as a virtue, which is, as far as I see it, roughly congruent with their description of responsivity.

These differences in meaning can be further clarified by analysing an assertion by designer Milton Glaser, who contends that given designers' access to production and manufacturing, they have a 'unique opportunity to have a different role than an average person, [because of this] there is more opportunity and more responsibility'[59]. Thorpe and Gamman might reply that such a view would look 'delusional and ultimately naive', as it ignores 'the power structure that informs design production'[60]. If we understand Glaser's usage of 'responsibility' in the sense of having a duty to fully control something, then the view I just attributed to Thorpe and Gamman would be correct.

However, if we understand it as a virtue, then responsibility can be retained without implying that a designer is necessarily responsible *for* an outcome in the sense of having full control over process and results, if this degree of control is possible at all.- Be that as it may, when responsibility is understood as the readiness and willingness to convert care into design, the possibility of responsibility does not seem farfetched and seems viable *even* in a market-led environment. Being responsible is hence a way of self-enactment. It is being invested in one's profession, caring for its internal goods, caring for other's flourishing because that is constitutive of pursuing one's life narrative quest. It is not about having full control of the process, nor about eschewing compromise.

That designers do not have full control over the outcomes of the processes they advocate, and that compromises are necessary, does not entail that they cannot be responsible in the sense intended here[61*]. What is more, responsivity can be said to be at the root of responsibility as a virtue. Virtuous behaviour is precisely about being able to juggle conflicting demands, to reach good compromises, seek to create opportunities for action, and to effectively do something.

At the same time, because alternative courses of action are presupposed, it is up to designers to produce a selective response. And what counts as virtuous behaviour is, with some exceptions, impossible to determine beforehand separately from a situation[62*]. Yet this does not mean that they can act arbitrarily; on the contrary, it means that they could be said to be acting virtuously only if they are moved by the right *reasons* for acting in a particular way.

Thorpe and Gamman are spot on when they claim that 'the notion that the designer should "have all the answers" or skills to resolve complex and "wicked" problems . . . is *unrealistic*'[63]. A responsible designer may have the readiness and willingness to deal with the plural demands made on them, but their actual ability to transform these demands into design is bounded by the actual opportunities they have to do so. The opportunities depend on internal factors, but also on external constraining (but also enabling) conditions that affect the designer's endeavours: macro socio-political issues, operational and technological concerns, availability of resources, and so on. Institutional dynamics is also a key factor, and I will touch on this topic in the conclusion of this book. Likewise, even though designers occupy a prominent role in how products, services, or environments get from abstract idea to concretion, there are many other professionals and technicians involved in this process; designers are never the sole actors in the design and development process at large.

However, that a designer's actual opportunities are constrained by economic, political, institutional, and social arrangements is not a sufficient invalidation of responsibility either. Although the notion of responsivity is useful to emphasise the readiness to respond that is characteristic of responsibility as a virtue, the arguments Thorpe and Gamman put forward against responsibility are not a bar to being responsible. It is precisely in difficult situations where the designer needs to respond to a plurality of normative demands where the virtue of responsibility is called for. It is in unclear, unstable situations where much is at stake that the designer develops the ability to navigate complexity and uncertainty. At the same time, it is precisely in a constrained situation where a designer can develop their practical wisdom as they decide what is good, what *needs* to be done, and what *can* be done.

Paternalism

Responsibility is presented here as a virtue that designers who are invested in their professions exhibit, which guides the reflection about how they can best pursue their professional purposes. In turn, this reflection is aided by regulative ideals concerning different design dimensions (technical, aesthetical, ethical, etc.). The overarching purpose that was proposed for professional design coalesces around contributing to others' wellbeing. The second objection has to do with this explicit concern, which could be deemed paternalistic as it could interfere with others' agency[64].

Paternalism is the interference with another person against their will, with the aim of promoting that person's good or wellbeing. A difference can be stipulated between 'soft' and 'hard' paternalism[65]. Soft paternalism restricts the freedom of a non-autonomous person because it judges that their decision making capacities are insufficient. For example, preventing a person from crossing the road when we realise there is an oncoming vehicle just out of the person's sight in order to protect the person from being run over. Soft paternalism is also directed to people assumed to lack autonomous decision making on the whole (such as children or people with Alzheimer's disease). Conversely, hard paternalism restricts the freedom of autonomous agents to protect them from their own voluntary choices. A classic example is the prohibition that exists in many countries concerning the use of recreational drugs.

The issue of paternalism is vast, and there are many other nuances that could be discussed, but I believe this rough characterisation provides a sufficient introduction to deal with the objection; I do not mean to solve the problem of paternalism in design, which is an issue that would need a more extensive treatment. I will only make the weaker claim that an explicit concern for others' wellbeing is not *necessarily* paternalistic.

You probably remember the example of the housefly etched inside of urinals to prevent spillage. This design, which nudges people to putatively good behaviour, could be seen as a prime example of paternalism. Defenders of this type of paternalism (called 'libertarian paternalism') argue that design is simply making things easier for people by facilitating making the socially beneficial choice. They argue that whereas a person is steered toward choices that are supposedly in their (or the public's) interests, the person retains the freedom to act otherwise. Like in all paternalism, however, a conception of the 'good' that is promoted is required. The obvious question that could be raised is: who gets to decide the good, and based on what criteria or standards?

This shows that paternalism, even in its soft variety, is indeed a legitimate ethical concern. It may be an effective tool to advance socially beneficial causes, but it does interfere with human agency; even defenders of paternalism share the view that because of that it needs to be justified by sound reasons[66].

Arguably, design can be operationalised for hard paternalism too, which strengthens the objection. Automated systems that allow you or deny access to services and goods can serve to illustrate how design can serve hard paternalism. For instance, take Twitter's moderation policy, which with the aim of curtailing potentially abusive and manipulative content about COVID-19 uses artificial intelligence to identify rule-breaking content and asks its users to delete tweets that were labelled as potentially harmful. Some accounts are 'challenged' automatically when the likelihood of abuse is deemed high, whereas other content that is harder to analyse is reviewed manually before the users are asked to remove it. This happens, for example, to tweets that include 'denial of global or local health authority'[67].

A fictional example: imagine in a not-so-distant future that you are no longer able to buy a 750-ml sugar-sweetened soft drink because your credit card is denied due to instructions from your wearable health monitor which did not authorise the transaction. The designers of such system clearly want what is best for you! And they act to realise that aim: they design a system that makes it impossible or very hard for you to do something that is bad for you[68*].

Designing to putatively promote others' wellbeing while neglecting that a necessary condition for a good human life is to be able to *decide for oneself what the good is* would be a misdirected endeavour. To negate the possibility to critically reflect on how one's life should be lived is to contribute to a negation of human agency. Shannon Vallor writes, 'It is essential for human agency that our moral practice . . . remains our *own conscious activity and achievement* rather than passive, unthinking submission'[69].

Justifying hard paternalism is thus more problematic because it restricts the agency of substantially autonomous persons in order to advance their wellbeing against their own will. The writer C. S. Lewis (1898–1963) persuasively illustrates what the worst-case scenario of hard paternalism can be:

> Of all tyrannies, a tyranny sincerely exercised for the good of its victims may
> be the most oppressive. It would be better to live under robber barons than
> under omnipotent moral busybodies. . . . [T]hose who torment us for our own
> good will torment us without end for they do so with the approval of their own
> conscience[70].

The charge of paternalism looks hard to overcome. A designer that embarks on paternalism (especially of the hard type) may be undermining themself methodologically: in order to promote flourishing, it undermines it by restricting agency.

But there is a way out of this tangle that does not require solving the problem of paternalism. As I advanced previously, an explicit concern for others' wellbeing is not necessarily paternalistic provided these others are involved in and have a sufficient say on the definition of the comprehensive goals that are pursued in a given design project. This is where the Capability Approach becomes especially relevant in a professional ethical sense. The approach can function both as a guide toward excellence and as safety net, which can help designers avoid the perils of paternalism, helping them do well as they pursue the good and preventing them from 'doing spectacularly bad things'[71].

Aiming to achieve the standards of excellence that are at the root of the practice of professional design requires designers to discover and understand what specific capabilities need to be extended through artefacts. It is, however, not really up to them to prescribe specific ways of being that are alien to people's life goals by camouflaging functionings in their designs. Doing this would be very similar to hard paternalism. Capabilities are about having real opportunities to live the life people have reason to value; that is, the reasons must be theirs.

A responsible designer would seek to promote and extend those capabilities that are relevant to the people that will be affected by and interacting with the artefacts they design. Comprehending what capabilities are worthy of being promoted is what enables the designer to grasp how they can truly contribute to others' flourishing. This comprehension can be best achieved by operating from a stance of 'receiving-not-projecting' and 'feeling with'. Participatory and open-ended cooperative co-design approaches that involve relevant actors and stakeholders are especially suitable for that purpose.

This does not mean that a designer needs to start a project with a clean slate with regard to what flourishing consists of. A virtuous designer would be able to adjust their pre-existing conception as they learn from others during the development of a design project. This is a by-product of 'feeling with' others, of engaging in moral imagination, and of exercising practical wisdom: by mentally simulating and playing out scenarios from the vantage point of others who may be radically different than us, and by balancing the different concerns that are relevant in a given situation, our conception of the good evolves, and so do our notions of what it means to flourish as a human being. This is clearly a process of learning and self-enactment, but it is also a transformational one. Thinking practically about what should be done and why, and ultimately committing to a course of action, is also a journey of self-growth. Needless to say, this is one of the most precious internal goods design can offer designers.

In this chapter we have explored the notion of responsibility as a virtue and what it might mean for a designer to be responsible, developing a view that focuses on care as a central element. In this account, responsibility as a virtue serves to structure a professional ethics for design that is centred on practices and internal goods. Without in the least eschewing the role that empathy plays in care, a view is proposed in which the importance a profession has for the designer *as a person* can provide a sufficient motive for acting responsibly. Practical wisdom is presented as the virtue that enables the designer to journey from a basic awareness about the ethical saliency of a feature in a design situation to the concrete realisation of the willingness to care, which is instantiated in artefacts that contribute to others' wellbeing.

NOTES

1. Debbie Millman, *How to Think Like a Great Graphic Designer* (New York: Allworth Press, 2007), 32.

2. Geoff Moore, *Virtue at Work: Ethics for Individuals, Managers, and Organizations* (Oxford: Oxford University Press, 2017), 80. Italics in the original.

3. Alasdair MacIntyre, *After Virtue: A Study in Moral Theory* (Notre Dame: University of Notre Dame Press, 2007), 220.

4. Jessica Nihlén Fahlquist, 'Responsibility as a Virtue and the Problem of Many Hands', in *Moral Responsibility and the Problem of Many Hands*, edited by Ibo Van de Poel, Lambèr Royakkers, and Sjoerd D. Zwart (New York: Routledge, 2015), 194.

5. Harry G. Frankfurt, *Necessity, Volition, and Love* (Cambridge: Cambridge University Press, 1999), 111. Italics in the original.

6. Dina Meyers cited in Larry May, *The Socially Resposive Self: Social Theory and Professional Ethics* (Chicago: The University of Chicago, 1996), 96.

7. Garrath Williams, 'Responsibility as a Virtue', *Ethical Theory and Moral Practice* 11, no. 4 (2008): 459.

8. Ibid., 462–69.

9. Ezio Manzini, *Politics of the Everyday* (London: Bloomsbury, 2019), 54.

10. Ibid., 55.

11. Although self-employed designers do in principle have the freedom to question the design brief, which is even expected by clients and is a standard methodological step, the question remains to what extent designers working within the consultancy model have the freedom to alter it *radically*. Designers embedded in organisations might have a better chance

at succeeding in this task. For a discussion of this issue, see Dan Hill, *Dark Matter and Trojan Horses: A Strategic Design Vocabulary* (Moscow: Strelka Press, 2012).

12. Nel Noddings cited in Nihlén Fahlquist, 'Responsibility as a Virtue and the Problem of Many Hands', 194.

13. Nel Noddings, *Caring: A Relational Approach to Ethics & Moral Education* (Berkeley: University of California Press, 2013), 18.

14. Ibid.

15. Nihlén Fahlquist, 'Responsibility as a Virtue and the Problem of Many Hands', 194; May, *The Socially Resposive Self*, 88.

16. Noddings, *Caring*, 24.

17. Ibid., 30.

18. Ibid.

19. Indi Young, *Practical Empathy: For Collaboration and Creativity in Your Work* (New York: Rosenfeld Media, 2015), vii. César Astudillo ignited this reflection and pointed her work out to me.

20. Ibid., 18.

21. Patricia Werhane, *Moral Imagination and Management Decision Making* (Washington: Business Roundtable Institute for Corporate Ethics, 2009), 4.

22. Ibid. Italics in the original.

23. Amy Coplan and Peter Goldie, *Empathy: Philosophical and Psychological Perspectives* (Oxford: Oxford University Press, 2011), xxvii.

24. Laura Forlano, 'The Danger of Intimate Algorithms',https://www.publicbooks.org/the-danger-of-intimate-algorithms/.

25. Martin Hoffman cited in Coplan and Goldie, *Empathy*, xxiii.

26. Forlano, 'The Danger of Intimate Algorithms'.

27. Jon Elster, *Sour Grapes: Studies in the Subversion of Rationality*, Cambridge Philosophy Classics edition (Cambridge: Cambridge University Press, 2016), 25.

28. Ingrid Robeyns, 'Sen's Capability Approach and Gender Inequality: Selecting Relevant Capabilities', *Feminist Economics* 9, no. 2-3 (2003): 63.

29. Ibid.

30. Johannes D. Hattula, Walter Herzog, Darren W. Dahl, and Sven Reinecke, 'Managerial Empathy Facilitates Egocentric Predictions of Consumer Preferences', *Journal of Marketing Research* 52, no. 2 (2015).

31. MacIntyre, *After Virtue*, 219.

32. Note that the phrase 'actual design' does not necessarily entail a design for a 'new' material or immaterial artefact that is simply *added* to the world. An actual design can also be the reconfiguration or the redesign of what already exists to make it more environmentally and socially sustainable. This idea is related to what Tony Fry has called 'elimination design' and Cameron Tonkinwise 'undesigning'. Tony Fry, *Design Futuring: Sustainability, Ethics and New Practice* (Oxford: Berg, 2009), 71–75; Cameron Tonkinwise, '"I Prefer Not To": Anti-Progressive Designing', in *Undesign: Critical Practices at the Intersection of Art and Design*, edited by Gretchen Coombs, Andrew McNamara, and Gavin Sade (Milton Park: Routledge, 2018).

33. Nihlén Fahlquist, 'Responsibility as a Virtue and the Problem of Many Hands', 196.

34. Adam Thorpe and Lorraine Gamman, 'Design with Society: Why Socially Responsive Design Is Good Enough', *CoDesign* 7, no. 3-4 (2011): 219.

35. May, *The Socially Resposive Self*, 88.

36. Roger Martin, *The Design of Business* (Brighton: Harvard Business School Press, 2009), 165.

37. Domènec Melé, *Business Ethics in Action* (London: Red Globe Press, 2020), 42.

38. All quotes from Stephen Kemmis, 'Phronēsis, Experience, and the Primacy of Praxis', in *Phronesis as Professional Knowledge: Practical Wisdom in the Professions*, edited by Elizabeth-Anne Kinsella and Allan Pitman (Rotterdam: Sense Publishers, 2012), 155. Italics in the original.

39. Gary Klein, Jennifer K. Phillips, Erica L. Rall, and Deborah A. Peluso, 'A Data–Frame Theory of Sensemaking', in *Expertise out of Context: Proceedings of the Sixth International*

Conference on Naturalistic Decision Making, edited by R Hoffman (Mahwah: Lawrence Erlbaum Associates, 2007).

40. Kemmis, 'Phronēsis, Experience, and the Primacy of Praxis', 155. Italics in the original.

41. Nihlén Fahlquist, 'Responsibility as a Virtue and the Problem of Many Hands', 196.

42. Vijay Kumar, *101 Design Methods: A Structured Approach for Driving Innovation in Your Organization* (Hoboken: John Wiley & Sons, 2012).

43. Frame creation focuses not on the generation of solutions but on the ability to create new approaches to the problem situation itself. Kees Dorst, *Frame Innovation: Create New Thinking by Design* (Cambridge: The MIT Press, 2015).

44. Elizabeth B. N. Sanders and Pieter Jan Stappers, 'Co-Creation and the New Landscapes of Design', *CoDesign* 4, no. 1 (2008).

45. Ezio Manzini, *Design, When Everybody Designs: An Introduction to Design for Social Innovation* (Cambridge: The MIT Press, 2015), 158.

46. Aristotle, *Nicomachean Ethics*, translated by Roger Crisp (Cambridge: Cambridge University Press, 2004), 30, 1106b.

47. To simplify, I concentrate here only on the two extremes (the reckless and the wise) of the practical wisdom continuum; there is thus a gradient in between the two, as practical wisdom is an ability that needs to be developed from absolute novice to expert.

48. David Casacuberta and Ariel Guersenzvaig, 'Using Dreyfus' Legacy to Understand Justice in Algorithm-Based Processes', *AI & Society* 34, no. 2 (2019): 315–16.

49. Daniel Hall cited in Moore, *Virtue at Work*, 177.

50. Kemmis, 'Phronēsis, Experience, and the Primacy of Praxis', 156.

51. Thorpe and Gamman, 'Design with Society', 220.

52. Ibid.

53. Ibid., 219.

54. Ibid., 227.

55. Manzini, *Politics of the Everyday*, 53.

56. Thorpe and Gamman, 'Design with Society', 219. Italics in the original.

57. Ibid., 220.

58. Ibid., 227.

59. Millman, *How to Think Like a Great Graphic Designer*, 39.

60. Thorpe and Gamman, 'Design with Society', 220.

61. Contra Thorpe and Gamman, I also believe that designers can be responsible in the other two senses as well (both as obligation and attribution), but this is not the argument I wanted to make here.

62. Exceptions to this would be murder or rape, which are always wrong regardless of the circumstances.

63. Thorpe and Gamman, 'Design with Society', 224.

64. Naturally, this objection has to do with the suggested overarching purpose itself as well.

65. Alan Wertheimer, *Rethinking the Ethics of Clinical Research: Widening the Lens* (Oxford: Oxford University Press, 2011), 24–28.

66. The issue of justification is discussed in ibid., 25–29.

67. Vijaya Gadde and Matt Derella, 'An Update on Our Continuity Strategy During Covid-19',https://blog.twitter.com/en_us/topics/company/2020/An-update-on-our-continuity-strategy-during-COVID-19.html.

68. One needs not to imagine that design could also be instrumentalised for mass surveillance and social control, as this is already happening with million ethnic Uyghurs and other Turkic Muslims in Xinjiang, China. Maya Wang, *China's Algorithms of Repression: Reverse Engineering a Xinjiang Police Mass Surveillance App* (New York: Human Rights Watch, 2019).

69. Shannon Vallor, *Technology and the Virtues: A Philosophical Guide to a Future Worth Wanting* (New York: Oxford University Press, 2016), 203. Italics in the original.

70. C. S. Lewis, *God in the Dock: Essays on Theology and Ethics* (Grand Rapids: Wm. B. Eerdmans Publishing Co., 1970), 292.

71. George DeMartino, 'Epistemic Aspects of Economic Practice and the Need for Professional Economic Ethics', *Review of Social Economy* 71, no. 2 (2013): 171.

Chapter Nine

Flourishing and Enduring
as a Designer

This closing chapter binds together and sums up the key themes that we have discussed in the book, considering in greater depth a topic that was left unexplored so far: the tension between internal and external goods. In doing so, it explores the virtues of constancy and integrity, which are necessary for maintaining an uncompartmentalised self. The chapter deals with the dynamics of current professional design, which is embedded in for-profit organisations. It closes by suggesting plausible ways in which designers can navigate the complexities they encounter in their working lives.

INSTITUTIONS AND EXTERNAL GOODS: TENSIONS AND CORRUPTION

Although we have so far concentrated on *practices*, there is an important complementary concept in the MacIntyrean account: 'institutions', which has a very specific meaning and should not be conflated with 'organisation'. MacIntyre's understanding of the notion is straightforward though; in contrast to practices, which are concerned with the pursuit of *internal* goods (which can only be attained through the practice that provides them), institutions are 'characteristically and necessarily concerned' with the acquisition of *external* goods such as money, material goods, power, and status (which can be attained in a myriad of ways and are not directly connected to a particular practice)[1]. Institutions are indispensable to sustain 'the practices of which they are the bearers. For no practices can survive for any length of time unsustained by institutions'[2]. An institution *hosts* the practice, enabling it to function and, eventually, to flourish. Practices *need* institutions and the external goods they provide for their continuance.

For instance, the management of a digital music service might prioritise measurable performance indicators (referrals, subscriptions to the service, direct product purchases, revenue, etc.) that can be seen as external goods. These external goods may be different from the goods (or goals) that are pursued by the design practitioners, who might be more interested in *internal* goods such as a beautiful and efficient interface, an elegant and intuitive interaction between the user and the service, the pleasure the user of the service gets out of experiencing a usable or accessible solution, the adequacy in the messages the interface displays, a clever workaround for getting a system to do something that it was not going to be able to do, and so on.

Insofar as the external goods are pursued in order to maintain the practice, the tension is a healthy one, but when priorities are reversed and external goods become ends in themselves, the tension endangers the core of the practice, which is the pursuit of internal goods. This danger is motivated by 'the corrupting power of institutions'[3], concerned, above all, with the acquisition of external goods. The conflicting interests generate an irresolvable tension between practices and institutions as the ideals, cooperation, and creativity of the practice are always vulnerable to the acquisitiveness and competitiveness of the institution[4].

Can't an institution be primarily concerned with internal goods? In the MacIntyrean understanding of the notion that is not possible: institutions primarily seek external goods and practices primarily seek internal goods. This seems rather counterintuitive if we understand 'institution' in its everyday sense (closely related to 'organisation'). We think it must be certainly possible that an institution is concerned primarily with internal goods; after all these institutions exist (for example, Doctors Without Borders). But this tangle is merely terminological, and Geoff Moore gets us out of it by using the notion of 'organisations' to refer to 'practice-institution combinations'[5] that are hospitable to practices, which cover the everyday sense of 'institution'. In the rest of this chapter, 'institution' must be understood in the narrower MacIntyrean sense, whereas the word 'organisation' will be used to include the notions of 'practice-institution combinations' and 'institution' in its everyday sense.

Let us review the opposition between internal and external goods with three short design situations, from extreme to everyday. In the first situation, an interface designer at a cab-hailing company is tasked with being part of a team that will design software that will make it more difficult for regulators to investigate their company. (Note: this case is not fictitious; it is based on 'Greyball', a software developed by Uber that was aimed at identifying and circumventing officials trying to investigate the company[6]). In the second situation, a designer uses 'dark patterns' and nudging techniques to boost revenue for their client. To start the check-out process, they use a big red button that has two bundled functions: it adds extra items to the shopping cart and it serves to continue onto the next screen. The button's label is 'Continue and take advantage of our current promotion'. They also include a much less conspicuous text link that says 'Continue without taking advantage of the promotion.' In the third situation, a designer is working on a feature like 'autoplay' or 'endless scrolling' in a digital service, with the main goal of increasing 'user engagement' (the time a person spends using the service).

These three situations exemplify how the designer is prioritising external goods above internal goods as their primary motive for their design, which boils down to pursuing money or power. The overall goal that is pursued in these cases is to instrumentalise design to give the project commissioner or your boss what they ask for (for example, selling more ads) and get paid for it. What is pursued has nothing to do with the purpose of promoting others' capabilities, nor with the primary pursuit of the countless internal goods that are associated with design as a practice. Clearly, all of these are instances of design, but they are far removed from what is a plausibly internal good of design.

Naturally, someone might argue that a person might still truly benefit from the promotion that was bundled with the red button or truly enjoy the content served thanks to the 'autoplay' feature, but this is beside the point. A good outcome is not the same as an internal good, which must be pursued for its own sake from the start, by aiming at it, by being the 'for-the-sake-of-which-we-act'. As I argued in chapter 7, a good outcome cannot *become* an internal good, because an external object cannot be internalised. Benefitting the user thanks to the design or the inclusion of the 'autoplay' feature *could have resulted* in an internal good if benefitting the user by extending their capabilities was not a mere instrument for the pursuit of external goods.

It would be a mistake, however, to neglect external goods or to disregard their key importance. After all, they sustain the practice. For a designer, external goods are embodied in the wages or fees they receive for their work (and also in social recognition, prestige, etc.) and enable them to sustain not only their livelihood, but also the different practices they engage with (design, family life, hobbies, arts, sports, and so on)[7*]. Yet internal goods are overall and across time more important to the practitioners than external goods, if only because it is in the pursuit of internal goods that they are able to flourish as human beings. Nevertheless, depending on individual and societal contextual factors (age, health issues, economic crises, and so on) a designer might, understandably, temporarily emphasise the acquisition of external goods that are considered indispensable to sustain other important practices the designer engages with (family life, education, and so on).

Still, however important, the designer is not always *directly* motivated or guided in their actions by internal goods no matter how central they are, as they also tend to be elusive. The view I put forward presents a scaffolded account of how these internal goods are attained. Designers are guided by regulative ideals, which have the internal goods of the practice and the practice itself at their foundation. Regulative ideals, as we saw in chapter 7, guide lower-level purposes that are converted into actions by responsible designers. Meanwhile, being a professional is constitutive of who they are: they care for promoting others' wellbeing through the design profession and in doing so they flourish as persons. Personal investment provides sufficient motivation for having a willingness to act according to those purposes and those regulative ideals.

In 1964, Dutch graphic designer Jan van Toorn defiantly asserted: 'It is not our job to please business'[8]. But this is not quite right; as professionals, designers arguably owe to clients the delivery of work that meets their requirements. It is thus their job to please business somehow. But pleasing business is not the designer's *only* job. Professional designers have many other purposes, among which contributing to others' flourishing is paramount. Pleasing business is problematic when it becomes the designer's main concern that *systematically* overrides internal goods as the main driver behind their actions. Van Toorn was onto something significant.

When pursuing external goods becomes the designer's habitual disposition, the practice is damaged. When the failure to pursue the true ends of design becomes prevalent, the professional practice becomes eventually debased. The consequences this has for the practice of design are evident: if this becomes the norm, the leniency it was granted to the design profession due to its youth and its claims to professional status become moot.

Just as responsibility as a virtue is essential in realising the professional's commitment to promoting others' wellbeing, the virtues also become crucial in protecting the practice from the pernicious and corrupting effects of the unbalanced pursuit of external goods. MacIntyre explains that 'The integrity of a practice causally requires the exercise of the virtues by at least some of the individuals who embody it in their activities'[9]. And to the virtues that are especially important we turn next.

CONSTANCY, INTEGRITY, AND COMPARTMENTALISATION

Safeguarding practices is crucial not only in order to protect the internal goods themselves that the practice offers, but also to protect the environments in which we can flourish as human beings. MacIntyre suggests two virtues that are especially relevant to shield the practice against the damaging effects of an unfettered pursuit of external goods: *constancy* and *integrity*.

Constancy

Constancy 'requires that those who possess it pursue the same goods through extended periods of time, not allowing the requirements of changing social contexts to distract them from their commitments or to redirect them'[10]. At the same time, 'the point of constancy is that it ties present action to our past and future pursuit of the good life'[11], ensuring a sense of unity or continuity for the self. Constancy is generally taken to limit the flexibility of our character. To me, its focus rests not necessarily on prescribing a rigid, unchangeable mindset toward the commitments one has, but on how it enables one to resist situational influences when one is conscientiously committed to the pursuit of determinate goods.

In chapter 8, I argued for the importance of negotiating the validity of a design solution in order to extend one's 'field of possibility'. Constancy is required to conduct this negotiation. Assuming their reasons are compelling, a designer that finds it important that the wellbeing of *all* footpath users is not disturbed or harmed by misbehaving scooter riders will show constancy if they stick to that view in the face of rejection or dismissal from other negotiating parties. Though the word does not come up explicitly, Paula Scher's quote in chapter 7 about the 'things that have to be held on to, things that have to be protected' is closely connected to constancy[12].

Business ethicist Angus Robson emphasises a related point, 'a justification must be forthcoming either explicitly in argument or implicitly in practice for why resistance is being offered, which answers the question, "What good was at risk?" or "What bad things were being proposed?"'[13] These questions connect constancy with practical wisdom, which is needed to answer them and to know what to do from there. Design's regulative ideals, and a shared conception of the goods that are pursued by the profession, inform the designer about the goods that are at stake and how their individual choices fit into the larger historical narrative of the tradition of design.

A tradition offers designers thus something secure and practical that they can hold fast to in order to resist situational influences; simultaneously, it enables them to see their own life as part of a communal narrative quest. A shared purpose is not only a common destination to aim at, but also a shared commitment to overcome a problem shared with others.

Integrity

Integrity is the second virtue that designers need in order to resist the corrupting influence of institutions and more generally to form 'our character in such a way that we can pursue good purposes both for ourselves and for the common good of the community'[14]. Just like constancy, integrity sets limits to our flexibility of character, and it aids us in avoiding the perils of compart-mentalisation.

Compartmentalisation, which we touched upon in chapter 5 and will re-visit again shortly, is about doing something as a professional (for instance, tricking people into buying things they do not need, or worse, designing interfaces to fool regulators) that we find ethically inadmissible, say, as a citizen or a parent. When compartmentalising, we tell ourselves things like 'design is just a job', 'I am not my job', 'I'm a designer, my job is to help companies with X, whether I like X or not', and so on and so forth. And thus we are able to carry on. But our narrative quest is harmed by compartmental-ising our existence and excluding a key practice (work) from that which we consider to be our 'real' life.

If constancy is about a continuity in time, integrity is about a continuity across the different spheres that constitute our lives. MacIntyre explains that:

> To have integrity is to refuse to be, to have educated oneself so that one is no longer able to be, one kind of person in one social context, while quite another in other contexts. It is to have set inflexible limits to one's adaptability to the roles that one may be called upon to play [15].

What MacIntyre tells us is that integrity is not only about being internally consistent and coherent in our role as designers; his point is that to have integrity is to be consistent and coherent *across all our roles*. Not only as designers but as friends, parents, citizens, and so on. This contrasts markedly with the way the expression 'behaving like a professional' is often under-stood, which is as setting aside personal considerations and adopting a tech-nical rationality that expunges our *personal* views. Integrity is the opposite of all that. Of course, one could be more relaxed at home than at work and make more jokes with a friend than with a client. Integrity is not about that, but about how we enact the virtues across the different spheres of our life. Moore summarises the idea, 'if we possess integrity, we would be no less just, no less caring, no less patient at work as at home' [16].

This seems to indicate that integrity is more about a dispositional mindset than about beliefs or principles. This also leaves room for choosing how a particular virtue is enacted in the different contexts and practices; so, being caring at home need not be enacted in the same way as being caring at work. The important part of integrity is maintaining the caring disposition. Along these lines, a lack of integrity could thus be seen as failing to make a suitable adaptation to a particular situation.

So, for example, imagine that the designer of the big red button that inadvertently adds unwanted items to the shopping cart (let us call them Taylor) realises that said feature would be manipulative, but chooses to not say anything about it and simply implements it because they believe that it their job is 'to design interface elements, not to be a complainer'. This would be a clear-cut case of compartmentalisation. We could find it also in many of the examples that we have dealt with: the scooter service, the field for gender, and the ads for the bookmaker. A designer who realises that ads for gambling are fraught with ethical dilemmas, but simply sets that aside because designing ads is 'their job' and it is 'not up to them to decide what the ads are about', is compartmentalising. In these situations, we find a conflict between institutional values on the one hand and the personal values on the other. We will explore this type of conflict shortly using the same red button example. But before we do that, I must hasten to add that not every designer that designs the red button feature or the ads for gambling is *necessarily* compartmentalising. This possibility will also be explored below and will serve to come full circle, integrating all the elements that constitute our account.

Role Morality and Compartmentalisation

Maintaining different roles for the different spheres of one's life (as a professional, a parent, a friend, etc.) need not be an ethical issue as long as these different roles are morally in unison, for which integrity is required. Maintaining different roles becomes more problematic when the role holder (the individual behind those roles) is unable to exhibit moral reasoning that goes beyond what is 'pre-established' in the roles they adopt. This means that the individual's moral reasoning is conditioned and defined by 'those considerations sanctioned in each context by the norms defining and governing' the roles they inhabit[17]. To illustrate, as a designer you think and act as a designer, as a friend you think and act as a friend; and so on. The problem occurs when one no longer feels connected *as an individual* to one's behaviour and reasoning as a role holder. Taylor, who did not want to be a 'complainer', exemplifies this; they dissociate the person they 'really' believe to be from the role they happen to take as designer.

Adopting role morality can have pernicious influence on the ethical reasoning of the person holding a particular role. As MacIntyre argues: 'Their lives express the social and cultural order that they inhabit in such a way that they have become unable to recognize, let alone to transcend its limitations'[18]. Taylor, in this case, views themself as not having any responsibility beyond what is assigned to them by their role and the social structure where they happen to find themselves. In summary, role morality is the temporary replacement of one's own values with the values and expectations that fit the role and the institution one works for.

When a person is disconnected from their own values and actions as a role holder, a precondition for moral agency is removed. A person can be a moral agent only insofar as they recognise themselves as such and have a sense of moral identity that is not fully determined by the social order in which they happen to find themselves[19]. After all, one would have a very fluid moral identity if one's moral identity was contingent on the role one has; what is more, a person who changes their moral identity only because they change jobs or roles would be lacking the virtue of constancy.

It is evident that strict adherence to role morality is fundamentally inconsistent with the account of professional ethics I have put forward. Somebody who compartmentalises has surrendered moral ownership over their own actions; claiming that 'I had no other choice because it's part of my job' is a frequent explanation one gives oneself and others. Those who do not feel personally attached to what they do could hardly develop the virtues of responsibility and care this account proposes.

This limitation in agency is, partly at least, self-assigned[20]*. Granted, one may find oneself in an environment that is not conducive to moral behaviour and where moral agency is challenged in different degrees. Hannah Arendt famously theorised how the erosion of agency and blind, unthinking obedience to norms can lead to the most unspeakable evils[21]. *Most* designers, however, do not work amid Nazi bureaucrats or for them. Be that as it may, it must be acknowledged that most of them do work within institutions that 'create expectations for individuals to conform their behaviours'[22]. At the same time, compartmentalisation is something the designer themself does to ward off the discomfort and anxiety caused by their conflicting thoughts and feelings[23]. Although it is absolutely related to the context in which it occurs, it is not *caused* by it. People do retain a reasonable degree of control over their mental life, which could allow them to react differently to the situation.

Four Scenarios Around Compartmentalisation

Let us revisit the red button situation by stipulating four scenarios or ways of dealing with it that might arise after Taylor recognises that the button feature 'is not quite right'.

1) In the first scenario, after an unproductive discussion with their boss, Taylor considers making a statement and quitting their job to uphold their professional values and maintain their integrity. Taylor avoids compartmentalisation, but, at first sight, this course of action seems to be unnecessary and precipitous. Nevertheless, quitting may be a necessary course of action in *some* cases; if these requests are constant we would concede that it may be a good idea to seek another job. Or, regardless the frequency, when 'the harm is great enough, we have to recognize that a professional is indeed required to risk loss of job to prevent harm from occurring'[24], as Larry May argues. Volkswagen's Dieselgate or Uber's Greyball would possibly fit this description. The question remains whether a designer tasked with designing a manipulative red button ought to behave like a moral saint. The harmful effects

of the red button are not comparable to those of Dieselgate, and provided that manipulative buttons are not a constant request, sacrificing themselves and quitting their jobs seems to be an exaggerated and unnecessary response.

The virtue ethical approach I am putting forward does not expect from a designer to become Don Quixote, fighting *every* design decision with ethical relevance as if it were a battle for life. After all, practical reason is about being critical and reflective, but also about choosing one's battles and administering one's own resources (time, energy, attention, and so on). What is more, while a designer that fights every decision could arguably function in some organisations, in most cases this would only result in the designer being fired. It would be unreasonable to advocate for such an ethic. Iris Murdoch brilliantly explains why such a quixotic *ethos* is not a reasonable expectation:

> It is a task to come to see the world as it is. A philosophy which leaves duty without a context and exalts the idea of freedom and power as a separate top level value ignores this task and obscures the relation between virtue and reality. We act rightly 'when the time comes' not out of strength of will but out of the quality of our usual attachments and with the kind of energy and discernment which we have available. And to this the whole activity of our consciousness is relevant[25].

2) In the second scenario, Taylor feels uneasy about the red button feature, but they simply shrug their shoulders and carry on with the design task because 'that's the job'. This is the simplest form of compartmentalisation, which was described in the previous section and needs no further treatment.

3) In the third scenario, Taylor initially behaves responsibly, communicating their concerns about the big red button, but fails to convince their boss of exploring alternative solutions. They are frustrated and reluctantly decide to go on with the design of the red button. May argues that 'loss of integrity is not a function of change of beliefs or values, but rather of unreflective change'[26]. At this juncture we cannot tell whether Taylor is compartmentalising. They would be if they go on with the design of the red button without adequate reflection; put differently, by coping with their ambivalence by rationalising their change of views ('I tried, but as a professional I've got to do what I am paid for, even if I do not like it'). Let us assume this is what they do. In this case, after their initial attempt to maintain integrity, Taylor becomes compartmentalised (their views are in a different compartment than their actions as role holder).

4) In the fourth scenario, Taylor ends up accepting the inclusion of the red button feature in the design, but does not compartmentalise their decision. In this scenario Taylor sets more modest conditions for maintaining integrity and professional value than the Taylor in scenario 1. What they seek to fight and resist in the first place is the internal process of compartmentalisation, because it prevents them from exercising responsibility, which they need to

adequately deal with a situation with conflicting normative demands. Taylor succeeds in maintaining an unfragmented self, but doing this does not generate a course of action. They engage in reflection on the harm that would be done in either case and come to realise that there may be no single truly good outcome to start with.

For the sake of argument, assume that our designer eventually concludes that the most reasonable course of action is to 'bite the bullet' and design the red button. After all, they want to keep the job; it is a job that they like, with a good portfolio of clients who commission interesting projects, and the pay is good too. Taylor sees all this but is still not compartmentalising: they do not attribute the actions they personally reject to the role holder they happen to embody. They 'own' their decisions. They might also realise and acknowledge that by placing the values of the organisation that employs them ahead of their own, they are not actually pursuing the internal goods of design, but external ones. Taylor does this without losing sight of what the direct and indirect effects of their actions might be on them and to others. Owning the decision allows them to maintain their integrity. Integrity is lost by avoiding reflection, not by changing one's view after reflecting on the matter. Aided by practical wisdom, Taylor can cope with the value conflict by becoming aware of what *good* they are safeguarding (in this case their job) and what their true possibilities for action are.

The upshot of the fourth scenario is that what is crucial for virtuous behaviour under difficult circumstances is not necessarily achieving a good outcome (the manipulative red button cannot possibly be a *good* outcome). The way Taylor has approached decision making is more crucial than the result they have obtained. By showing moral and emotional maturity and confronting a situation without shutting off unwanted emotional experiences, Taylor's virtues and character are further developed and strengthened. The virtues of practical wisdom, honesty, integrity, and courage enable Taylor to balance the multiple normative demands they face as a whole person (having a job *and* designing well are both important things to them). Constancy aids them in not changing their views too often: they understand that if they design too many 'big red buttons' to keep their job, they cannot become the type of designer they want to be. And Taylor could not bear to become a designer that manipulates for a living.

Encountering conflicts such as having to design something that we disapprove of because we consider it for instance manipulative, unfair, or discriminatory is not infrequent; this conflict places the designer in a situation where they have to choose between alternatives whereby none of them are truly good options. We would like to be able to choose the 'lesser evil', but sometimes it is not possible to do so, because the alternatives are incommensurable. Unresolvable dilemmas, contradictions, and uncertainties are central to professional design practice, and they need to be accepted and embraced as a part of it. Much like we have accepted and embraced the 'wickedness' of design problems.

It is these types of unresolvable situations that require a selective response to multiple demands, which enable us to develop the virtues. Sometimes situations are too messy, and we barely manage to muddle through, but it is through honestly seeking to produce responsible, uncompartmentalised responses where we learn about ourselves. We learn to maintain integrity and gain better understanding of the goods that we pursue, as well as what is at stake, and what is lost. Even when the outcome we reach is a *rotten* one that does not make us proud.

CLOSING REMARKS AND CONNECTIONS

It is key to emphasise that even though the scenarios here are based on an individual, it would be a mistake to view this account as individualistic and professional designers as selfish, preference-maximising decision makers who define their own standards in terms of what they want *individually*. In the communitarian account I have put forward, there is a shared conception of the good, which informs designers about what they *ought* to desire and what design is for[27]. The practice of professional design (and any practice for that matter) is a 'socially established cooperative human activity'[28] that transcends and connects practitioners to one another; this makes everyone's self-enactment and flourishing an individual as well as a collective quest.

By connecting their individual goals and action to the purposes of the profession, the designer is not a lonesome decision maker that confronts the challenges on their own. They act and decide within a tradition that provides a sense of collective purpose in the form of a shared understanding of the kind of goods that are worth pursuing. At the same time, every individual designer shares with the other practitioner a common commitment to the pursuit of those goods; what is more, being aware of this connection to others pushes them to 'generalize beyond a particular context for action, but also invites [them] to generalize beyond [their] own experience'[29] when engaged in ethical reflections. A designer that is aware of the interconnected nature of their practice will necessitate more than having individual *motives*. They will demand of themselves to have *reasons* for acting in a particular way. Individual motives, it is worth reiterating, only become good reasons when aligned to shared professional purposes.

The internal goods of design are the goods that designers, individually and collectively, are primarily committed to *by virtue of being designers*, but these goods are not the only ones that designers can obtain. MacIntyre asserts that the virtues are connected to three types of goods: 'those internal to practices, those which are the goods of an individual life and those which are the goods of the community'[30]. By providing others with the instrumental conditions for their own flourishing (in the form of material and immaterial artefacts), designers are fortunate enough to be connected to the three kinds of goods.

External goods such as money or power are true goods, but they are pursued for the sake of one of the other three types of goods. So how can we reconcile the focus on internal goods that professional design has with the focus on external goods that institutions have?

Formally seen, there is no opposition; the goal of an institution is to procure external goods so that the practice it hosts can flourish. When an institution is hospitable to the practice it hosts, we have an 'organisation' as we discussed previously. A business firm can be an example of an organisation in which internal and external goods are in balance; Amartya Sen eloquently argued that the 'success of the firm can itself be fruitfully seen as a public good' if they go beyond mere economic growth or profits and in the direction of obtaining meaning and social legitimacy by contributing to the good of society. For-profit businesses could and do make a plausible claim to making an essential contribution to everyone's flourishing by serving important human needs in all areas of life. So, take for instance Herman Miller; their salespeople need to procure external good *in order* to make great chairs that enable people to work comfortably and safely. Designers need to design chairs that are marketable, but meeting marketing goals cannot be their primary purpose; their primary concern must reside in designing good chairs that are conducive to the internal goods of design. Whether or not a chair sells well is contingently attached to chairs by 'the accidents of social circumstance'[31].

But this account of the fit between design and business may be unsatisfying to some and be dismissed as idealistic. Especially in the last forty years, business interests have come to impress their values upon the whole world. It would be then extremely naive to expect that the goal of capability expansion and human flourishing will be easily aligned with the objectives of a company driven solely to maximize shareholder wealth. It is no news that in today's modern capitalism, value extraction is generally rewarded more highly than value creation[32]. Everyone who has a job knows that market forces affect and reduce all spheres of human activity to that which can be exploited for profitability; even those social realms that were protected from it such as health and education are increasingly concerned with the metrics of profit making and other notions that are alien to the core of practices.

The open challenge remains, then, to reflect on how for-profit business fits into the account of professional ethics I have provided. Naturally, the reflection can be extended to also include non-commercial organisations, which are also increasingly governed with an utmost concern for 'the bottom line'. This is not the place to scrutinise the relationship between capitalism and society at large or to argue that the market system by itself is not a viable recipe for human flourishing. My goal for the remaining pages is far more modest: to continue with the review of the key ideas of the account I have put forward and to conceptually reconcile the daily work of design practitioners, who largely work for business organisations, with the pursuit of the internal goods of design and the ideals of professionalism.

Throughout the book I have emphasised that designers may have to endure a reduction in their agency as they are often not fully decisive in determining a design outcome. To put it differently, designers do not undertake the whole of design by themselves. They are not able to generate results in the exact way they intend, and many other stakeholders are involved during design and in the ulterior adoption process. And fortunately so! In chapter 4, I mentioned Don Ihde's 'designer fallacy', which covers this indeterminacy.

The account I offered, however, is not contingent on having full control over the outcome. In this last chapter, we have also discussed how institutions can have a corrosive effect on the practice when external goods crowd out internal ones. And have also touched upon the loss of individual integrity and the problem of compartmentalisation. My account is more centrally connected to resisting these institutional pressures than with actually reaching excellent results, which may be contingent on factors that lie outside the designer. This account is about how the designer can develop the virtues and, enabled by a set of them (practical wisdom, integrity, and courage are some of them), can pursue the purposes of design. Seeking excellence in the outcome is important in design, but the process is just as crucial.

In a furious attack on design's professional and educational worlds, published in 1975, Victor Papanek listed ten ways of 'bringing design back into the mainstream of life', the first one of which is:

> Some designers will be able to connect themselves differently in the future: why do thousands of us work for industry, but almost none of us for trade unions? Why do we work directly for cigarette companies or carmakers, but almost never for cancer clinics or autonomous groups or pedestrians or bicyclists?[33]

The scenario envisioned by Papanek is already a reality, as the examples included in this book illustrate, designers participate in the design of everything from hospitals to improved urban grids for cyclists. Notwithstanding this, designers are still working for 'industry' to use Papanek's term. (But as I just argued, as long as the external goods serve the internal goods of design that need not be troublesome.) The point I try to make is the following: if the set of possibilities for design that was proposed by Papanek sounded implausible considering the business order back then, but eventually became realities, why should we accept the present limitations imposed by the business establishment as if they were the whole set of possibilities that actually exist now?

Although this is a rhetorical question that needs no explicit answer, I will nonetheless try to first answer it with a strong claim: although it is not easy, there is space for designers to act in ways that are consistent with one's narrative quest and with the purpose of design. Even if it seems impossible or even utopian to do so. Even within a business environment. A weaker claim can also be advanced: it is not impossible to act, as Larry May puts it, 'in a way he or she can live with'[34]. Acting in a way one can live with means avoiding compartmentalisation by making it personal and owning the decisions one makes. It means reflecting on which important things have to be balanced, musing on what is at stake, and what is lost. Besides moral maturity, this reflection requires emotional maturity, especially in the form of emotional regulation skills, which can help us avoid being overwhelmed without having to resort to compartmentalising ourselves in order to cope.

It is not controversial to claim that some companies might be more hospitable to the pursuit of the internal goods of design, for instance because they already are truly engaged in the practice of design (even some banks do). Conversely, other companies may use design only instrumentally, with total disregard for designs' internal goods. Surely, a designer working for companies like these will have a more difficult time to pursue worthy ends when there is an overemphasis on external goods. What is more, without the possibility of pursuing some sort of ethical engagement with others with and for whom one works, the designer will hardly be able to flourish within such a company.

Business organisations, however impersonal and bureaucratic, are not perfect machines. Metaphors can only go so far: organisations seldom fully eradicate human agency. Even a social determinist like Marshall McLuhan argued that 'there is absolutely no inevitability as long as there is a willingness to contemplate what is happening'[35]. Designers are not just a cog in the wheel; they are persons who together with others shape and co-shape those organisations to a greater or lesser extent. Donella Meadows asserted that 'Systems can't be controlled, but they can be designed and redesigned'[36]. Organisations are systems, not machines.

Along these lines, design and designers can have an important influence in organisations—they can ignite internal cultural change; drive strategy; envision future scenarios not only to prepare for 'whatever is coming', but to bring organisations closer to a more desirable, just way of living; they frame and reframe challenges; foster communication and creativity; and bring new ideas to reality. In sum, they enable and provide a way for organisations to extend 'the field of possibility' contesting its current limits (which are often embodied in the 'validity' of a design solution). Again, whether these initiatives can blossom will greatly depend on the soil in which they are planted; that is, the organisation. But it is worth bearing in mind that grass and flowers can grow almost everywhere, even in asphalt.

Nevertheless, things do go south, and designers might find themselves involved in certain affairs that they find more morally questionable they are able to bear. But even inside the largest multinational corporations, designers can influence the course of events when they engage in examine the ethical challenges they face and act upon the conclusions of that examination. To exemplify, in June 2018, following a campaign among Google's staff, its management announced that the company would not design or deploy artificial intelligence systems in areas that are likely to cause great overall harm such as autonomous weapons[37]. Conceivably, this collective action started small, with individuals expressing their uneasiness to each other and sharing thoughts in private, over coffee, and then it grew to a structured and collective deliberation resulting in a coherent discourse articulated and aided by the virtues of practical wisdom, courage, and integrity.

So, to truly prosper, good initiatives need to be able to transcend the constraining focus on external goods that institutions that host them inevitably seek to impose. But is it ever possible to go beyond cost-benefit analysis and the 'bottom line'? Of course it is, as the work of contemporary economists like Kate Raworth and Mariana Mazzucato indicates. What these and other progressive thinkers argue for is to place growth and profit making *within* the boundaries of what communities and individuals deem desirable. They maintain that societal values and common goals can and must regain a central place in economic thinking, but they do not argue for an eradication of profit, only for its domestication so that humanity can regain an 'ecologically safe and socially just space'[38].

Raworth proposes a conceptual framework for economics known as 'the Doughnut', which visualises social and planetary boundaries as two concentric rings of different size (hence, the doughnut); shortfalls in human wellbeing is the inner ring and an overshoot of pressure on Earth's life-giving systems is everything beyond the outer ring. In between those two boundaries marked by the rings lies a sweet spot; a safe and just space where humanity can flourish. This is not a fringe theory; in early April 2020 the city of Amsterdam in The Netherlands became the first city in the world to adopt the Doughnut Model to guide its development. Hopefully by the time this book is published other cities will have followed suit[39].

Repairing the economic system so that it serves the public interest may still seem impossible to some, yet this very sense of impossibility arises, according to writer George Monbiot, out of 'the loss of a common purpose, which in turn leads to a loss of belief in ourselves as a force for change'[40]. In the account I have defended, the professional is not an isolated individual but a full member of a community of practice that is guided by a shared purpose and rewarded with goods that are common to all. Bearing this in mind can enable us to overcome defeatism, cynicism, and fatalism.

In line with McLuhan and Meadows, who were quoted previously, Mazzucato asserts: 'structural forces are results of decision-making inside organisations. There is nothing inevitable or deterministic about it'[41]. To be able to adequately contribute to a positive transformation, however modestly, designers need to discover and learn about alternatives to the neoliberal fixation with profit maximisation and the laws of supply and demand. Consider, for instance, the 'mission-driven' approach that Mazzucato advocates to steer economic growth, which could be an alternative way to rethink the role that organisations and others at the value chain could play in addressing broad societal challenges for the public good[42]. Unsurprisingly, by providing a clear direction for economic agents to work on (such as challenges around cancer, soil health, or food), Mazzucato's approach nicely fits with the purpose-driven ethical account I have defended here.

Designers working for companies and organisations that embrace a mission-driven approach (whether it is a non-profit or a business) would have more opportunities to promote others' wellbeing and flourish as designers. But there is more; as they exercise their profession, they interact with other practitioners who also are engaged in developing their own narrative quest by pursuing the *telos* of the practices they belong to in an occupational setting (in the roles of executives or managers, financial controllers, or others working in operations, manufacturing, sales, procurements, and so on). They too can be guided by their own regulative ideals and be assisted by practical wisdom, courage, and integrity in conceiving new and better ways of doing things. Where better is not necessarily bigger, faster, or cheaper, but more sustainable and equitable for all.

These other stakeholders themselves are also finding or trying to find ways to act with which they can live with. They too are trying to become the kind of person they want to be, and they also consider and adjust their life goals as they go along and learn more about themselves and the world. My modest suggestion is that designers engage in the pursuit of the internal goods of their practice, trusting the others with whom they interact will do the same from their vantage point in relation to their own practices[43].

Achieving design's internal goods is also a matter of seeking interpersonal engagement with other stakeholders in order to find recognise, understand, and appreciate the purposes these other stakeholders pursue. If it is true that practices contribute to the common good, designers may be able to find common ground with other practitioners who collaboratively pursue the internal goods of their own practices just as designers pursue theirs. A shared focus on the flourishing and the common good of the community provides a shared directionality and enables members of different professions to pursue a joint purpose and collaborate with each other.

This virtue ethical account of design is practice-centred and practical in the sense of being 'practicable'. Admittedly, it is not practical in the sense of offering easy-to-follow rules, but design professional ethics cannot be about providing a set of rules to be followed blindly. It needs to be an open-ended inquiry into how we want to live as designers and as persons. Seeing things this way means going beyond deciding what is right in a particular situation, but also reflecting on the very ends that are pursued in professional action.

With a focus on designing responsibly and promoting others' flourishing and capabilities the framework for professional ethics I have put forward can enable and empower designers to discover and understand what can and needs to be otherwise, what factors and conditions might constrain their plans, and what real possibilities open up for designing new ways of living. Although it is not designers alone who get to reorganise everyday life, *designerly* ethical reflection becomes indispensable to imagine and help delineate futures that are less wasteful and carbon-intensive in all spheres of life, but also more just, fair, and equitable. Thoughtlessness will get us nowhere.

For most of the designers I encountered through the years (many of whom were first my students), design is not just a job they do for a living but a central part of their life. The words of designer Milton Glaser eloquently illustrate how designing can be constitutive of one's life: 'For me, work was about survival. I had to work in order to have any sense of being human. . . . For whatever reason, work is what I do'[44]. This centrality of design to the designers' lives is more frequent than not. Even if not all designers depend on design alone so crucially and exclusively, Glaser's words may resonate with most of us: design is what we do. What is more, what we do, the professional practice of design, is at its best when we seek the goods of design. The goods of design are ends in themselves for us: we flourish and grow as persons as we pursue them. Meanwhile, the results of our work are the means that enable others to achieve their larger purposes in life.

Although I believe, naturally, that my arguments are compelling, the reader might find themself agreeing with some elements and rejecting some others. The account of design professional ethics I have put forward is certainly not intended to be the last word, but a part of what I hope may become an emergent area of inquiry within design ethics around the design profession. Also, if my arguments so far have been correct, this account does provide ways in which professional designers can realistically deal with ethically challenging demands from clients and bosses while being able to maintain their integrity. It enables designers to pursue their professional journey toward the good without having to make unreasonable sacrifices to uphold their standards; in other words, without ethics becoming the 'duty without context' that Iris Murdoch warned against. At the same time, this account shows a plausible way in which contributing to the flourishing of others can go beyond being a remote aspirational ideal for the design profession, but a purpose worth pursuing in practice.

NOTES

1. Alasdair MacIntyre, *After Virtue: A Study in Moral Theory* (Notre Dame: University of Notre Dame Press, 2007), 194.

2. Ibid.

3. Ibid.

4. Ibid.

5. Geoff Moore, *Virtue at Work: Ethics for Individuals, Managers, and Organizations* (Oxford: Oxford University Press, 2017), 10.

6. Mike Isaac, 'How Uber Deceives the Authorities Worldwide',https://www.nytimes.com/2017/03/03/technology/uber-greyball-program-evade-authorities.html.

7. A similar account can be provided for a design firm. Pursuing and achieving external goods such as winning projects and getting paid for them is what enables a firm to sustain their core practice; that is, providing design services. For a MacIntyrean account of virtuous organisations, see Moore, *Virtue at Work*.

8. Wim Crouwel and Jan Van Toorn, *The Debate: The Legendary Contest of Two Giants of Graphic Design* (New York: The Monacelli Press, 2015), 55.

9. MacIntyre, *After Virtue*, 195.

10. Alasdair MacIntyre, 'Social Structures and Their Threats to Moral Agency', *Philosophy* 74, no. 289 (1999): 318.

11. Angus Robson, 'Constancy and Integrity: (Un)Measurable Virtues?' *Business Ethics: A European Review* 24, no. S2 (2015): 121.

12. Debbie Millman, *How to Think Like a Great Graphic Designer* (New York: Allworth Press, 2007), 50–51.

13. Robson, 'Constancy and Integrity', 123.

14. Moore, *Virtue at Work*, 80.

15. MacIntyre, 'Social Structures and Their Threats to Moral Agency', 317.

16. Moore, *Virtue at Work*, 81.

17. MacIntyre, 'Social Structures and Their Threats to Moral Agency', 327.

18. Ibid.

19. Ibid., 320.

20. Ibid., 327. MacIntyre uses much harsher language: 'The divided self is complicit with others in bringing about its own divided states and so can be justly regarded as their co-author'.

21. Hannah Arendt, *Eichmann in Jerusalem: A Report on the Banality of Evil* (London: Penguin Books, 2006).

22. Larry May, *The Socially Resposive Self: Social Theory and Professional Ethics* (Chicago: The University of Chicago, 1996), 72.

23. Gary R. VandenBos, ed. *APA Dictionary of Psychology* (Washington: American Psychological Association, 2015), 269.

24. May, *The Socially Resposive Self*, 110.

25. Iris Murdoch, *The Sovereignty of Good* (London: Routledge, 2001), 91–92.

26. May, *The Socially Resposive Self*, 24.

27. This is loosely based on Charles Taylor, *The Malaise of Modernity* (Toronto: House of Anansi Press, 1991), 16.

28. MacIntyre, *After Virtue*, 187.

29. Chris Higgins, *The Good Life of Teaching: An Ethics of Professional Practice* (West Sussex: Wiley-Blackwell, 2011), 23.

30. Alasdair MacIntyre, 'A Partial Response to My Critics', in *After Macintyre: Critical Perspectives on the Work of Alasdair Macintyre*, edited by John P. Horton and Susan Mendus (Notre Dame: University of Notre Dame Press, 1994), 284.

31. MacIntyre, *After Virtue*, 188.

32. Mariana Mazzucato, *The Value of Everything: Making and Taking in the Global Economy* (London: Allen Lane, 2018).

33. Victor Papanek, 'Edugraphology: The Myths of Design and the Design of Myths', in *Looking Claser 3: Classic Writings on Graphic Design*, edited by Michael Bierut, Jessica Helfand, Steven Heller and Rick Poynor (New York: Alworth Press, 1999), 254.

34. May, *The Socially Resposive Self*, 25.

35. Marshall McLuhan and Quentin Fiore, *The Medium Is the Massage* (London: Penguin Books, 1967), 25.

36. Donella Meadows, *Thinking in Systems: A Primer* (White River Junction: Chelsea Green Publishing, 2008), 169.

37. Phoebe Braithwaite, 'Google's Artificial Intelligence Ethics Won't Curb War by Algorithm',https://www.wired.co.uk/article/google-project-maven-drone-warfare-artificial-intelligence; Sundar Pichai, 'AI at Google: Our Principles',https://blog.google/technology/ai/ai-principles/.

38. Kate Raworth, *Doughnut Economics: Seven Ways to Think Like a 21st-Century Economist* (London: Random House Business Books, 2017), 45; Mazzucato, *The Value of Everything*.

39. Adele Peters, 'Amsterdam Is Now Using the "Doughnut" Model of Economics: What Does That Mean?',https://www.fastcompany.com/90497442/amsterdam-is-now-using-the-doughnut-model-of-economics-what-does-that-mean.

40. George Monbiot, *Out of the Wreckage: A New Politics for an Age of Crisis* (London: Verso, 2018), 22.

41. Mazzucato, *The Value of Everything*, 280.

42. Mariana Mazzucato, *Mission-Oriented Research & Innovation in the European Union: A Problem-Solving Approach to Fuel Innovation-Led Growth* (Brussels: European Commission, 2018).

43. This is related to and inspired by a point made by Ezio Manzini: whoever acts under complexity 'must assume responsibility for everything he can see and do from his point of observation and action on the world'. Ezio Manzini, *Politics of the Everyday* (London: Bloomsbury, 2019), 53.

44. Debbie Millman, *How to Think Like a Great Graphic Designer* (New York: Allworth Press, 2007), 32.

Coda:
Teaching Design Professional Ethics

Here, I will include some reflections and suggestions on several aspects that arise from the account of design professional ethics that were presented over the previous pages. They are primarily intended for design educators at the upper-level undergraduate and graduate levels; nevertheless, they might be of interest to other readers as well. I must hasten to add that this is not an exhaustive overview; my aim is to delineate at a general level how a design educator could possibly connect the normative content that I presented here with their daily practice and foster the development of ethical expertise in students.

My assumption from the point of departure is that these educators teach at institutions that seek to prepare students to be oriented toward the public good (broadly, toward something that is a good thing for society as whole). It goes without saying that this is not necessarily opposed to having a friendly disposition toward business organisations, as long as these have goals that go beyond mere profit[1].

HOW TO FOSTER THE DEVELOPMENT OF ETHICAL EXPERTISE

The practice-centred outlook on ethics that was adopted here requires, as a basic premise, a point of departure that takes ethical behaviour to be more than following rules or calculating the pros and cons of an outcome. The goal of a virtuous decision maker is to develop a strong practical wisdom that enables them to do what is required in a given situation; in other words, it aids one in determining what is *good* in an ethical sense. As I argued in chapter 8, practical wisdom does not require applying rules and principles to

257

come to a good judgement; those with well-developed practical wisdom can often intuitively see what it is good.

This is an important point that the design educator should keep in mind when educating students, but this is not to say that teaching explicit rules and general principles is wrong. On the contrary, defenders of virtue ethics frequently argue that principles and rules codify accumulated practical wisdom. For example, respecting others' dignity, the injunctions against lying or gratuitously hurting, or the so-called Silver Rule ('That which is hateful to you do not do to another'[2]) are good rules because they highlight precisely what a virtuous person typically would do or avoid doing. But they would do so or avoid doing so out of virtue, not out of duty or utility calculation.

General principles are useful for beginners and intermediate decision makers who lack a strong practical wisdom; they can provide a guideline for moral action, but do not replace deliberation and judgement. Following rules blindly would be self-defeating, as there is no learning without reflection. Also, rules and principles can be useful for wise people too; they can guide experts encountering novel situations and assist them (especially in situations with there is much at stake) in verifying their own judgement to see if it is flawed or sound.

Consequently, learning 'the rules of ethics' should not be the ultimate learning objective when teaching ethics to designers. The goal must be to first and foremost enable the student to develop ethical know-how. Reflection in-action and reflection on-action must be the locus of teaching; in this context, the value of declarative knowledge is contingent to the extent that it aids reflection. No longer is design simply taught by proclaiming rules about form, colour, interaction, material, sustainability, and so on to students in the hope that they will internalise them and design with those rules in mind. Naturally, many rules are taught in the formative years (it would be absurd, for example, to deny that truly understanding the Gestalt principles is essential for a student of design), but every design instructor knows that learning context-independent rules is only the first step of the process to becoming a designer.

To be clear, I do not oppose teaching the basic ethical theories and their vocabularies. Neither am I against showcasing and discussing some out of the myriad of ethical principles formulated by designers and design scholars (such as Milton Glaser's 'Road to Hell in 12 Steps' or William McDonough's 'Hannover Principles'). Nor am I against examining and discussing paradigmatic design cases that contain features that might be interesting from an ethical perspective; indeed, many an ethical insight can be gained from analysing cases. An instructor could, for instance, discuss the role designers played in the design of software that companies have used to attempt to flout laws and deceive the public and regulators[3]. But, much in the same way that knowing the history of typography and what a 'stem' is does not prepare one to space type properly, knowing the differences between Kant and Mill does not prepare designers for making moral judgements.

What is needed is to adopt an approach for teaching design professional ethics that considers it *a constitutive part of the reflective practice that is professional design*. This means that teaching ethics is about teaching how to make sound moral judgements *in practice*. This relates, on the one hand, to the need to foreground the importance that shared conceptions of the good that is pursued have for the practice and its practitioners; the goal of which is to avoid seeing ethical decision making as a purely personal matter but as a *socially and historically situated endeavour*.

On the other hand, the focus on practice entails that teaching ethics need not greatly differ from the way design is taught nowadays. But this is too tentative; I will make a bolder claim: we should teach ethics much in the same way we teach the many other skills, abilities, and mindsets that are necessary to be a *good* designer.

The studio class is possibly the key teaching environment in contemporary design education; I believe it offers the opportunities for reflective participation from students and mentoring from teachers that are necessary for skill acquisition and development. I propose to integrate a concern for ethics in the studio class by integrating it into decision making in the same way that decision making integrates concerns regarding, for instance, form, interaction, material, and sustainability. The instructor is not simply teaching how to decide—they are teaching how to make decisions about material, form, and so on. Decision making is *always* about something.

Does that mean that every design instructor should be formally trained in ethics? Of course not; just like not every studio course is taught by a person that is an expert in *all* facets of design. Every studio course has its special focus, which determines the requirements for the type of instructors that may be necessary. I will explore three different scenarios and explore the type of instructors they might require.

In the first scenario, the topic of the design studios does not appear at first sight to be ethically problematic; a studio course on colour, for example. For courses like this it would seem desirable that the instructor is *sufficiently* aware of the importance and relevance of ethics in design and can accompany their students whenever they encounter a situation with ethical saliency, acting more as a co-discoverer than as a source of knowledge. One can expect that most experienced design instructors would be able to intuitively recognise situations like this, even if they are not able to say much about it; that is, to produce declarative knowledge about it.

Some areas of design seem to be more fraught with ethical dilemmas than others, so in the second scenario, a studio course on, say, e-commerce, might pose serious ethical challenges to the students. These challenges need to be seen as prime opportunities to address ethics beyond only acknowledging *that* there is something ethically important at stake, as in the first scenario. The instructor here would need to also be able to aid the student in understanding *what* is at stake and what is going on in a given situation (though not necessarily using formal ethical vocabulary), guiding the student to explore the topic in further detail. This presupposes from the instructor, next to their pedagogical abilities, a sufficiently strong practical wisdom as well as some familiarity with design ethics.

The third scenario shows a studio course that has a strong focus on developing ethical expertise as a learning outcome. For example, a studio course on the topic of health services might be an excellent opportunity to integrate advanced issues into a traditional studio course. The goal here is to embed ethical reflection in the application of methods and techniques already in use (in chapter 8, I showed how co-creation or SWOT analysis can serve to develop practical wisdom). This is not to say that we should eschew methods and techniques that are especially focussed on the ethical side of design decisions; on the contrary, these could prove to be very useful, and I will mention a few of them shortly.

Depending on the duration and scope, a course like this could be led by a designer with extensive knowledge of ethics or by a team of people combining advanced knowledge of design and ethics. At ELISAVA, where I teach, a multidisciplinary team is the preferred approach for an intensive ten-week studio course; the team of instructors typically include instructors trained in design, the humanities, philosophy, or social sciences, as well as specialised technicians and guest professional designers[4].

The upshot of all this is that we should not separate ethics from other aspects of designing. At some point during development, a student might think exclusively in terms of expressive form; at other times, this exclusive attention could be dedicated to the materials that would be used. But eventually, to produce a good result, the student will need to reflect on the design *as a whole*. At some point in the process, thus, ethics will need to be reflected upon as the constitutive part of the whole it is. The point that needs to be emphasised further is that the goal of ethical education should not be internalising rules, but developing ethical expertise *in practical action*.

Although the specific topics of ethical concern that must be addressed during a studio course (or a regular lecture-based class, for that matter) need to be decided case by case[5*], I propose a basic set of central abilities and dispositions that need to be promoted so that students can acquire and develop the moral know-how that supports and enables responsibility as a virtue. For descriptive purposes, they are presented here as separate entities, but they are closely related to one another. At the same time, the list can be seen as the minimum requirements for enabling the student to develop ethical expertise. The items on the list are congruent with the notions covered in chapter 8. The basic set of abilities and dispositions that need to be developed are:

- *Practitioner's stance:* developing a sense of being a practitioner inscribed in a socially and historically situated practice; at the same time, developing an understanding and appreciation of the internal goods of design; grasping that the standards of excellence that govern the practice of design, include not only the technical or aesthetical aspects of design to which the students are often exposed, but also ethical ones; being able to assess, question, and rethink the standards.
- *Professional stance:* developing professionalism as a 'thick' ethical notion, rather than merely as competent technical performance; linking the *telos* of design to professionalism and to the student's own narrative quest; being able to connect the student's personal motives for a particular design decision with the purposes and ethical aims of the profession in terms of regulative ideals. Both the practitioner and the professional stance contribute to developing character and virtue in the form of a personal investment in design that endows the designer with moral motivation to act ethically.
- *Moral sense-making:* developing the part of practical wisdom that enables the student to gain an initial understanding of a situation, its context, and to properly make sense of it from an ethical perspective; being able to find ways to connect the purpose of design with the project that one is working on (using the *telos* of design as a frame to understand the design problem).
- *Care:* fostering the transformation of caring about something into a willingness to act; avoiding paternalism by gaining a deep understanding of others and their goals, values, and which capabilities need to be specially promoted through design.
- *Moral imagination:* envisioning possible courses of action and assessing their potential consequences; being aware of and able to balance the conflicting interests of stakeholders.
- *Moral reasoning and judgement:* judging and choosing which of the courses of action is the most ethically justified, integrating analysis and intuition; being able to explain decisions and produce reasons.
- *Enacting courses of action:* the part of practical wisdom that brings it all together, converting care into actual design artefacts.
- *Communication skills:* related to all of the previous points, being able to persuasively explain one's motives and justify one's decisions in terms of how these are conducive to others' flourishing, producing thus *reasons* for acting in one way or another.

At the generic level, some of these abilities are already present in the regular 'toolbox' of design; in this way, much of the instrumental knowledge students already have can be appropriately applied to design professional ethics (for instance, the reasoning capacities that are required to make sense of a design brief can also be used in moral sense-making; similarly, moral imagination is not that different from scenario-based design).

Design students are starting practitioners who enter a practice and engage *indirectly* with other practitioners (for instance, through standards, vocabulary, methods, etc.), but also *directly* with other practitioners and stakeholders with whom they interact. Although the abilities included in the previous list are directed to the individual agent, it would be a mistake to neglect the defining social character of design activity. Along these lines, teamwork skills such as listening, rapport building, and conflict resolution are indispensable, and the methods used should contemplate this necessity.

There are also behaviours that undermine the development of ethical know-how. Business ethicist Domènec Melé enumerates some of them: focussing solely on one's interests, lacking concern for others, analysing problems with superficiality, making hasty decisions, being self-complacent, and being inconstant[6].

To close this initial reflection on ethics education, if we want to teach a 'practice-based', situated ethics, we need to put rules and norms in their proper context, not favouring them over the non-declarative intuitive behaviour that expert decision makers exhibit in practice. The design educator should help the student gain expertise in situations that mimic real practice, because to effectively develop moral expertise a student needs an environment that allows them to deliberately *practice* ethical decision making, engaging its many dimensions: deliberation, judgement, emotion, motivation, and, most importantly, action.

CONNECTIONS WITH DESIGN METHODOLOGY

There is a large spectrum of approaches, methods, techniques, and practical tools that seek to enable and aid designers to integrate ethics and values in their design process. In the following, I will briefly introduce a few of them, encompassing from full-fledged approaches to practical tools. Although these methods and approaches originate in different epistemic traditions and perspectives, at least as I see it they fit well with the account of design professional ethics that I introduced in this book.

1. Value-sensitive design: an approach to the design of technology that, in the words of design scholars Batya Friedman and David Hendry, 'provides theory, method, and practice to account for human values in a principled and systematic manner throughout the technical design process'[7]. Originally developed in the early 1990s by Friedman and Peter Kahn for the design of information systems, the current formulation of the key tenets of the approach are analyses of stakeholders (direct and indirect, and at various levels: individual, group, and societal); distinctions among designer values, values explicitly supported by the technology, and stakeholder values; integrative and iterative investigations; and a commitment to progress (not perfection).

Human values can be understood as 'what is important to people in their lives, with a focus on ethics and morality', without spelling out what is important and thus retaining space for plural interpretations[8]. Some of these human values are human welfare, privacy, trust, autonomy, and environmental sustainability. The approach is guided by several methods that codify and operationalise the approach[9].

The approach is open to foreground the wellbeing of the natural world, for example by emphasising that also non-human entities (organisations, non-human species, or natural objects) can be considered stakeholders in the design process. At the same time (and much like in my account), value-sensitive design 'privileges the perspectives and values of human beings [as the approach concerns] the process of technology that is carried out by human beings'[10].

2. Transition design: according to Terry Irwin, one of its pioneers, it is an approach 'for addressing "wicked" problems . . . and catalysing societal transitions toward more sustainable and desirable futures'[11]. Transition design is structured around the following activities: 1) visualising and mapping complex problems, 2) contextualising them, 3) identifying conflicts and aligning, 4) facilitating the co-creation of desirable futures, and 5) identifying leverage points for situating design interventions. The approach sees the involvement of all stakeholders as crucial in resolving wicked problems and designing for systems-level change. Design educators Stacie Rohrbach and Molly Steenson argue for the thoughtful integration of transition design into design education as the approach can empower designers to 'seed and catalyse' positive systemic change[12].

3. Ethical cycle: ethicists and educators Ibo Van de Poel and Lambèr Royakkers describe it as a tool that helps students and professional designers 'to make a systematic and thorough analysis of the moral problem and to justify your final decisions in moral terms'[13]. The tool can aid the development and acquisition of practical reasoning skills by guiding a sound analysis and providing opportunities for reflection. Van de Poel and Royakkers, the method's creators, hold the view that reaching a moral judgment 'is not a straightforward or linear process in which you simply apply ethical theories to find out what to do'[14]. Consequently, they consider that 'the formulation of possible "solutions", and the ethical judging of these solutions go hand in hand'[15]. This highlights what I take to be a basic requirement any method must fulfil to be adequately paired to the account of both design professional ethics and the reflective nature of design that were presented in this volume.

4. Mepham's ethical matrix: it is a framework for rational ethical analysis developed by philosopher Ben Mepham[16]. The matrix supports non-philosophers in the analysis of the ethical impacts of a design from the perspective of the different groups affected by it. Originally developed for assessing the impact of genetically modified organisms, my colleague David Casacuberta and I have successfully been using an adapted version of the original matrix with engineering and design students as well as with professionals, working individually and in teams. The working of the framework is straightforward: the different stakeholders that may be affected by a design are located vertically along the first column of the matrix table, while different ethical theories (for example, Kantianism, utilitarianism, or feminist ethics) or principles (for example, wellbeing, autonomy, or justice) are located horizontally in the first row. The remaining empty cells are used to include a description of the expected outcome that a design might potentially have for a particular stakeholder according to an ethical principle or theory. Assuming that most ethical theories contain some truth about morality, Mepham's matrix allows us to gain ethical insights by analysing from multiple ethical perspectives what a design (or a design feature) might mean for different stakeholders.

5. Pre-mortem: it is technique developed by decision theorist Gary Klein on the basis of the 'post-mortem' technique, which is frequently used to assess a project after its conclusion (which, in turn, is a technique based on the medical autopsy). Conversely, in the pre-mortem the goal is to assess a future state of affairs (the outcome of one's project) in order to identify and mitigate threats, risks, and likely failures at the outset *before* they occur. The exercise starts by assuming the project has failed completely; it then focusses on the development of possible failure scenarios and works backwards from there to the present to try to establish plausible explanations for those imaginary failures. The technique is especially useful to mitigate 'optimism bias' (thinking that everything will go according to plan), to challenge key assumptions made, and anticipate unintended effects. It is a combination of individual and collective reasoning that stimulates divergent thinking and internal dissent by requiring participants to envision ways in which things might go wrong in order to prevent this from happening as much as one is able to, which is crucial for practical wisdom[17].

6. Playful methods: under this label I group tools that are less analytical (than, for example, the ethical cycle or Mepham's matrix) and rely on play and creativity to foster moral reasoning. Designer Jet Gispen created the 'design noir' technique[18], which resembles the pre-mortem technique but uses role-playing and humour to arrive at ethical insights about dystopian scenarios in order to improve a design. It starts by envisioning and developing two extreme situations involving the design, it follows by acting them out, and it ends with a reflexion on the first-hand experience. This tool is part of the 'Ethics for Designers toolkit', which includes analytical and non-analytical tools with the goal of developing three dimensions of moral reasoning: sensitivity, creativity, and advocacy[19]. Another example would be the 'Envisioning Cards' created by the Value Sensitive Design Research Lab[20], which are designed to evoke consideration and discussion of the long-term influence of new technology across four dimensions: stakeholders, time, values, and pervasiveness. These cards are aimed for educational and industry practice; although they also support analytical uses, they support a playful approach too: for example, by selecting a random card for getting unstuck by performing the activity that is indicated on the card or by using them to engage clients to discuss their concerns.

7. *Asking questions:* when relevant and sharp, questions constitute a powerful tool than can be used during all types of courses: studio, discussion, and seminar. Asking questions is a frequent method to ignite ethical reflection. Many scholars have proposed and compiled sets of questions that can be readily (or with minor adaptations) used in class. To enumerate a few, media theorist Neil Postman proposed a set of questions to be asked about a new technology (for example, 'What new problems might be created because we have solved this problem?')[21]. Philosopher Bruno Latour proposed a set of questions during the early stage of the COVID-19 pandemic; though I have not yet used these in class, they look promising for an exercise (for example, 'What are the activities now suspended that you would like to see not resumed?')[22]. In my classes to prompt awareness, deliberation, and discussion, I use a set of cards with questions about ethical aspects of design and technology. The set of cards is largely an adaptation of questions posed by philosopher L. M. Sacasas[23] that are very useful to explore the wide-ranging moral dimension the technologies we use and design (for example, 'What feelings does the use of this technology generate in me toward others?'). Besides some questions of my own, the set also includes some from a list of questions commonly attributed to philosopher Jacques Ellul: '76 Reasonable Questions to Ask About Any Technology' (for example, 'How does it affect our way of seeing and experiencing the world?')[24].

Again, this short overview is not meant to be exhaustive; its goal is only to illustrate the type of available methods that could serve to operationalise the normative content presented in this volume, which is an issue that, together with a more extensive reflection on design education and ethics, would require a whole new book to be carried out comprehensively.

NOTES

1. Amartya Sen, 'Does Business Ethics Make Economic Sense?' *Business Ethics Quarterly* 3, no. 1 (1993): 50.
2. Babylonian Talmud, Shabbat 31a. A similar maxim can be found in the writings of Confucius: 'Do not impose on others what you would not choose for yourself'; The Analects XV.24.
3. Such as the already mentioned Volkswagen's 'Dieselgate' and Uber's 'Greyball' cases.
4. For a description of how we work at ELISAVA on a studio course involving generative and participatory research for socially responsible design practice, see Ariel Guersenzvaig, 'Llagostera Youth Center', in *Developing Citizen Designers*, edited by Elizabeth Resnick (New York: Bloomsbury, 2016).

5. Some capabilitarian scholars have argued that it is important to develop an inclusive deliberative process for defining curricula content 'in which all voices (students, teachers, university management and staff, politicians, and society at large) can be heard'. Alejandra Boni-Aristizábal and Carola Calabuig-Tormo, 'Enhancing Pro-Public-Good Professionalism in Technical Studies', *Higher Education* 71, no. 6 (2016).

6. Domènec Melé, *Business Ethics in Action* (London: Red Globe Press, 2020), 89.

7. Batya Friedman and David Hendry, *Value Sensitive Design: Shaping Technology with Moral Imagination* (Cambridge: The MIT Press, 2019), 3–4.

8. Ibid., 24.

9. For a detailed discussion of value sensitive methods, see ibid., 59–103.

10. Ibid., 27–29.

11. Terry Irwin, 'The Emerging Transition Design Approach' (paper presented at the DRS2018 Conference, University of Limerick, 2018).

12. See their paper for curricula, methods, and case studies. Stacie Rohrbach and Molly Steenson, 'Transition Design: Teaching and Learning', *Cuadernos del Centro de Estudios en Diseño y Comunicación* 19, no. 73 (2019).

13. Ibo Van de Poel and Lambèr Royakkers, *Ethics, Technology, and Engineering: An Introduction* (Malden: Wiley-Blackwell, 2011), 137.

14. Ibid., 135.

15. Ibid.

16. Ben Mepham, 'A Framework for the Ethical Analysis of Novel Foods: The Ethical Matrix', *Journal of Agricultural and Environmental Ethics* 12, no. 2 (2000); Ben Mepham, Matthias Kaiser, Erik Thorstensen, Sandy Tomkins, and Kate Millar, *Ethical Matrix Manual* (The Hague: LEI, Wageningen UR, 2006).

17. Gary Klein, 'Performing a Project Premortem', *Harvard Business Review*, September 2007 (2007).

18. Jet Gispen, 'Design Noir',https://www.ethicsfordesigners.com/design-noir.

19. Ibid.

20. Batya Friedman, Lisa Nathan, Shaun Kane, and John Lin, 'Envisioning Cards', https://www.envisioningcards.com/.

21. Neil Postman, *Building a Bridge to the 18th Century: How the Past Can Improve the Present* (New York: Alfred A. Knopf, 2000), 42–57.

22. Bruno Latour, 'Where to Land after the Pandemic? A Paper and Now a Platform',http://www.bruno-latour.fr/node/852.html.

23. L. M. Sacasas, 'Do Artifacts Have Ethics?' https://thefrailestthing.com/2014/11/29/do-artifacts-have-ethics/.

24. Jacques Ellul, '76 Reasonable Questions to Ask About Any Technology', http://www.thewords.com/articles/ellul76quest.htm.

Further Reading

This list of books and journal articles consists of a selection of volumes included in the main bibliography and some additional ones. These sources for further reading are categorised into topics that roughly follow the order in which they appear in the book. Though some of the items may belong to multiple topics, I choose not to repeat them across categories. The list does not aim to be exhaustive.

DESIGN AND POLICY MAKING

- Gordon, Eric, and Gabriel Mugar. *Meaningful Inefficiencies: Civic Design in an Age of Digital Expediency*. Oxford: Oxford University Press, 2020.
- Kimbell, Lucy. *Applying Design Approaches to Policy Making: Discovering Policy Lab*. Brighton: University of Brighton, 2014.
- Schaminée, André. *Designing With and Within Public Organizations: Building Bridges between Public Sector Innovators and Designers*. Amsterdam: Bis Publishers, 2018.
- Vandenbroeck, Philippe. *Working with Wicked Problems*. Brussels: King Baudouin Foundation, 2012.

PARTICIPATORY DESIGN AND EMERGING ROLES FOR DESIGN

- Dorst, Kees. *Frame Innovation: Create New Thinking by Design*. Cambridge: The MIT Press, 2015.
- Nelson, Harold G., and Erilk Stolterman. *The Design Way: Intentional Change in an Unpredictable World*. Cambridge: The MIT Press, 2014.

- Manzini, Ezio. *Design, When Everybody Designs: An Introduction to Design for Social Innovation*. Cambridge: The MIT Press, 2015.
- Meroni, Anna, and Daniela Sangiorgi. *Design for Services*. Surrey: Gower Publishing Limited, 2011.
- Yee, Joyce, Emma Jefferies, and Lauren Tan. *Design Transitions: Inspiring Stories. Global Viewpoints. How Design Is Changing*. Amsterdam: Bis Publishers, 2013.

CONSUMPTION AND THE ANTHROPOCENE

- Bonneuil, Christophe, and Jean-Baptiste Fressoz. *The Shock of the Anthropocene: The Earth, History and Us*. London: Verso, 2017.
- Dauvergne, Peter. *The Shadows of Consumption: Consequences for the Global Environment*. Cambridge: MIT Press, 2008.
- Lodziak, Conrad. *The Myth of Consumerism*. London: Pluto Press, 2002.
- McNeill, J. R., and Peter Engelke. *The Great Acceleration*. Cambridge: The Belknap Press of Harvard University Press, 2014.

PHILOSOPHY AND STUDIES OF TECHNOLOGY

- Bijker, Wiebe E., Thomas Parke Hughes, Trevor Pinch, and Deborah G. Douglas, eds. *The Social Construction of Technological Systems: New Directions in the Sociology and History of Technology*. Anniversary edition. Cambridge: The MIT Press, 2012.
- Franklin, Ursula. *The Real World of Technology*. Toronto: House of Anansi Press, 1990.
- Latour, Bruno. *Pandora's Hope*. Cambridge: Harvard University Press, 1999.
- Miller, Daniel. *The Comfort of Things*. Cambridge: Polity Press, 2008.
- Mitcham, Carl. *Thinking through Technology: The Path between Engineering and Philosophy*. Chicago: The University of Chicago Press, 1994.
- Verkerk, Maarten, Jan Hoogland, Jan Van der Stoep, and Marc De Vries. *Philosophy of Technology: An Introduction for Technology and Business Students*. Oxon: Routledge, 2016.
- Winner, Langdon. *The Whale and the Reactor: A Search for Limits in an Age of Technology*. Chicago: The University of Chicago Press, 1986.

DESIGN ETHICS

- Buchanan, Richard. 'Design Ethics'. In *Encyclopedia of Science, Technology, and Ethics*, edited by Carl Mitcham, 504–10. Detroit: Macmillan Reference, 2005.

- Chan, Jeffrey K. H. 'Design Ethics: Reflecting on the Ethical Dimensions of Technology, Sustainability, and Responsibility in the Anthropocene'. *Design Studies* 54 (2018): 184–200.
- Costanza-Chock, Sasha. *Design Justice: Community-Led Practices to Build the Worlds We Need*. Cambridge: The MIT Press, 2020.
- Dilnot, Clive. 'Clive Dilnot, Ethics in Design: 10 Questions'. In *Design Studies: A Reader*, edited by Hazel Clark and David Brody, 180–90. Oxford: Bloosmbury, 2009.
- Findeli, Alain. 'Ethics, Aesthetics, and Design'. *Design Issues* 10, no. 2 (1994): 49–68.
- Jasanoff, Sheila. *The Ethics of Invention: Technology and the Human Future*. New York: W.W. Norton & Company, 2016.
- Manzini, Ezio. *Politics of the Everyday*. London: Bloomsbury, 2019.
- Margolin, Victor. 'Design, the Future and the Human Spirit'. *Design Issues* 23, no. 3 (2007): 4–15.
- Mitcham, Carl. 'Ethics into Design'. In *Discovering Design: Explorations in Design Studies*, edited by Richard Buchanan and Victor Margolin, 173–89. Chicago: The University of Chicago Press, 1995.
- Parsons, Glenn. *The Philosophy of Design*. Cambridge: Polity, 2016.
- Papanek, Victor. *Design for the Real World*. Second edition. Chicago: Academy Chicago Publishers, 1984.
- Scherling, Laura, and Andrew DeRosa, eds. *Ethics in Design and Communication: Critical Perspectives*. London: Bloosmbury, 2020.
- Tonkinwise, Cameron. 'Ethics by Design, or the Ethos of Design'. *Design Philosophy Papers* 2, no. 2 (2004): 129–44.
- Van de Poel, Ibo, and Lambèr Royakkers. *Ethics, Technology, and Engineering: An Introduction*. Malden: Wiley-Blackwell, 2011.
- Van den Hoven, Jeroen, Pieter E. Vermaas, and Ibo Van de Poel. *Handbook of Ethics, Values, and Technological Design*. Dordrecht: Springer Reference, 2015.
- Verbeek, Peter-Paul. *Moralizing Technology: Understanding and Designing the Morality of Things*. Chicago: The University of Chicago Press, 2011.
- Vermaas, Pieter E., and Stéphane Vial. *Advancementes in the Philosophy of Design*. Dordrecht: Springer, 2018.

GENERAL INTRODUCTIONS TO ETHICS

- Rachels, James, and Stuart Rachels. *The Elements of Moral Philosophy*. Ninth edition. New York: McGraw-Hill Education, 2018.
- Shafer-Landau, Russ. *The Fundamentals of Ethics*. Fourth edition. Oxford: Oxford University Press, 2017.

VIRTUE ETHICS

- Annas, Julia. *Intelligent Virtue*. Oxford: Oxford University Press, 2011.
- Hursthouse, Rosalind. *On Virtue Ethics*. Oxford: Oxford University Press, 1999.
- Lutz, Christopher. *Reading Alasdair Macintyre's after Virtue*. London: Continuum, 2012.
- MacIntyre, Alasdair. *After Virtue: A Study in Moral Theory*. Notre Dame: University of Notre Dame Press, 2007.

APPLIED TOPICS ANALYSED FROM A VIRTUE ETHICAL PERSPECTIVE

- Higgins, Chris. *The Good Life of Teaching: An Ethics of Professional Practice*. West Sussex: Wiley-Blackwell, 2011.
- Moore, Geoff. *Virtue at Work: Ethics for Individuals, Managers, and Organizations*. Oxford: Oxford University Press, 2017.
- Vallor, Shannon. *Technology and the Virtues: A Philosophical Guide to a Future Worth Wanting*. New York: Oxford University Press, 2016.

THE CAPABILITY APPROACH

- Nussbaum, Martha. *Creating Capabilities: The Human Development Approach*. Cambridge: The Belknap Press of Harvard University Press, 2011.
- Robeyns, Ingrid. *Wellbeing, Freedom and Social Justice: The Capability Approach Re-Examined*. Cambridge: Open Book Publishers, 2017.

CAPABILITIES AND DESIGN

- Cipolla, Carla. 'Sustainable Freedoms, Dialogical Capabilities and Design'. In *Cumulus Working Papers Nantes 16/06*, 59–65. Helsinki: University of Art and Design Helsinki, 2006.
- Dong, Andy. 'The Policy of Design: A Capabilities Approach'. *Design Issues* 24, no. 4 (2008): 76–87.
- Manzini, Ezio. 'Design, Ethics and Sustainability. Guidelines for a Transition Phase'. *Cumulus Working Papers Nantes 16/06*, 9–15. Helsinki: University of Art and Design Helsinki, 2006.
- Oosterlaken, Ilse. 'Design for Development: A Capability Approach'. *Design Issues* 25, no. 4 (2009): 91–102.
- Oosterlaken, Ilse. 'Human Capabilities in Design for Values'. In *Handbook of Ethics, Values, and Technological Design: Sources, Theory, Val-*

ues and Application Domains, edited by Jeroen van den Hoven, Pieter E. Vermaas, and Ibo van de Poel, 221–50. Dordrecht: Springer Netherlands, 2015.

- Oosterlaken, Ilse, and Jeroen Van den Hoven, eds. *The Capability Approach, Technology and Design*. Dordrecht: Springer, 2012.

Bibliography

Ackoff, Russell L. *Re-Creating the Corporation: A Design of Organizations for the 21st Century*. New York: Oxford University Press, 1999.

Agencia de Ecología Urbana de Barcelona. 'Superblocks'. n/a. Accessed February 24, 2020, http://www.bcnecologia.net/en/conceptual-model/superblocks.

Ahern, John J. *An Historical Study of the Professions and Professional Education in the United States*. Chicago: Loyola University, 1971.

AIGA. 'About AIGA'. 2020. Accessed February 24, 2020,https://www.aiga.org/about/.

———. 'AIGA Standards of Professional Practice'. 2010. Accessed February 24, 2020,https://www.aiga.org/standards-professional-practice.

———. *Design Business and Ethics*. Third edition. New York: AIGA | the professional association for design, 2009.

Airaksinen, Timo. 'Professional Ethics'. In *Encyclopedia of Applied Ethics (Second Edition)*, edited by Ruth Chadwick, 616–23. San Diego: Academic Press, 2012.

Ajuntament de Barcelona. 'Superblocks'. n/a. Accessed February 24, 2020,https://ajuntament.barcelona.cat/ecologiaurbana/en/what-we-do-and-why/quality-public-space/superblocks.

Allen, James. 'Why There Are Ends of Both Goods and Evils in Ancient Ethical Theory'. In *Strategies of Argument Essays in Ancient Ethics, Epistemology, and Logic*, edited by Mi-Kyoung Lee, 231–54. Oxford: Oxford University Press, 2014.

Alter, Adam. *Irresistible: Why Are You Addicted to Technology and How to Set Yourself Free*. London: Vintage, 2017.

Alvarez, Maria. 'Reasons for Action: Justification, Motivation, Explanation'. *The Stanford Encyclopedia of Philosophy* (Winter 2017 Edition), Stanford University, 2017. Accessed March 2, 2020,https://plato.stanford.edu/archives/win2017/entries/reasons-just-vs-expl.

Alzola, Miguel. 'Virtuous Persons and Virtuous Actions in Business Ethics and Organizational Research'. *Business Ethics Quarterly* 25, no. 3 (2015): 287–318.

Amabile, Teresa M. 'Componential Theory of Creativity'. In *Encyclopedia of Management Theory*, edited by Eric H. Kessler, 134–38. Los Angeles: SAGE, 2013.

———. 'Social Psychology of Creativity: A Consensual Assessment Technique'. *Journal of Personality and Social Psychology* 43, no. 5 (1982): 997–1013.

Annas, Julia. *Intelligent Virtue*. Oxford: Oxford University Press, 2011.

Antonelli, Paola. *Humble Masterpieces: Everyday Marvels of Design*. New York: Regan Books, 2005.

Arendt, Hannah. *Eichmann in Jerusalem: A Report on the Banality of Evil*. London: Penguin Books, 2006.

Aristotle. *Nicomachean Ethics*. Translated and edited by Roger Crisp. Cambridge: Cambridge University Press, 2004.

————. *Politics*. Translated by Ernest Barker. Oxford: Oxford University Press, 1995.
Armstrong, Alan E. *Nursing Ethics: A Virtue-Based Approach*. London: Palgrave Macmillan, 2007.
Attfield, Robin. *Environmental Ethics: An Overview for the Twenty-First Century*. Second edition. Cambridge: Polity Press, 2014.
Babich, Nick. 'The Power of Defaults'. UX Planet, 2017. Accessed February 24, 2020,https://uxplanet.org/the-power-of-defaults-992d50b73968.
Badwan, Basil, Roshit Bothara, Mieke Latijnhouwers, Alisdair Smithies, and John Sandars. 'The Importance of Design Thinking in Medical Education'. *Medical Teacher* 40, no. 4 (2018/04/03 2018): 425–26.
Bakker, Conny, Marcel Den Hollander, Ed Van Hinte, and Yvo Zijlstra. *Products That Last Product Design for Circular Business Models*. Amsterdam: Bis Publishers, 2014.
Banks, Sarah, and Ann Gallagher. *Ethics in Professional Life: Virtues for Health and Social Care*. London: Palgrave Macmillan, 2009.
Barber, Bernard. 'Some Problems in the Sociology of the Professions'. *Daedalus* 92, no. 4 (1963): 669–88.
Bayles, Michael D. *Professional Ethics*. Belmont: Wadsworth Publishing Company, 1981.
Berghoff, Hartmut. '"Organised Irresponsibility'? The Siemens Corruption Scandal of the 1990s and 2000s'. *Business History* 60, no. 3 (2018/04/03 2018): 423–45.
Bernard, Gert, and Gert Joshua. 'The Definition of Morality'. *The Stanford Encyclopedia of Philosophy* (Fall 2017 Edition), Stanford University, 2017. Accessed February 24, 2020,https://plato.stanford.edu/archives/fall2017/entries/morality-definition/.
BEUC, The European Consumer Organisation. *Dynamic Currency Conversion: When Paying Abroad Costs You More Than It Should*. Brussels: The European Consumer Organisation, 2017.
Bijker, Wiebe E., Thomas Parke Hughes, Trevor Pinch, and Deborah G. Douglas, eds. *The Social Construction of Technological Systems: New Directions in the Sociology and History of Technology*. Anniversary edition. Cambridge: The MIT Press, 2012.
Bilbao International. 'Exposición: "Objetos Imposibles" De Jacques Carelman'. 2011. Accessed February 24, 2020,http://www.bilbaointernational.com/en/exhibition-impossible-objects-by-jacques-carelman/.
Blackwell, Lewis. *The End of Print: The Graphic Design of David Carson*. London: Lawrence King Publishing, 1995.
Boatright, John R. 'Swearing to Be Virtuous: The Prospects of a Banker's Oath'. *Review of Social Economy* 71, no. 2 (2013/06/01 2013): 140–65.
Boehnert, Joanna. *Design, Ecology, Politics: Towards the Ecocene*. London: Bloosmbury, 2018.
Boni-Aristizábal, Alejandra, and Carola Calabuig-Tormo. 'Enhancing Pro-Public-Good Professionalism in Technical Studies'. *Higher Education* 71, no. 6 (2016/06/01 2016): 791–804.
Bonneuil, Christophe, and Jean-Baptiste Fressoz. *The Shock of the Anthropocene: The Earth, History and Us*. London: Verso, 2017.
Borden, Sandra L. *Journalism as Practice: Macintyre, Virtue Ethics and the Press*. New York: Routledge, 2010.
Boulanin, Vincent, and Maaike Verbruggen. *Mapping the Development of Autonomy in Weapon Systems*. Solna: Stockholm International Peace Research Institute, 2017.
Bournemouth Borough Council. 'Bins and Recycling'. 2017. Accessed February 24, 2020,https://www.bournemouth.gov.uk/BinsRecycling/BinsandRecycling.aspx.
Bovens, Luc. 'The Ethics of Dieselgate'. *Midwest Studies In Philosophy* 40, no. 1 (2016): 262–83.
Bowie, Norman E. 'The Profit Seeking Paradox'. In *Papers on the Ethics of Administration*, edited by N. Dale Wright, 97–120. Provo, Utah: Brigham Young University, 1988.
Braithwaite, Phoebe. 'Google's Artificial Intelligence Ethics Won't Curb War by Algorithm'. Wired Magazine, 2018. Accessed July 22, 2018,https://www.wired.co.uk/article/google-project-maven-drone-warfare-artificial-intelligence.
Brand, Stephan, Maximilian Petri, Philipp Haas, Christian Krettek, and Carl Haasper. 'Hybrid and Electric Low-Noise Cars Cause an Increase in Traffic Accidents Involving Vulnerable

Road Users in Urban Areas'. *International Journal of Injury Control and Safety Promotion* 20, no. 4 (2013/12/01 2013): 339–41.

Brignull, Harry. 'Dark Patterns'. 2017. Accessed February 24, 2020,https://darkpatterns.org/.

Bucciarelli, Louis L. *Designing Engineers*. Cambridge: The MIT Press, 1994.

Buchanan, Richard. 'Design Ethics'. In *Encyclopedia of Science, Technology, and Ethics*, edited by Carl Mitcham, 504–10. Detroit: Macmillan Reference, 2005.

———. 'Design Research and the New Learning'. *Design Issues* 17, no. 4 (2001): 3–23.

———. 'Human Dignity and Human Rights: Thoughts on the Principles of Human-Centered Design'. *Design Issues* 17, no. 3 (2001): 35–39.

———. 'Wicked Problems in Design Thinking'. *Design Issues* 8, no. 2 (1992): 5–21.

Bürdek, Bernhard E. *Design: History, Theory and Practice of Product Design*. Basel: Birkhäuser, 2005.

Butler, David. 'Consumers Union Praises Senator's Call for Ftc Investigation of Airline "Dynamic Pricing"'. News release, 2018.

CABE. *The Value of Good Design*. London: Commission for Architecture and the Built Environment, 2002.

CABE Space. *The Value of Public Space: How High Quality Parks and Public Spaces Create Economic, Social and Environmental Value*. London: Commission for Architecture and the Built Environment, 2014.

Carr, David. *Professionalism and Ethics in Teaching*. London: Routledge, 2000.

Casacuberta, David, and Ariel Guersenzvaig. 'Using Dreyfus' Legacy to Understand Justice in Algorithm-Based Processes'. *AI & Society* 34, no. 2 (2019): 313–19.

Castillo, Greg. *Cold War and the Home Front: The Soft Power of Midcentury Design*. Minneapolis: The University of Minnesota Press, 2010.

Chan, Jeffrey K. H. 'Design Ethics: Reflecting on the Ethical Dimensions of Technology, Sustainability, and Responsibility in the Anthropocene'. *Design Studies* 54 (2018): 184–200.

Chin, Elizabeth. *Purchasing Power: Black Kids and American Consumer Culture*. Minneapolis: University of Minnesota Press, 2001.

Cipolla, Carla. 'Sustainable Freedoms, Dialogical Capabilities and Design'. In *Cumulus Working Papers Nantes 16/06*, 59–65. Helsinki: University of Art and Design Helsinki, 2006.

Committee on Ethics of the American College of Obstetricians and Gynecologists. 'Informed Consent'. 2009. Accessed February 24, 2020,https://www.acog.org/Clinical-Guidance-and-Publications/Committee-Opinions/Committee-on-Ethics/Informed-Consent.

Cooper, Rachel. 'Design Research – Its 50-Year Transformation'. *Design Studies* 65, (2019): 6–17.

Cooper, Tim. 'Planned Obsolescence'. In *Encyclopedia of Consumer Culture*, edited by Dale Southerton, 1094–96. Thousand Oaks: SAGE, 2011.

———. 'Slower Consumption: Reflections on Product Life Spans and the "Throwaway Society"'. *Journal of Industrial Ecology* 9, no. 1-2 (2005): 51–67.

Coplan, Amy, and Peter Goldie. *Empathy: Philosophical and Psychological Perspectives*. Oxford: Oxford University Press, 2011.

Cortina, Adela. 'El Sentido De Las Profesiones'. In *10 Palabras Clave En Ética De Las Profesiones*, edited by Adela Cortina and Jesus Conill, 13–28. Pamplona: Verbo Divino, 2000.

———. *Hasta un Pueblo de Demonios: Ética Pública y Sociedad*. Madrid: Taurus, 1998.

Costanza-Chock, Sasha. *Design Justice: Community-Led Practices to Build the Worlds We Need*. Cambridge: The MIT Press, 2020.

Cox, A., P. Oladimeji, and H. Thimbleby. 'Number Entry Interfaces and Their Effects on Errors and Number Perception'. Paper presented at the Proceedings of the IFIP Conference on Human-Computer Interaction—Interact 2011, Lisbon, 2011.

Crandall, Beth, Gary Klein, and Robert R. Hoffman. *Working Minds: A Practitioner's Guide to Cognitive Task Analysis*. Cambridge: The MIT Press, 2006.

Criado Perez, Caroline. *Invisible Women: Exposing Data Bias in a World Designed for Men*. London: Vintage, 2019.

Crisp, Roger. 'Well-Being'. *The Stanford Encyclopedia of Philosophy* (Fall 2017 Edition), Stanford University, 2017. Accessed February 24, 2020,https://plato.stanford.edu/archives/fall2017/entries/well-being/.

Cross, Nigel. *Designerly Ways of Knowing*. Basel: Birkhäuser, 2007.

———. 'Editorial: Design as a Discipline'. *Design Studies* 65 (2019): 1–5.

———. *Engineering Design Methods: Strategies for Product Design*. Fourth edition. Chichester: Wiley, 2008.

Crouwel, Wim, and Jan van Toorn. *The Debate: The Legendary Contest of Two Giants of Graphic Design*. New York: The Monacelli Press, 2015.

Csikszentmihaly, Mihaly, and Eugene Rochberg-Halton. *The Meaning of Things: Domestic Symbols and the Self*. Cambridge: Cambridge University Press, 1981.

Damasio, Antonio. *Descartes' Error: Emotion, Reason, and the Human Brain*. London: Vintage, 2006.

Davis, Michael. 'Profession as a Lens for Studying Technology'. In *The Ethics of Technology: Methods and Approaches*, edited by Sven Ove Hansson, 83–96. London: Rowman & Littlefield International, 2017.

———. *Thinking Like an Engineer: Studies in the Ethics of a Profession*. Oxford: Oxford University Press, 1998.

———. 'Thinking Like an Engineer: The Place of a Code of Ethics in the Practice of a Profession'. *Philosophy and Public Affairs* 20, no. 2 (1991): 150–67.

DeMartino, George. ''Econogenic Harm': On the Nature of and Responsibility for the Harm Economists Do as They Try to Do Good'. In *The Oxford Handbook of Professional Economics Ethics*, edited by George DeMartino and Deirdre McCloskey, 71–100. New York: Oxford University Press, 2016.

———. 'Epistemic Aspects of Economic Practice and the Need for Professional Economic Ethics'. *Review of Social Economy* 71, no. 2 (2013): 166–86.

Dempsey, Samantha, and Clara Taylor. 'Designer's Oath: Collaboratively Defining a Code of Ethics for Design'. *Touchpoint: The Journal of Service Design* 7, no. 1 (2015): 29–31.

Department of Health of the United Kingdom. *Design Bugs Out–Product Evaluation*. Runcorn: Department of Health, 2011.

Design Council. *Design for Public Good*. London: Design Council, 2013.

Dong, Andy. 'The Policy of Design: A Capabilities Approach'. *Design Issues* 24, no. 4 (2008): 76–87.

Doris, John M. 'Persons, Situations, and Virtue Ethics'. *Noûs* 32, no. 4 (1998/12/01 1998): 504–30.

Dorst, Kees. *Frame Innovation: Create New Thinking by Design*. Cambridge: The MIT Press, 2015.

Drozynski, Kate. 'The 10 Stages of Falling Down a Youtube Rabbit Hole',http://www.mtv.com/news/2283473/youtube-rabit-hole/.

Dubberly, Hugh. 'How Do You Design: A Comparison of Models'. Dubberly Design Office, 2005. Accessed February 24, 2020,http://www.dubberly.com/articles/how-do-you-design.html.

Dzur, Albert W. *Democratic Professionalism: Citizen Participation and the Reconstruction of Professional Ethics, Identity, and Practice*. University Park: The Pennsilvanya State University Press, 2008.

Eban, Katherine. *Bottle of Lies: The inside Story of the Generic Drug Boom*. New York: Ecco Press, 2019.

Edgar, Andrew. 'Professionalism in Health Care'. In *Handbook of the Philosophy of Medicine*, edited by Thomas Schramme and Steven Edwards, 677–97. Dordrecht: Springer, 2017.

Editorial Board. 'Strengthening the Credibility of Clinical Research'. *The Lancet* 375, no. 9722 (2010): 1225.

Ehmer, Josef. 'Artisans and Guilds, History Of'. In *International Encyclopedia of the Social & Behavioral Sciences (Second Edition)*, edited by James D. Wright, 46–51. Oxford: Elsevier, 2015.

Ehrenfeld, John. *Sustainability by Design*. New Haven: Yale University Press, 2009.

Eisend, Martin. 'Explaining the Impact of Scarcity Appeals in Advertising: The Mediating Role of Perceptions of Susceptibility'. *Journal of Advertising* 37, no. 3 (2008): 33–40.

Ellul, Jacques. '76 Reasonable Questions to Ask About Any Technology'. n/a. Accessed April 21, 2020, http://www.thewords.com/articles/ellul76quest.htm.

Elster, Jon. *Sour Grapes: Studies in the Subversion of Rationality.* Cambridge Philosophy Classics edition. Cambridge: Cambridge University Press, 2016.

Escobar, Arturo. *Designs for the Pluriverse: Radical Interdependence, Autonomy, and the Making of Worlds.* Durham: Duke University Press, 2017.

Etzioni, Amitai. 'Communitarianism Revisited'. *Journal of Political Ideologies* 19, no. 3 (2014): 241–60.

———. *Happiness Is the Wrong Metric: A Liberal Communitarian Response to Populism.* Cham: Springer, 2018.

European Commission. 'Antitrust: Commission Fines Google €4.34 Billion for Illegal Practices Regarding Android Mobile Devices to Strengthen Dominance of Google's Search Engine'. European Commission, 2018. Accessed February 24, 2020, http://europa.eu/rapid/press-release_IP-18-4581_en.htm.

———. 'Commission Welcomes Parliament Vote on Decreasing Vehicle Noise'. News release, 2014, https://ec.europa.eu/commission/presscorner/detail/en/IP_14_363.

———. 'New Rules Make Household Appliances More Sustainable'. News release, 2019.

———. 'Poverty, Middle Class and Purchasing Power'. Knowledge for Policy, n/a. Accessed February 24, 2020, https://ec.europa.eu/knowledge4policy/foresight/topic/growing-consumerism/poverty-middle-class-purchasing-power_en.

Eyal, Nir. *Hooked: How to Build Habit-Forming Products.* New York: Portfolio Penguin, 2014.

Feng, Patrick, and Andrew Feenberg. 'Thinking About Design: Critical Theory of Technology and the Design Process'. In *Philosophy and Design: From Engineering to Architecture*, edited by Pieter E. Vermaas, Peter Kroes, Andrew Light, and Steven A Moore, 105–18. Dordrecht: Springer, 2008.

Ferrero, Ignacio, and Alejo José G. Sison. 'Aristotle and Macintyre on the Virtues in Finance'. In *Handbook of Virtue Ethics in Business and Management*, edited by Alejo José G. Sison, Gregory R. Beabout, and Ignacio Ferrero, 1153–61. Dordrecht: Springer Netherlands, 2017.

Findeli, Alain. 'Ethics, Aesthetics, and Design'. *Design Issues* 10, no. 2 (1994): 49–68.

Fogelin, Robert. 'The Logic of Deep Disagreements'. *Informal Logic* 25, no. 1 (2005): 3–11.

Foot, Philippa. 'The Problem of Abortion and the Doctrine of Double Effect'. *Oxford Review* 5 (1967): 5–15.

———. *Virtues and Vices and Other Essays in Moral Philosophy.* Oxford: Oxford University Press, 2002.

Forbrukerrådet. *Deceived by Design.* Oslo: Forbrukerrådet, 2018.

Forlano, Laura. 'The Danger of Intimate Algorithms'. Public Books, 2020, https://www.publicbooks.org/the-danger-of-intimate-algorithms/.

Frankfurt, Harry G. *Necessity, Volition, and Love.* Cambridge: Cambridge University Press, 1999.

Freud, Sigmund. *Civilization and Its Discontents.* Reprint edition. New York: W. W. Norton & Company, 2010.

Friedman, Batya, and David Hendry. *Value Sensitive Design: Shaping Technology with Moral Imagination.* Cambridge: The MIT Press, 2019.

Friedman, Batya, Lisa Nathan, Shaun Kane, and John Lin. 'Envisioning Cards'. 2011. Accessed April 25, 2020, https://www.envisioningcards.com/.

Frog Design. *The Business Value of Design.* San Francisco: Frog Design, 2017.

Fuad-Luke, Alastair. 'Adjusting Our Metabolism: Slowness and Nourishing Rituals of Delay in Anticipation of a Post-Consumer Age'. In *Longer Lasting Products: Alternatives to the Throwaway Society*, edited by Tim Cooper, 133–55. Surrey: Gower Publishing Limited, 2010.

Fuelfor. 'Standalone Imaging Suite Creating a Patient-Centric Clinic Experience', 2018. Accessed February 24, 2020, https://www.fuelfor.net/standalone-imaging-suite-.

Fukasawa, Naoto, and Jasper Morrison. *Super Normal: Sensations of the Ordinary*. Baden: Lars Müller Publishers, 2006.

Fry, Tony. *Design Futuring: Sustainability, Ethics and New Practice*. Oxford: Berg, 2009.

Gadde, Vijaya, and Matt Derella. 'An Update on Our Continuity Strategy During Covid-19'. Twitter Blog, 2020. Accessed April 22, 2020,https://blog.twitter.com/en_us/topics/company/2020/An-update-on-our-continuity-strategy-during-COVID-19.html.

Gardner, Howard, Mihaly Csikszentmihaly, and William Damon. *Good Work: When Excellence and Ethics Meet*. New York: Basic Books, 2001.

Garland, Ken. 'First Things First'. 1964. Accessed February 24, 2020,http://www.kengarland.co.uk/KG-published-writing/first-things-first/.

Garvey, James. *The Ethics of Climate Change: Right and Wrong in a Warming World*. London: Continuum, 2008.

George, Stephen L., and Marc Buyse. 'Fraud in Clinical Trials'. *Clinical Investigation* 5, no. 2 (2015): 161–73.

Giachritsis, Christos. *Generating Simulations to Enable Testing of Alternative Routes to Improve Wayfinding in Evacuation of over-Ground and Underground Terminals*. Teddington: BMT Group Ltd, 2014.

Gidlow, D. A. 'Lead Toxicity'. *Occupational Medicine* 54, no. 2 (2004): 76–81.

Gilbert, Adrian. *Waffen-SS: Hitler's Army at War*. New York: Da Capo Press, 2019.

Gispen, Jet. 'Design Noir'. 2017. Accessed April 25, 2020,https://www.ethicsfordesigners.com/design-noir.

Goldschmidt, Gabriela, and William L. Porter, eds. *Design Representation*. London: Springer-Verlag, 2004.

Gottesfeld, Perry. 'The West's Toxic Hypocrisy over Lead Paint'. New Scientist, 2013. Accessed February 24, 2020,https://www.newscientist.com/article/mg21829190-200-the-wests-toxic-hypocrisy-over-lead-paint/.

Giachritsis, Christos. *Generating Simulations to Enable Testing of Alternative Routes to Improve Wayfinding in Evacuation of over-Ground and Underground Terminals*. Teddington: BMT Group Ltd, 2014, 2.

Groth, Olaf J., Mark J. Nitzberg, and Stuart J. Russell. 'AI Algorithms Need FDA-Style Drug Trials'. *Wired Opinion*, 2019. Accessed February 24, 2020,https://www.wired.com/story/ai-algorithms-need-drug-trials/.

Guersenzvaig, Ariel. 'Book Review'. *Journal of Design Research* 17, no. 1 (2019): 87–91.

———. 'Llagostera Youth Center'. In *Developing Citizen Designers*, edited by Elizabeth Resnick, 218–21. New York: Bloomsbury, 2016.

Harman, Gilbert. 'No Character or Personality'. *Business Ethics Quarterly* 13, no. 1 (2003): 87–94.

Harrabin, Roger. 'EU Brings in "Right to Repair" Rules for Appliances'. BBC, 2019. Accessed February 24, 2020,https://www.bbc.com/news/business-49884827.

Hartmann, Marie, Kaja Misvær Kistorp, and Emilie Strømmen Olsen. 'Using Prototyping and Co-Creation to Create Ownership and Close Collaboration: Reducing the Waiting Time for Breast Cancer Patients'. In *This Is Service Design Doing*, edited by Marc Stickdorn, Adam Lawrence, Markus Hormess, and Jakob Schneider, 252–55. Sebastopol: O'Reilly Media, 2018.

Hattula, Johannes D., Walter Herzog, Darren W. Dahl, and Sven Reinecke. 'Managerial Empathy Facilitates Egocentric Predictions of Consumer Preferences'. *Journal of Marketing Research* 52, no. 2 (2015): 235–52.

Heskett, John. *Design: A Very Short Introduction*. Oxford: Oxford University Press, 2002.

Higgins, Chris. *The Good Life of Teaching: An Ethics of Professional Practice*. West Sussex: Wiley-Blackwell, 2011.

Hill, Dan. *Dark Matter and Trojan Horses: A Strategic Design Vocabulary*. Moscow: Strelka Press, 2012.

Hoser, Paul. 'Nationalsozialistische Deutsche Arbeiterpartei (NSDAP), 1920-1923/1925-1945'. Historische Lexikon Bayerns, n/a. Accessed February 24, 2020,https://www.historisches-lexikon-bayerns.de/Lexikon/Nationalsozialistische_Deutsche_Arbeiterpartei_(NSDAP),_1920-1923/1925-1945.

Hursthouse, Rosalind. *On Virtue Ethics*. Oxford: Oxford University Press, 1999.

Hursthouse, Rosalind, and Glen Pettigrove. 'Virtue Ethics'. *The Stanford Encyclopedia of Philosophy* (Winter 2018 Edition), Stanford University, 2018. Accessed February 24, 2020,https://plato.stanford.edu/archives/win2018/entries/ethics-virtue/.

IEEE Global Initiative on Ethics of Autonomous and Intelligent Systems. *Ethically Aligned Design: A Vision for Prioritizing Human Well-Being with Autonomous and Intelligent Systems*, Version 2. Piscataway: IEEE, 2018.

Ihde, Don. 'The Designer Fallacy and Technological Imagination'. In *Philosophy and Design: From Engineering to Architecture*, edited by Pieter E. Vermaas, Peter Kroes, Andrew Light, and Steven A Moore, 51–59. Dordrecht: Springer Netherlands, 2008.

———. *Postphenomenology and Technoscience: The Peking University Lectures*. Albany: State University of New York Press, 2009.

———. 'Technology and Prognostic Predicaments'. *AI & Soc* 13, no. 1-2 (1999): 44–51.

Illich, Ivan. *Deschooling Society*. London: Marion Boyars, 2002.

———. 'Disabling Professions'. In *Disabling Professions*, edited by Ivan Illich, John McKnight, Irving K. Zola, Jonathan Caplan, and Harley Shaiken, 11–39. London: Marion Boyars, 1977.

———. *Medical Nemesis: The Expropriation of Health*. New York: Pantheon Books, 1976.

———. *Tools for Conviviality*. New York: Fontana/Collins, 1973.

Irwin, Terry. 'The Emerging Transition Design Approach'. Paper presented at the DRS2018 Conference, University of Limerick, 2018.

———. 'Transition Design: A Proposal for a New Area of Design Practice, Study, and Research'. *Design and Culture* 7, no. 2 (2015/04/03 2015): 229–46.

Isaac, Mike. 'How Uber Deceives the Authorities Worldwide'. The New York Times, 2017,https://www.nytimes.com/2017/03/03/technology/uber-greyball-program-evade-authorities.html.

Jackson, Cecil W. *Detecting Accounting Fraud: Analysis and Ethics*. Essex: Pearson Education, 2015.

Jacobs, Chip, and William J. Kelly. *Smogtown: The Lung-Burning History of Pollution in Los Angeles*. Woodstock: The Overlook Press, 2008.

Jacobs, Jane. *The Death and Life of Great American Cities*. New York: Random House, 1961.

Jasanoff, Sheila. *The Ethics of Invention: Technology and the Human Future*. New York: W.W. Norton & Company, 2016.

Jenness, James W., Jeremiah Singer, Jeremy Walrath, and Elisha Lubar. *Fuel Economy Driver Interfaces: Design Range and Driver Opinions*. Washington, DC: U.S. National Highway Traffic Safety Administration, 2009.

Jones, John Chris. *Design Methods*. Second edition. Chichester: John Wiley and Sons, 1992.

Joyce, Tang. 'Professionalization'. In *International Encyclopedia of the Social Sciences*, edited by William A. Darity, 515–17. Farmington Hills: Macmillan Reference, 2008.

Kahneman, Daniel, and Amos Tversky. 'The Simulation Heuristic'. In *Judgment under Uncertainty: Heuristics and Biases*, edited by Daniel Kahneman, Paul Slovic, and Amos Tversky, 201–08. New York: Cambridge University Press, 1982.

Kaiser, Brittany. *Targeted: The Cambridge Analytica Whistleblower's inside Story of How Big Data, Trump, and Facebook Broke Democracy and How It Can Happen Again*. New York: Harper, 2019.

Kane, Charles, Walter Bender, Jody Cornish, and Neal Donahue. *Learning to Change the World: The Social Impact of One Laptop Per Child*. New York: Palgrave Macmillan, 2012.

Kanso, Heba. 'Amid Egypt's Anti-Gay Crackdown, Gay Dating Apps Send Tips to Stop Entrapment'. Reuters, 2017. Accessed February 24, 2020,https://reut.rs/2itGG1d.

Kant, Immanuel. *Groundwork of the Metaphysics of Morals*. Cambridge: Cambridge University Press, 1997.

Kasser, Tim. *The High Price of Materialism*. Cambridge: The MIT Press, 2002.

Kemmis, Stephen. 'Phronēsis, Experience, and the Primacy of Praxis'. In *Phronesis as Professional Knowledge: Practical Wisdom in the Professions*, edited by Elizabeth-Anne Kinsella and Allan Pitman, 147–61. Rotterdam: Sense Publishers, 2012.

Keulartz, Jozef, and Jac A. A. Swart. 'Animal Flourishing and Capabilities in an Era of Global Change'. In *Ethical Adaptation to Climate Change: Human Virtues of the Future*, edited by Allen Thompson and Jeremy Bendik-Keymer. Cambridge: The MIT Press, 2012.

Kirby, Emma Jane. 'The Map That Saved the London Underground'. 2014. Accessed February 24, 2020,https://www.bbc.com/news/magazine-25551751.

Klein, Gary. 'Performing a Project Premortem'. *Harvard Business Review*, September (2007): 18–19.

———. *Sources of Power*. Cambridge: The MIT Press, 1998.

Klein, Gary, Jennifer K. Phillips, Erica L. Rall, and Deborah A. Peluso. 'A Data–Frame Theory of Sensemaking'. In *Expertise out of Context: Proceedings of the Sixth International Conference on Naturalistic Decision Making*, edited by R Hoffman, 113–58. Mahwah: Lawrence Erlbaum Associates, 2007.

Kolko, Jon. *Exposing the Magic of Design: A Practitioner's Guide to the Methods and Theory of Synthesis*. Oxford: Oxford University Press, 2010.

Kraemer, Kenneth L., Jason Dedrick, and Prakul Sharma. 'One Laptop Per Child: Vision Vs. Reality'. *Communications of the ACM* 52, no. 6 (2009): 66–73.

Krippendorff, Klaus. *The Semantic Turn: A New Foundation for Design*. Boca Raton: CRC Press, 2006.

Kumar, Vijay. *101 Design Methods: A Structured Approach for Driving Innovation in Your Organization*. Hoboken: John Wiley & Sons, 2012.

Kunda, Gideon. *Engineering Culture: Control and Commitment in a High Tech Culture*. Philadelphia: Temple University Press, 1993.

Ladd, John. 'The Quest for a Code of Professional Ethics: An Intellectual and Moral Confusion'. In *Engineering, Ethics, and the Environment*, edited by P. Aarne Vesilind and Alastair S. Gunn, 210–18. Cambridge: Cambridge University Press, 1998.

Lang, Steven, and Julia Rothenberg. 'Neoliberal Urbanism, Public Space, and the Greening of the Growth Machine: New York City's High Line Park'. *Environment and Planning A: Economy and Space* 49, no. 8 (2016): 1743–61.

Larmore, Charles. 'Right and Good'. (1998).https://www.rep.routledge.com/articles/thematic/right-and-good/v-1.

Latour, Bruno. 'Where to Land after the Pandemic? A Paper and Now a Platform'. 2020. Accessed April 29, 2020,http://www.bruno-latour.fr/node/852.html.

———. *Pandora's Hope*. Cambridge: Harvard University Press, 1999.

Lawson, Bryan. *How Designers Think: The Design Process Demystified*. Fourth edition. Oxford: Architectural Press, 2006.

Lawson, Bryan, and Kees Dorst. *Design Expertise*. Burlington: Architectural Press, 2009.

Lee, Dave. 'Microsoft Staff: Do Not Use Hololens for War'. BBC News, 2019. Accessed February 24, 2020,https://www.bbc.com/news/technology-47339774.

Lesko, Jim. 'Industrial Design at Carnegie Institute of Technology, 1934-1967'. *Journal of Design History* 10, no. 3 (1997): 269–92.

Lewis, C. S. *God in the Dock: Essays on Theology and Ethics*. Grand Rapids: Wm. B. Eerdmans Publishing Co., 1970.

Litman-Navarro, Kevin. 'We Read 150 Privacy Policies. They Were an Incomprehensible Disaster'. *New York Times*, 2019. Accessed May 12, 2020,https://www.nytimes.com/interactive/2019/06/12/opinion/facebook-google-privacy-policies.html.

Lodziak, Conrad. *The Myth of Consumerism*. London: Pluto Press, 2002.

Lombard, Kara-Jane, ed. *Skateboarding: Subcultures, Sites and Shifts*. London: Routledge, 2016.

Lopez, Gerry. *Surf Is Where You Find It*. Venture: Patagonia Books, 2008.

Lopez, Tomas. 'Poor Ballot Design Hurts New York's Minor Parties . . . Again'. 2014. Accessed February 24, 2020,https://www.brennancenter.org/our-work/analysis-opinion/poor-ballot-design-hurts-new-yorks-minor-partiesagain.

Loughran, Kevin. 'Parks for Profit: The High Line, Growth Machines, and the Uneven Development of Urban Public Spaces'. *City & Community* 13, no. 1 (2014): 49–68.

Lutz, Christopher. *Reading Alasdair Macintyre's After Virtue*. London: Continuum, 2012.

MacIntyre, Alasdair. *After Virtue: A Study in Moral Theory*. Notre Dame: University of Notre Dame Press, 2007.

———. *Ethics in the Conflicts of Modernity: An Essay on Desire, Practical Reasoning, and Narrative*. Cambridge: Cambridge University Press, 2016.

———. 'A Partial Response to My Critics'. In *After Macintyre: Critical Perspectives on the Work of Alasdair Macintyre*, edited by John P. Horton and Susan Mendus, 283–304. Notre Dame: University of Notre Dame Press, 1994.

———. 'Social Structures and Their Threats to Moral Agency'. *Philosophy* 74, no. 289 (1999): 311–29.

Man, Hongjie. 'Informed Consent and Medical Law'. In *Legal and Forensic Medicine*, edited by Roy G. Beran, 865–79. Berlin, Heidelberg: Springer Berlin Heidelberg, 2013.

Manzini, Ezio. 'Design, Ethics and Sustainability. Guidelines for a Transition Phase'. In *Cumulus Working Papers Nantes 16/06*, 9–15. Helsinki: University of Art and Design Helsinki, 2006.

———. *Design, When Everybody Designs: An Introduction to Design for Social Innovation*. Cambridge: The MIT Press, 2015.

———. *Politics of the Everyday*. London: Bloomsbury, 2019.

———. 'Social Innovation and Design—Enabling, Replicating and Synergizing'. In *Changing Paradigms: Designing for a Sustainable Future*, edited by Peter Stebbing and Ursula Tischner, 328–37. Aalto: Aalto University School of Arts, Design and Architecture, 2015.

Margolin, Victor. 'Design, the Future and the Human Spirit'. *Design Issues* 23, no. 3 (2007): 4–15.

Martin, Roger. *The Design of Business*. Brighton: Harvard Business School Press, 2009.

Mason, Andrew. 'Macintyre on Liberalism and Its Critics: Tradition, Incommensurability and Disagreement'. In *After Macintyre: Critical Perspectives on the Work of Alasdair Macintyre*, edited by Susan Mendus and John P. Horton. Cambridge: Polity, 1994.

———. 'Macintyre on Modernity and How It Has Marginalized the Virtues'. In *How Should One Live?: Essays on the Virtues*, edited by Roger Crisp. Oxford: Oxford University Press, 1996.

May, Larry. *The Socially Resposive Self: Social Theory and Professional Ethics*. Chicago: The University of Chicago, 1996.

Mazzucato, Mariana. *Mission-Oriented Research & Innovation in the European Union: A Problem-Solving Approach to Fuel Innovation-Led Growth*. Brussels: European Commission, 2018.

———. *The Value of Everything: Making and Taking in the Global Economy*. London: Allen Lane, 2018.

McBride, Dawn M., and J. Cooper Cutting. *Cognitive Psychology: Theory, Process, and Methodology*. Second edition. Thousand Oaks: SAGE, 2019.

McCoy, Katherine. 'Good Citizenship: Design as a Social and Political Force'. In *The Social Design Reader*, edited by Elizabeth Resnick, 137–44. London: Bloomsbury, 2019.

McLaughlin, Jacqueline E., Michael D. Wolcott, Devin Hubbard, Kelly Umstead, and Traci R. Rider. 'A Qualitative Review of the Design Thinking Framework in Health Professions Education'. *BMC Medical Education* 19, no. 1 (2019): 98.

McLuhan, Marshall, and Quentin Fiore. *The Medium Is the Massage*. London: Penguin Books, 1967.

McMullan, John L., and Delthia Miller. 'Advertising the 'New Fun-Tier': Selling Casinos to Consumers'. *International Journal of Mental Health and Addiction* 8, no. 1 (2010/01/01 2010): 35–50.

McNamara, Andrew, Justin Smith, and Emerson Murphy-Hill. 'Does ACM's Code of Ethics Change Ethical Decision Making in Software Development?' Paper presented at the Proceedings of the 2018 26th ACM Joint Meeting on European Software Engineering Conference and Symposium on the Foundations of Software Engineering, Lake Buena Vista, FL, USA, 2018.

McNeill, J. R. *Something New under the Sun: An Environmental History of the Twentieth-Century World*. New York: W. W. Norton & Company, 2000.

———— and Peter Engelke. *The Great Acceleration*. Cambridge: The Belknap Press of Harvard University Press, 2014.

Meadows, Donella. *Thinking in Systems: A Primer*. White River Junction: Chelsea Green Publishing, 2008.

Melé, Domènec. *Business Ethics in Action*. London: Red Globe Press, 2020.

Mepham, Ben. 'A Framework for the Ethical Analysis of Novel Foods: The Ethical Matrix'. *Journal of Agricultural and Environmental Ethics* 12, no. 2 (2000): 165–76.

Mepham, Ben, Matthias Kaiser, Erik Thorstensen, Sandy Tomkins, and Kate Millar. *Ethical Matrix Manual*. The Hague: LEI, Wageningen UR, 2006.

Meroni, Anna, and Daniela Sangiorgi. *Design for Services*. Surrey: Gower Publishing Limited, 2011.

Meyers, Christopher. *The Professional Ethics Toolkit*. Hoboken: Willey Blackwell, 2018.

Mica, Adriana. 'Unintended Consequences: History of the Concept'. In *International Encyclopedia of the Social & Behavioral Sciences*, edited by James D Wright, 744–49. Amsterdam: Elsevier, 2015.

Mill, John Stuart. *Utilitarianism and on Liberty*. Malden: Blackwell Publishing, 2003.

Miller, Daniel. *The Comfort of Things*. Cambridge: Polity Press, 2008.

————. *Material Culture and Mass Consumption*. Oxford: Basil Blackwell, 1987.

Millman, Debbie. *How to Think Like a Great Graphic Designer*. New York: Allworth Press, 2007.

Mitcham, Carl. 'Ethics into Design'. In *Discovering Design: Explorations in Design Studies*, edited by Richard Buchanan and Victor Margolin, 173–89. Chicago: The University of Chicago Press, 1995.

Monbiot, George. *Out of the Wreckage: A New Politics for an Age of Crisis*. London: Verso, 2018.

Monteiro, Mike. *How Designers Destroyed the World, and What We Can Do to Fix It*. San Francisco: Mule Design, 2019.

Moore, Geoff. *Virtue at Work: Ethics for Individuals, Managers, and Organizations*. Oxford: Oxford University Press, 2017.

Moy, Dianne, and Chris Ryan. 'Using Scenarios to Explore System Change: Veil, Local Food Depot'. In *Design for Services*, edited by Anna Meroni and Daniela Sangiorgi, 161–71. Surrey: Gower Publishing Limited, 2011.

Murdoch, Iris. *The Sovereignty of Good*. London: Routledge, 2001.

Nelson, Harold G., and Erilk Stolterman. *The Design Way: Intentional Change in an Unpredictable World*. Cambridge: The MIT Press, 2014.

Newport, Cal. *Digital Minimalism: On Living Better with Less Technology*. London: Penguin Business, 2019.

Newton, Lisa. 'Professionalization: The Intractable Plurality of Values'. In *Profits and Professions: Essays in Business and Professional Ethics*, edited by Wade L. Robison, Michael S. Pritchard and Joseph Ellin, 23–36. Clifton: Humana Press, 1983.

Nihlén Fahlquist, Jessica. 'Responsibility as a Virtue and the Problem of Many Hands'. In *Moral Responsibility and the Problem of Many Hands*, edited by Ibo Van de Poel, Lambèr Royakkers, and Sjoerd D. Zwart, 187–208. New York: Routledge, 2015.

Noble, David. *America by Design: Science, Technology, and the Rise of Corporate Capitalism*. New York: Alfred A. Knopf, 1977.

Noddings, Nel. *Caring: A Relational Approach to Ethics & Moral Education*. Berkeley: University of California Press, 2013.

Noggle, Robert. 'The Ethics of Manipulation'. *The Stanford Encyclopedia of Philosophy* (Summer 2018 Edition), Stanford University 2018. Accessed February 24, 2020,https://plato.stanford.edu/archives/sum2018/entries/ethics-manipulation/.

Norden, Lawrence, David Kimball, Whitney Quesenbery, and Margaret Chen. *Better Ballots*. New York: Brennan Center for Justice at New York University School of Law, 2008.

Norman, Donald. *The Design of Everyday Things*. Revised and expanded edition. New York: Basic Books, 2013.

Norman, Donald A. *Emotional Design: Why We Love (or Hate) Everyday Things*. New York: Basic Books, 2004.

Nussbaum, Martha. *Creating Capabilities: The Human Development Approach*. Cambridge: The Belknap Press of Harvard University Press, 2011.

Nussbaum, Martha C. 'Virtue Ethics: A Misleading Category?'. *The Journal of Ethics* 3, no. 3 (1999): 163–201.

Oakley, Justin, and Dean Cocking. *Virtue Ethics and Professional Roles*. Cambridge: Cambridge University Press, 2001.

Oosterlaken, Ilse. 'Design for Development: A Capability Approach'. *Design Issues* 25, no. 4 (2009): 91–102.

Ortega y Gasset, José. *Toward a Philosophy of History*. Translated by Helene Weyl. New York: W. W. Norton & Company, 1941.

Pan, Bing. 'The Power of Search Engine Ranking for Tourist Destinations'. *Tourism Management* 47 (2014): 79–87.

Papanek, Victor. *Design for the Real World*. Second revised edition. Chicago: Academy Chicago Publishers, 1984.

———. 'Edugraphology: The Myths of Design and the Design of Myths'. In *Looking Claser 3: Classic Writings on Graphic Design*, edited by Michael Bierut, Jessica Helfand, Steven Heller, and Rick Poynor. New York: Alworth Press, 1999.

Parisier, Eli. *The Filter Bubble: What the Internet Is Hiding from You*. London: Penguin Books, 2011.

Parsons, Glenn. *The Philosophy of Design*. Cambridge: Polity, 2016.

Parsons, Talcott. 'Professions'. In *International Encyclopedia of the Social Sciences*, edited by David L. Sills, 536–47. New York: Macmillan, 1968.

Pellegrino, Edmund D., and David C. Thomasma. *The Virtues in Medical Practice*. New York: Oxford University Press, 1993.

Pendall, Rolf, Christopher Hayes, Arthur George, Zach McDade, Casey Dawkins, Jae Sik Jeon, Eli Knaap, et al. *Driving to Opportunity: Understanding the Links among Transportation Access, Residential Outcomes, and Economic Opportunity for Housing Voucher Recipients*. Washington, DC: The Urban Institute, 2014.

Peters, Adele. 'Amsterdam Is Now Using the "Doughnut" Model of Economics: What Does That Mean?' *Fast Company*, 2020. Accessed May 2, 2020,https://www.fastcompany.com/90497442/amsterdam-is-now-using-the-doughnut-model-of-economics-what-does-that-mean.

Petit, Victor, and Bertrand Guillaume. 'Scales of Design: Ecodesign and the Anthropocene'. In *Advancements in the Philosophy of Design*, edited by Pieter E. Vermaas and Stéphane Vial, 473–94. Cham: Springer International Publishing, 2018.

Phillips, Michael J. *Ethics and Manipulation in Advertising*. Westport: Quorum, 1997.

Pichai, Sundar. 'AI at Google: Our Principles'. 2018. Accessed July 22, 2018,https://blog.google/technology/ai/ai-principles/.

Pincoffs, Edmund. 'Quandary Ethics'. *Mind* 80, no. 320 (1971): 552–71.

Pollack, Sydney. 'Sketches of Frank Gehry'. Artificial Eye, 2007.

Postman, Neil. *Building a Bridge to the 18th Century: How the Past Can Improve the Present*. New York: Alfred A. Knopf, 2000.

Preece, Jenny. *Human-Computer Interaction*. Harlow: Pearson Education, 1994.

Press, Mike, and Rachel Cooper. *The Design Experience: The Role of Design and Designers in the Twenty-First Century*. London: Routledge, 2003.

Proudfoot, Michael, and A. R. Lacey. 'Normative', in *The Routledge Dictionary of Philosophy*. Fourth edition. London: Routledge, 2010.

Quito, Anne. 'Branding the World's Newest Country'. Works That Work, 2014. Accessed February 24, 2020,https://worksthatwork.com/4/branding-south-sudan.

Rachels, James, and Stuart Rachels. *The Elements of Moral Philosophy*. Ninth edition. New York: McGraw-Hill Education, 2018.

Radder, Hans. 'Why Technologies Are Inherently Normative'. In *Handbook of the Philosophy of Science*, edited by Dov Gabbay, Paul Thagard, and John Woods, 887–921. Amsterdam: Elsevier, 2009.

Rae, Jeneanne. 'The Power & Value of Design Continues to Grow across the S&P 500'. *DMI Journal* 27, no. 4 (2015): 4–11.

Rams, Dieter. *Less but Better / Weniger, Aber Besser*. Berlin: Die Gestalten Verlag, 2014.

Raworth, Kate. *Doughnut Economics: Seven Ways to Think Like a 21st-Century Economist*. London: Random House Business Books, 2017.

Resnick, Elizabeth. *The Social Design Reader*. London: Bloomsbury, 2019.

Rittel, Herb, and Melvin M. Webber. 'Dilemmas in a General Theory of Planning'. *Policy Sciences* 4, no. 2 (1973): 155–69.

Robertson, Adi. 'OLPC's $100 Laptop Was Going to Change the World—Then It All Went Wrong'. 2018. Accessed February 24, 2020,https://www.theverge.com/2018/4/16/17233946/olpcs-100-laptop-education-where-is-it-now.

Robeyns, Ingrid. 'Sen's Capability Approach and Gender Inequality: Selecting Relevant Capabilities'. *Feminist Economics* 9, no. 2-3 (2003): 61–92.

———. *Wellbeing, Freedom and Social Justice: The Capability Approach Re-Examined*. Cambridge: Open Book Publishers, 2017.

Robison, Wade L. *Ethics within Engineering: An Introduction*. London: Bloomsbury, 2017.

Robson, Angus. 'Constancy and Integrity: (Un)Measurable Virtues?' *Business Ethics: A European Review* 24, no. S2 (2015): S115–S29.

Rodríguez-Hidalgo, A., J. I. Morales, A. Cebrià, L. A. Courtenay, J. L. Fernández-Marchena, G. García-Argudo, J. Marín, et al. 'The Châtelperronian Neanderthals of Cova Foradada (Calafell, Spain) Used Imperial Eagle Phalanges for Symbolic Purposes'. *Science Advances* 5, no. 11 (2019): eaax1984.

Rohrbach, Stacie, and Molly Steenson. 'Transition Design: Teaching and Learning'. *Cuadernos del Centro de Estudios en Diseño y Comunicación* 19, no. 73 (2019): 235–63.

Rosenberger, Robert. *Callous Objects: Designs against the Homeless*. Third edition. Minneapolis: University of Minnesota Press, 2017.

Royal College of Art. 'Our History'. n/a. Accessed February 24, 2020,https://www.rca.ac.uk/more/about-rca/our-history/.

Russ, Jean. *Sustainability and Design Ethics*. Second edition. Boca Raton: CRC Press, 2019.

Sacasas, L. M. 'Do Artifacts Have Ethics?' 2014. Accessed April 25, 2020,https://thefrailestthing.com/2014/11/29/do-artifacts-have-ethics/.

Sanders, Elizabeth B. N., and Pieter Jan Stappers. 'Co-Creation and the New Landscapes of Design'. *CoDesign* 4, no. 1 (2008/03/01 2008): 5–18.

Sanders, Liz. 'Is Sustainable Innovation an Oxymoron?' In *Changing Paradigms: Designing for a Sustainable Future*, edited by Peter Stebbing and Ursula Tischner, 296–301. Aalto: Aalto University School of Arts, Design and Architecture, 2015.

Sangiorgi, Daniela, and Kakee Scott. 'Conducting Design Research in and for a Complex World'. In *The Routledge Companion to Design Research*, edited by Paul A. Rodgers and Joyce Yee, 114–31. London: Routledge, 2015.

Sapoznik, Harry. *Klezmer!: Jewish Music from Old World to Our World*. New York: Schirmer Trade Books, 1999.

Sarfatti Larson, Magali. 'Looking Back and a Little Forward: Reflections on Professionalism and Teaching as a Profession'. *Radical Teacher* 99, no. Spring 2014 (2014): 7–17.

———. *The Rise of Professionalism: A Sociological Analysis*. Berkeley: University of California Press, 1977.

Schaminée, André. *Designing With and Within Public Organizations: Building Bridges between Public Sector Innovators and Designers*. Amsterdam: Bis Publishers, 2018.

Schlosser, Markus. 'Agency'. *The Stanford Encyclopedia of Philosophy* (Fall 2015 Edition), Stanford University, 2015. Accessed February 24, 2020,https://plato.stanford.edu/archives/fall2015/entries/agency/.

Schneller, Annina. *Scratching the Surface: 'Appearance' as a Bridging Concept between Design Ontology and Design Aesthetics*. Advancementes in the Philosophy of Design. Edited by Pieter E. Vermaas and Stéphane Vial. Dordrecht: Springer, 2018.

Schön, Donald. 'Problems, Frames and Perspectives on Designing'. *Design Studies* 5, no. 3 (1984): 132–36.

———. *The Reflective Practitioner: How Professionals Think in Action*. New York: York: Basic Books, 1983.

Schwartz Cowan, Ruth. *More Work for Mother: The Ironies of Household Technology from the Open Hearth to the Microwave*. London: Free Association Books, 1989.

Semuels, Alana. 'No Driver's License, No Job'. *The Atlantic*, 2016. Accessed February 24, 2020,https://www.theatlantic.com/business/archive/2016/06/no-drivers-license-no-job/486653/.

Sen, Amartya. *Development as Freedom*. New York: Alfred A. Knopf, 2000.

———. 'Does Business Ethics Make Economic Sense?'. *Business Ethics Quarterly* 3, no. 1 (1993).

Shafer-Landau, Russ. *The Fundamentals of Ethics*. Fourth edition. Oxford: Oxford University Press, 2017.

Shaughnessy, Adrian. *How to Be a Graphic Designer without Losing Your Soul*. New expanded edition. New York: Princeton Architectural Press, 2010.

Sheppard, Benedict, John Edson, and Garen Kouyoumjian. 'More Than a Feeling: Ten Design Practices to Deliver Business Value'. McKinsey Design, updated December 12, 2019, 2017. Accessed February 24, 2020,https://www.mckinsey.com/business-functions/mckinsey-design/our-insights/more-than-a-feeling-ten-design-practices-to-deliver-business-value.

Sheppard, Benedict, Garen Kouyoumjian, Hugo Sarrazin, and Fabricio Dore. 'The Business Value of Design'. *McKinsey Quarterly*, no. 4 (2018): 58–72.

Simon, Herbert. *The Sciences of the Artificial*. Third edition. Cambridge: The MIT Press, 1996.

Sims, Ronald R., and Johannes Brinkmann. 'Enron Ethics (Or: Culture Matters More Than Codes)'. *Journal of Business Ethics* 45, no. 3 (2003): 243–56.

Singer, Peter. *Animal Liberation: The Definitive Classic of the Animal Movement*. Fortieth anniversary edition. New York: Open Road Media, 2015.

———. 'Famine, Affluence, and Morality'. *Philosophy & Public Affairs* 1, no. 3 (1972): 229–43.

Sinnott-Armstrong, Walter. 'Consequentialism'. *The Stanford Encyclopedia of Philosophy* (Summer 2019 Edition), Stanford University, 2018. Accessed February 24, 2020,https://plato.stanford.edu/archives/sum2019/entries/consequentialism/.

Sison, Alejo José G. 'Revisiting the Common Good of the Firm'. In *The Challenges of Capitalism for Virtue Ethics and the Common Good*, edited by Kleio Akrivou and Alejo José G. Sison, 93–120. Cheltenham: Edward Elgar Publishing, 2016.

SITEAL. 'Programa Conectar Igualdad'. n/a. Accessed February 24, 2020,http://www.tic.siteal.iipe.unesco.org/politicas/859/programa-conectar-igualdad.

Solomon, Robert C. *Ethics and Excellence: Cooperation and Integrity in Business*. Oxford: Oxford University Press, 1992.

Spool, Jared. 'Do Users Change Their Settings?' UIE, 2011. Accessed February 24, 2020,https://archive.uie.com/brainsparks/2011/09/14/do-users-change-their-settings/.

Stanford Law School. 'The Legal Design Lab'. Stanford Law School, n/a. Accessed February 24, 2020,https://law.stanford.edu/organizations/pages/legal-design-lab/.

Sunstein, Carl. 'The Ethics of Nudging'. *Yale Journal on Regulation* 32, no. 2 (2015): 413–50.

Suri, Jane Fulton. 'The Experience of Evolution: Developments in Design Practice'. *The Design Journal* 6, no. 2 (2003): 39–48.

Taylor, Charles. *The Malaise of Modernity*. Toronto: House of Anansi Press, 1991.

———. *Sources of the Self: The Making of the Modern Identity*. Cambridge: Harvard University Press, 1989.

Thaler, Richard, and Carl Sunstein. *Nudge: Improving Decisions About Health, Wealth, and Happiness*. New Haven: Yale University Press, 2008.

Thaler, Richard H., Cass R. Sunstein, and John P. Balz. 'Choice Architecture'. SSRN, 2010. Accessed February 24, 2020,https://ssrn.com/abstract=1583509.

Thorpe, Adam, and Lorraine Gamman. 'Design with Society: Why Socially Responsive Design Is Good Enough'. *CoDesign* 7, no. 3-4 (2011): 217–30.

Tonkinwise, Cameron. '"I Prefer Not To": Anti-Progressive Designing'. In *Undesign: Critical Practices at the Intersection of Art and Design*, edited by Gretchen Coombs, Andrew McNamara, and Gavin Sade, 74–84. Milton Park: Routledge, 2018.

Tovey, Michael. 'The Passport to Practice'. In *Design and Designing*, edited by Steve Garner and Chris Evans, 5–19. London, 2012.

Tromp, Nynke, Paul Hekkert, and Peter-Paul Verbeek. 'Design for Socially Responsible Behavior: A Classification of Influence Based on Intended User Experience'. *Design Issues* 21, no. 3 (2011): 3–19.

Tufekci, Zeynep. 'Youtube, the Great Radicalizer',https://www.nytimes.com/2018/03/10/opinion/sunday/youtube-politics-radical.html.

U.S. National Highway Traffic Safety Administration. *Federal Motor Vehicle Safety Standard No. 141, Minimum Sound Requirements for Hybrid and Electric Vehicles.* Washington, DC: Federal Register, 2019.

Urbina, Dante A., and Alberto Ruiz-Villaverde. 'A Critical Review of Homo Economicus from Five Approaches'. *American Journal of Economics and Sociology* 78, no. 1 (2019/01/01 2019): 63–93.

Vallor, Shannon. *Technology and the Virtues: A Philosophical Guide to a Future Worth Wanting.* New York: Oxford University Press, 2016.

Van de Poel, Ibo. 'Moral Responsibility'. In *Moral Responsibility and the Problem of Many Hands*, edited by Ibo Van de Poel, Lambèr Royakkers, and Sjoerd D. Zwart, 12–49. New York: Routledge, 2015.

———. 'The Problem of Many Hands'. In *Moral Responsibility and the Problem of Many Hands*, edited by Ibo Van de Poel, Lambèr Royakkers, and Sjoerd D. Zwart, 50–92. New York: Routledge, 2015.

Van de Poel, Ibo, and Peter Kroes. 'Can Technology Embody Values?'. In *The Moral Status of Technical Artefacts*, edited by Peter Kroes and Peter-Paul Verbeek, 103–24. Dordrecht: Springer, 2014.

Van de Poel, Ibo, and Lambèr Royakkers. *Ethics, Technology, and Engineering: An Introduction.* Malden: Wiley-Blackwell, 2011.

Van den Hoven, Jeroen. 'Human Capabilities and Technology'. In *The Capability Approach, Technology and Design*, edited by Ilse Oosterlaken and Jeroen Van den Hoven, 27–36. Dordrecht: Springer, 2012.

Van den Hoven, Jeroen, Pieter E. Vermaas, and Ibo Van de Poel. *Handbook of Ethics, Values, and Technological Design.* Dordrecht: Springer, 2015.

Van Wynsberghe, Aimee, and Scott Robbins. 'Ethicist as Designer: A Pragmatic Approach to Ethics in the Lab'. *Science and Engineering Ethics* 20, no. 4 (2014): 947–61.

VandenBos, Gary R., ed. *APA Dictionary of Psychology.* Washington: American Psychological Association, 2015.

Venturi, Robert. *Complexity and Contradiction in Architecture.* Second edition. New York: The Museum of Modern Art, 1977.

Verbeek, Peter-Paul. *Moralizing Technology: Understanding and Designing the Morality of Things.* Chicago: The University of Chicago Press, 2011.

Waheed, Hussain. 'The Common Good'. *The Stanford Encyclopedia of Philosophy* (Spring 2018 Edition), Stanford University, 2018. Accessed February 24, 2020,https://plato.stanford.edu/archives/spr2018/entries/common-good/.

Walker, Melanie, and Monica McLean. *Professional Education, Capabilities and the Public Good: The Role of Universities in Promoting Human Development.* Oxon: Routledge, 2013.

Walker, Stuart, and Jacques Giard, eds. *The Handbook of Design for Sustainability.* London: Bloomsbury, 2013.

Wand, Jonathan N., Kenneth W. Shotts, Jasjeet S. Sekhon, Walter R. Mebane, Michael C. Herron, and Henry E. Brady. 'The Butterfly Did It: The Aberrant Vote for Buchanan in Palm Beach County, Florida'. *American Political Science Review* 95, no. 4 (2001): 793–810.

Wang, Maya. *China's Algorithms of Repression: Reverse Engineering a Xinjiang Police Mass Surveillance App.* New York: Human Rights Watch, 2019.

Weber, Max. *The Protestant Ethic and the Spirit of Capitalism.* London: Routledge, 1992.

Werhane, Patricia. *Moral Imagination and Management Decision Making.* Washington: Business Roundtable Institute for Corporate Ethics, 2009.

Wertheimer, Alan. *Rethinking the Ethics of Clinical Research: Widening the Lens.* Oxford: Oxford University Press, 2011.

Whitbeck, Caroline. *Ethics in Engineering Practice and Research*. Second edition. Cambridge: Cambridge University Press, 2011.

White, Mark D. *The Manipulation of Choice: Ethics and Libertarian Paternalism*. New York: Palgrave Macmillan, 2013.

Whitehead, Jake, Robin Smit, and Simon Washington. 'Where Are We Heading with Electric Vehicles?' *Air Quality and Climate Change* 52, no. 3 (2018): 18–27.

Wicks, Peter. 'Secondary Literature'. International Society for MacIntyrean Enquiry, 2020. Accessed February 24, 2020,https://www.macintyreanenquiry.org/secondary-chronological.

Widdows, Heather. *Global Ethics: An Introduction*. Durham: Acumen, 2011.

Wilkinson, Richard G., and Kate Pickett. *The Spirit Level: Why Equality Is Better for Everyone*. London: Penguin Books, 2010.

Williams, Bernard. *Ethics and the Limits of Philosophy*. Abingdon: Routledge, 2006.

Williams, Garrath. 'Responsibility as a Virtue'. *Ethical Theory and Moral Practice* 11, no. 4 (2008): 455–70.

Wood, Greg, and Malcolm Rimmer. 'Codes of Ethics: What Are They Really and What Should They Be?' *International Journal of Value - Based Management* 16, no. 2 (2003): 181–95.

Woodham, Jonathan 'Central School of Art and Design'. In *A Dictionary of Modern Design*. Oxford: Oxford University Press, 2016.

World Bank. *Poverty and Shared Prosperity 2018: Piecing Together the Poverty Puzzle*. Washington, DC: World Bank, 2018.

World Design Organisation. 'Code of Professional Ethics'. Montreal: World Design Organization, n/a.

———. 'Who We Are'. World Design Organization, 2020. Accessed February 24, 2020,https://wdo.org/.

Wueste, Daniel E., ed. *Professional Ethics and Social Responsibility*. Lanham: Rowman & Littlefield Publishers, 1994.

York, Richard. 'Ecological Paradoxes: William Stanley Jevons and the Paperless Office'. *Human Ecology Review* 13, no. 2 (2006): 143–47.

Young, Indi. *Practical Empathy: For Collaboration and Creativity in Your Work*. New York: Rosenfeld Media, 2015.

Zizzo, Natalie, Emily Bell, and Eric Racine. 'What Is Everyday Ethics? A Review and a Proposal for an Integrative Concept'. *Journal of Clinical Ethics* 27, no. 2 (2016): 117–28.

Index

Ackoff, Russel L., 81

After Virtue, 3, 159. *See also* goods; practices; MacIntyre, Alasdair virtue(s); virtue ethics;

Alzola, Miguel, 146

Amabile, Teresa, 22

Annas, Julia, 145, 147

anthropocentrism, 199, 207n47; challenge of, 199; defence against the charge of, 199–200

anticipation of effects, 110, 261; limitations of, 111; principles and strategies for, 111, 262. *See also* mediation analysis; moral imagination; Value Sensitive Design

Arendt, Hannah, 244

Aristotle, 144, 147, 148, 177, 193, 226; and flourishing, 144

Astudillo, César, 87n77, 115n3, 234n19

Barber, Bernard, 43, 49

Bayles, Michael, 42, 43, 62

Benini, Julia, 37n18

Bentham, Jeremy, 153

Buchanan, Richard, 5, 7, 23, 71, 124, 125, 179

Cambridge Analytica data scandal, 52, 53, 59n56

capabilities, 191–193; and artefacts, 195; capability and design scholarship, 206n25; contrasted with functionings, 192; conversion factors and design, 195–197; defined, 192; design as amplifier of, 187, 193; as opportunities, 192, 193; and the overarching purpose of design, 194

capability approach, 192

capitalism, 15, 20, 21, 248. *See also* internal goods of design, tension with external goods; goods, tension between internal and external

care, 212, 214, 220–221, 222; and the designer's stance, 218; others as objects of, 191; professions as an object of, 212, 218. *See also* responsibility as a virtue

Carelman, Jacques, 98

Carr, David, 45, 50, 140n4

cars, 105, 106, 183, 196

Casacuberta, David, 140n20, 263

Chan, Jeffrey K. H., 123, 200

character, 147, 150, 151, 203, 204, 211. *See also* compartmentalisation; constancy; integrity; regulative ideals; virtue

choice architecture. *See* nudges

Cocking, Dean, 44, 151, 190

codes of ethics, 1, 4, 54–55; in design, 55, 128; discussion of, 55–57

cognitive landscape, 150, 158

www.ingramcontent.com/pod-product-compliance
Lightning Source LLC
Chambersburg PA
CBHW021810270326
41932CB00007B/121